OECD Science, Technology and Innovation Outlook 2018

ADAPTING TO TECHNOLOGICAL AND SOCIETAL DISRUPTION

This work is published under the responsibility of the Secretary-General of the OECD. The opinions expressed and arguments employed herein do not necessarily reflect the official views of OECD member countries.

This document, as well as any data and any map included herein, are without prejudice to the status of or sovereignty over any territory, to the delimitation of international frontiers and boundaries and to the name of any territory, city or area.

Please cite this publication as:
OECD (2018), *OECD Science, Technology and Innovation Outlook 2018: Adapting to Technological and Societal Disruption*, OECD Publishing, Paris.
https://doi.org/10.1787/sti_in_outlook-2018-en

ISBN 978-92-64-30756-8 (print)
ISBN 978-92-64-30757-5 (pdf)

Annual: OECD Science, Technology and Innovation Outlook
ISSN 2518-6272 (print)
ISSN 2518-6167 (online)

The statistical data for Israel are supplied by and under the responsibility of the relevant Israeli authorities. The use of such data by the OECD is without prejudice to the status of the Golan Heights, East Jerusalem and Israeli settlements in the West Bank under the terms of international law.

Photo credits: Cover © Sylvain Fraccola

Corrigenda to OECD publications may be found on line at: *www.oecd.org/publishing/corrigenda*.
© OECD 2018

You can copy, download or print OECD content for your own use, and you can include excerpts from OECD publications, databases and multimedia products in your own documents, presentations, blogs, websites and teaching materials, provided that suitable acknowledgement of OECD as source and copyright owner is given. All requests for public or commercial use and translation rights should be submitted to *rights@oecd.org*. Requests for permission to photocopy portions of this material for public or commercial use shall be addressed directly to the Copyright Clearance Center (CCC) at *info@copyright.com* or the Centre français d'exploitation du droit de copie (CFC) at *contact@cfcopies.com*.

Preface

Innovation enables countries to be more competitive, more adaptable to change and to support higher living standards. It provides the foundation for new businesses and new jobs and helps address pressing social and global challenges, such as health, climate change, and food and energy security.

While the opportunities for innovation are immense, they are not automatic. New realities are reshaping innovation, and policymakers should reflect on whether science, technology and innovation (STI) policies remain "fit for purpose" in driving sustainable growth and supporting societal well-being. The 2018 edition of the *OECD Science, Technology and Innovation Outlook* aims to help countries understand how the disruptive trends and issues we see are affecting our science practices, technology developments, innovation processes and STI policies.

A number of "game-changers" stand out, notably the rise of artificial intelligence (AI), accompanied by an unprecedented growth in data, and the fast expanding role of economies, such as China, that are leading developments in some emerging technologies. AI holds the potential for revolutionising the scientific process and new poles of STI activity are taking root, opening up new opportunities for countries to benefit from science and innovation. At the same time, issues of privacy, digital security, safety, transparency and competition have all risen up the policy agenda, defying quick solutions and demanding new and coordinated policy responses.

There are also growing demands on innovation, not only to support growth and job creation, but also to address a wide range of social and global challenges that are reflected in the Sustainable Development Goals (SDGs). The focus on the SDGs highlights the importance of linking innovation more closely to people's needs. In this respect, the digital transformation can also help to engage more people in innovation – democratise it, even – and make it more inclusive. Yet today, too few research and innovation funding programmes are explicitly linked to the SDGs.

A major challenge is to implement new governance and steering mechanisms that can deal with the public concerns and risks that come with some emerging technologies, e.g. AI or gene editing, so that the outcomes serve society. The speed and uncertainty of technological change makes it difficult for policymakers to exert oversight of emerging technologies. Preventing, correcting or mitigating potential negative effects, while still allowing entrepreneurial activity to flourish, is a balancing act facing all policymakers today.

If we are to harness the full promise of innovation for our economies and societies, it is essential to better understand these evolutions in the innovation landscape. New opportunities coincide with a growing divergence in productivity growth across businesses, as well as with innovation performance disparities between regions and countries. There is untapped potential for innovation to contribute to social inclusion and environmental goals, as well as a need to ensure that people are better prepared to participate in, and adapt to, the sometimes disruptive processes of innovation in their lives.

Governments need to become more agile, more responsive, more open to stakeholder participation and better informed of the potential opportunities and challenges of new technologies. Given the scale of such challenges, international co-operation has an essential role to play, but is threatened by the erosion of multilateralism in other areas. We must maintain a global mindset, strive for openness, and support multilateral co-operation to boost innovation for growth and well-being and manage its risks for the benefit of all. It is our responsibility to work together towards better STI policies, at national and international levels, to ensure that society as a whole shares the benefits of innovation for better lives, now and for the generations to come. The OECD is determined to play its role in bringing this goal to fruition.

Angel Gurría

Secretary-General

OECD

Foreword

The *OECD Science, Technology and Innovation Outlook 2018* is the twelfth edition in a biennial series that reviews key trends in science, technology and innovation (STI) policy in OECD countries and several major partner economies. The 14 chapters in this edition look at a range of topics, notably the opportunities and challenges related to enhanced data access, the impacts of artificial intelligence on science and manufacturing, and the influence of digitalisation on research and innovation. The report also discusses the shortcomings of current policy measures, how the Sustainable Development Goals are re-shaping STI policy agendas, and the need for more flexible and agile approaches to technology governance and policy design. While these disruptive changes pose a number of challenges for policymakers, the digital revolution under way also provides solutions for better policy targeting, implementation and monitoring.

This report relies on the latest academic work in the field, as well as research and innovation statistical data, and data on wider trends and issues. It makes extensive use of country responses to the 2017 European Commission/OECD International Survey on Science, Technology and Innovation Policy (https://stip.oecd.org). It also features contributions by renowned experts and academics to broaden the debate and provide more personal – and sometimes controversial – angles to it.

A common denominator across the chapters is the need for more adaptive policies that can better respond to disruptive scientific, technological and societal developments. This, in turn, creates new challenges: governments need to become more agile, while still ensuring policy coherence and maintaining public trust. During this necessary transition, emerging and fast-changing digital technologies both challenge policymakers, and provide them with solutions to better target, implement and monitor their interventions.

All the STI Outlook 2018 chapters feature concrete examples of national policy initiatives in order to contribute to the process of international policy learning. Complexity and uncertainty characterise the relationship between developments in STI and the economic and social challenges facing countries at all income levels. Consequently, an ever-greater need exists for exchanging information on existing policies, as well as the factors underlying their successes and failures.

Acknowledgements

The *OECD Science, Technology and Innovation Outlook* (STI Outlook) is prepared under the aegis of the OECD Committee for Scientific and Technological Policy (CSTP), with input from its working parties. CSTP Delegates contributed significantly through their responses to the joint European Commission/OECD International Survey on Science, Technology and Innovation Policy (STIP Compass). Renowned experts provided valuable contributions ("In my view") to broaden and deepen the debate.

The 2018 STI Outlook is a collective effort, co-ordinated by the Science and Technology Policy (STP) Division of the OECD Directorate for Science, Technology and Innovation (DSTI). It is produced under the guidance of Dominique Guellec. Michael Keenan and Philippe Larrue served as overall co-ordinators, and Sylvain Fraccola as the administrative co-ordinator.

Chapter 1, "An introduction to the 2018 edition of the STI Outlook", was prepared by Michael Keenan and Philippe Larrue, with inputs from Dominique Guellec.

Chapter 2, "Artificial intelligence and the technologies of the Next Production Revolution", was prepared by Alistair Nolan (DSTI). It draws in parts on the first chapter ("The next production revolution: Key issues and policy proposals") of the 2017 OECD publication, *The Next Production Revolution: Implications for Governments and Business*. This chapter benefited from the observations of David Rosenfeld of the DSTI Economic Analysis and Statistics Division, Luis Aranda, of the DSTI Division for Digital Economy Policy, Luke Mackle, of the OECD's Global Relations Secretariat, as well as inputs from Linde Wester, who is currently completing a PhD on quantum computing at Oxford University. This chapter also benefited from the contribution of Greg Ameyugo of CEA Tech List.

Chapter 3, "Perspectives on innovation policies in the digital age", was prepared by Dominique Guellec and Caroline Paunov (DSTI). It is based on the OECD Working Party on Innovation and Technology Policy (TIP) "Digital and open innovation" project. This chapter benefited from contributions by Sandra Planes Satorra (DSTI), Erik Brynjolfsson and Avinash Collis of the Massachusetts Institute of Technology, and Luc Soete of the University of Maastricht. Detailed feedback from TIP and CSTP experts and delegates is gratefully acknowledged.

Chapter 4, "STI policies for delivering on the Sustainable Development Goals", was prepared by Mario Cervantes and Soon Jeong Hong (DSTI). This chapter benefited from contributions by Ian Hughes of University College Cork and Alfred Watkins of the Global Solutions Summit.

Chapter 5, "Artificial intelligence and machine learning in science", was prepared by Ross D. King of the University of Manchester and Stephen Roberts of the Alan Turing Institute at the University of Oxford, under the guidance of Alistair Nolan (DSTI). This chapter also benefited from the contribution of Gary Marcus of New York University.

Chapter 6, "Enhanced access to data for science, technology and innovation", was prepared by Alan Paic and Carthage Smith (DSTI). It is based on recent activities of the CSTP and OECD Global Science Forum (GSF). This chapter benefited from contributions by Hon Michael Keenan MP of the Australian Government and Michelle Willmers of the University of Cape Town.

Chapter 7, "Gender in a changing context for STI", was prepared by Elizabeth Pollitzer of the Portia Organisation, Carthage Smith (DSTI) and Claartje Vinkenburg of Vrije Universiteit Amsterdam.

Chapter 8, "Public research funding: New funding approaches and instruments", was prepared by Dominique Guellec, Philippe Larrue and Frédéric Sgard (DSTI). It is based on recent CSTP and GSF activities. This chapter benefited from the contribution of Erik Arnold of the KTH Royal Institute of Technology (KTH).

Chapter 9, "The governance of public research policy across OECD countries", was prepared by Martin Borowiecki and Caroline Paunov (DSTI). It is based on TIP work on knowledge transfer between industry and science. Cynthia Lavison, Andrés Barreneche, Diogo Machado, Evgeny Moiseichev, Tadanori Moriguchi, Sandra Planes Satorra, Akira Tachibana and Malte Tötzke helped develop the OECD Database on Governance of Public Research (RESGOV). Detailed feedback on the policy questionnaire from TIP and CSTP experts and delegates is gratefully acknowledged.

Chapter 10, "Technology governance and the innovation process", was prepared by Sebastian Pfotenhauer of the Technical University of Munich and David Winickoff (DSTI). It is based on recent OECD Working Party on Biotechnology, Nanotechnology and Converging Technologies activities. This chapter benefited from the contribution of David Guston of Arizona State University.

Chapter 11, "New approaches in policy design and experimentation", was prepared by Piret Tõnurist of the OECD Directorate of Public Governance and Territorial Development. It is based on recent OECD Observatory of Public Sector Innovation activities.

Chapter 12, "The digitalisation of science and innovation policy", was prepared by Fernando Galindo-Rueda, Michael Keenan, Daniel Ker and Dmitry Plekhanov (DSTI), based on work conducted by the CSTP and its Working Party of National Experts on Science and Technology Indicators (NESTI). This chapter benefited from the contribution of Clinton Watson of the New Zealand Ministry of Business, Innovation and Employment.

Chapter 13, "Mixing experimentation and targeting: Innovative entrepreneurship policy in a digitised world", was prepared by Carlo Menon of the DSTI Structural Policy Division, based on work conducted by the OECD Committee on Industry, Innovation and Entrepreneurship. This chapter benefited from the contribution of Marco Cantamessa of Politecnico di Torino University.

Chapter 14, "Blue Sky perspectives towards the next generation of data and indicators on science and innovation", was prepared by Fernando Galindo-Rueda (Economic Analysis and Statistics Division in DSTI) developing further a previous summary of the main messages arising from the OECD Blue Sky Forum held in Ghent 2016, an event organised by the OECD Working Party of National Experts on Science and Technology Indicators (NESTI).

All the chapters of the 2018 STI Outlook were reviewed by Sarah Box, Dominique Guellec, Dirk Pilat and Andrew Wyckoff of the DSTI. The team thanks them for their valuable comments and guidance.

The overall publication owes much to Sylvain Fraccola who designed the infographics, and Blandine Serve for her statistical support. Thanks to Fernando Galindo-Rueda, Silvia Appelt, Hélène Dernis and Brigitte Van Beuzekom (DSTI) for their helpful statistical inputs. Giulia Ajmone Marsan (OECD Centre for Entrepreneurship), Andres Barreneche (DSTI) and Gernot Hutschenreiter (DSTI) also contributed.

The authors are grateful to Florence Hourtouat and Beatrice Jeffries for their secretarial assistance. Special thanks to Janine Treves of the OECD Public Affairs and Communication Directorate for her guidance, and to Romy de Courtay for her editorial contributions and bibliographic research. Their engagement had a significant effect on the publication's overall quality.

Table of contents

Preface 3

Foreword 5

Acknowledgements 7

Abbreviations and acronyms 19

Executive summary 21
    Digitalisation is changing innovation and science practices 21
    STI policy and governance are becoming more mission-oriented 22

Infographic: Adapting to technological and societal disruption 23

Chapter 1. An introduction to the STI Outlook 2018 25
    Introduction 26
    What are the economic, societal and technological drivers of STI policy changes? 26
    How are technological and societal change transforming innovation processes? 29
    How is science evolving to become more open, automated and gender-friendly? 33
    How is STI policy responding to societal and technological disruptions? 38
    How is STI governance adapting to a fast-changing context? 44
    Conclusion 48
    Note 49
    References 49

Chapter 2. Artificial intelligence and the technologies of the Next Production Revolution 51
    Introduction 52
    Production technologies: Recent developments and policy implications 52
    Selected cross-cutting policy issues 64
    Conclusion 69
    Notes 69
    References 70

Chapter 3. Perspectives on innovation policies in the digital age 75
    Introduction 76
    Changes in innovation characteristics induced by the digital transformation 76
    Changes in market structures and dynamics 80
    Implications for innovation policies 83
    The future of innovation policies in the digital context 90
    Notes 92
    References 92

Chapter 4. STI policies for delivering on the Sustainable Development Goals 95
    Introduction 96

The need to reset overarching STI policy frameworks 97
The strategic orientation of research towards the SDGs 99
Interdisciplinarity and greater inclusivity 105
The international STI co-operation imperative 106
Linking development aid and STI policies 107
Changing STI governance for sustainability transitions 110
The promise of digitalisation 114
Future outlook 116
Notes 117
References 117

Chapter 5. Artificial intelligence and machine learning in science 121

Introduction 122
Technological drivers are behind the recent rise of AI 123
Why AI in science matters 124
Human-AI interaction 127
AI across scientific domains 128
Using AI to select experiments 129
Explainability: What does it imply in the context of science? 130
A key policy concern: Gaps in education and training 131
A vision of AI and the future of science 131
Conclusion 133
Notes 134
References 134

Chapter 6. Enhanced access to publicly funded data for STI 137

Introduction 138
Public data for STI: An overview 138
Policy challenges to promoting enhanced access to data 145
Future outlook 157
Notes 158
References 159

Chapter 7. Gender in a changing context for STI 163

Gender equity: A persistent science, technology and innovation (STI) policy imperative 164
Childhood and gender stereotypes 167
Higher education 169
Careers in the research system 172
The changing context for STI and the importance of diversity 176
Future vision and how to achieve it 178
Notes 180
References 181

Chapter 8. New trends in public research funding 185

Introduction 186
Recent changes in research-funding 187
An analytical framework of funding instruments 189
The purpose fit of research-funding instruments 194
Advancing the research-funding agenda 198
A forward-looking view on research funding 201

Notes 202
References 202

Chapter 9. The governance of public research policy across OECD countries 205

Introduction 206
HEIs and PRIs in national STI strategies 207
Institutions allocating funding and evaluating performance 209
Autonomy of HEIs and PRIs 214
Stakeholder involvement in policy decision-making 215
Future outlook 218
Notes 219
References 219

Chapter 10. Technology governance and the innovation process 221

Embedding governance in innovation processes 222
Reframing governance as integral to the innovation process 223
Three instruments for process governance in innovation 227
Policy implications 233
Notes 235
References 236

Chapter 11. New approaches in policy design and experimentation 241

Introduction 242
Reaping the benefits of design thinking 243
Creating collective intelligence 245
Exploring the promises of behavioural insights 247
Experimenting with new STI policy approaches 248
Building government platforms 250
Anticipating disruptive change 251
Adopting systems thinking in STI policy making 252
Embracing new skills and capacities 254
Future outlook for STI policy design 255
Notes 256
References 257

Chapter 12. The digitalisation of science and innovation policy 261

Introduction 262
The DSIP landscape: A brief overview 262
The promises of DSIP 265
Main policy challenges 268
Future outlook 276
Notes 277
References 277

Chapter 13. Mixing experimentation and targeting: innovative entrepreneurship policy in a digitised world 279

Balancing targeting and experimentation: new developments in policy support to innovative
entrepreneurship and venture capital 280
How has innovative and high-growth entrepreneurship evolved in recent years? 281
Why and how should innovative and high-growth entrepreneurship be publicly supported? 282

Do big data and machine learning applications open new avenues to identify high growth potential start-ups? 285
How to target policies toward the identified high growth potential start-ups? 287
Is government venture capital an effective instrument to select and support high growth start-ups? 288
Future outlook: toward new balances between targeted and non-targeted policies? 292
Notes 293
References 294

Chapter 14. Blue Sky perspectives towards the next generation of data and indicators on science and innovation 297

Introduction 298
New perspectives on the policy use of STI data, statistics and analysis 300
Changing demands for STI data from a more sophisticated user base 302
Digitalisation: The expanding frontier for STI data and statistics 305
Perspectives for producers of STI data and statistics 307
Policy and governance perspectives for STI measurement 310
Conclusions and future outlook 311
References 313

Tables

Table 4.1. How the digital transformation can help achieve the SDGs: some examples 115
Table 6.1. Overview of FAIR principles 142
Table 7.1. Gender issues and STI 165
Table 8.1. Funding-instrument design parameters to fit specific purposes 195
Table 9.1. Common characteristics across OECD countries 207
Table 9.2. Performance contracts in Austria, Finland and Scotland 213
Table 10.1. Process governance in three policy instruments 228
Table 11.1. Traditional public-sector context versus design thinking 244
Table 12.1. Examples of interoperability enablers in DSIP and related systems 271

Figures

Figure 1.1. Trends in business R&D financed by businesses and government 32
Figure 1.2. Direct government funding and tax support for business R&D, 2015 and 2006 33
Figure 1.3. Government budget allocations for civil R&D, 2000-08 and 2008-17 41
Figure 1.4. Average annual growth of total government budgets and GBARD, 2009-16 43
Figure 1.5. Change in the share of government in the direct funding of gross domestic expenditure on R&D, 2009-16 (or latest year available) 44
Figure 2.1. An overview of policies affecting advanced production 65
Figure 3.1. Characteristics of innovation in the digital age 77
Figure 3.2. Policy issues and instruments requiring change to be effective in the digital age 84
Figure 3.3. Eight principles for innovation policies in the digital age 85
Figure 4.1. The SDGs 96
Figure 4.2. Growing societal concerns are changing balances in public R&D budgets 100
Figure 4.3. Main criteria for funding – competitive research grants 102
Figure 4.4. Breakdown of STI initiatives by targeted societal challenges, 2018 103
Figure 4.5. Percentage of STI policy instruments directed towards societal challenges 104

Figure 4.6. Promoting social and technological innovation in developing countries: The approach of Grand Challenges Canada 108
Figure 4.7. STI inputs to the SDG process 113
Figure 5.1. Hypothesis-driven closed-loop learning 126
Figure 6.1. An assessment of the relevance of the OECD principles concerning access to research data from public funding 141
Figure 6.2. Creating a value proposition for data repositories 154
Figure 7.1. Percentage of new female students entering tertiary education, by selected fields of education, 2015 170
Figure 7.2. Doctorate holders in the working-age population, 2016 171
Figure 7.3. Women researchers as a percentage of total researchers in each sector (headcount) 174
Figure 7.4. Gender inequality in research careers: A system-dynamic model 179
Figure 8.1. Components of GERD financed by government, OECD, 2005-2015 188
Figure 8.2. GUF as a percentage of civil GBARD, 2016 191
Figure 8.3. Classification of research-funding instruments by intensity, granularity and assessment type 193
Figure 8.4. Most frequently stated desired effects of research funding 194
Figure 8.5. Research funding in a policy context 200
Figure 9.1. Four core dimensions that shape the policy mix 206
Figure 9.2. Quantitative targets included in national STI strategies 208
Figure 9.3. Number of public agencies in charge of project-based funding allocations in countries with agencies in place 210
Figure 9.4. Year of introduction of performance contracts and shares of HEI institutional block funding involved 212
Figure 9.5. Autonomy of HEIs across the OECD-34 215
Figure 9.6. Who formally participates in the research and innovation council? 216
Figure 9.7. Who formally participates in public university boards? 217
Figure 10.1. Three imperatives of a process-based approach to governance 224
Figure 10.2. Upstream governance in the innovation process in three instruments 234
Figure 11.1. Predictiv's Approach 248
Figure 11.2. Determinants of innovation in the public sector 254
Figure 11.3. Six core skills for public-sector innovation 255
Figure 12.1. A stylised conceptual view of a DSIP initiative and its possible main components 264
Figure 12.2. Main challenges facing DSIP initiatives 268
Figure 12.3. Challenges in implementing and using DSIP systems 269
Figure 12.4. Types of information harnessed for DSIP systems 269
Figure 12.5. Use of interoperability enablers in DSIP systems 271
Figure 13.1. Venture capital investments over time in Europe and in the United States 282
Figure 13.2. The two approaches to encourage innovative start-ups 284
Figure 13.3. Investments in rapidly expanding technologies: public-mixed and private VC 290

Boxes

Box 1.1. Selected megatrends affecting STI 27
Box 1.2. Key science systems trends and issues 34
Box 1.3. Targeting entrepreneurship support on high-growth potential firms 48
Box 2.1. Recent Applications of AI in Production 53
Box 2.2. In my view: AI and digitalisation for workforce training and assistance 54
Box 2.3. Blockchain : Potential applications in production 59
Box 2.4. Getting supercomputing to industry: Possible policy actions 66

Box 2.5. A new computing regime: The race for quantum computing..67
Box 3.1. In my view: GDP and well-being in the digital economy 80
Box 3.2. STI strategies aiming to achieve digital transformation..89
Box 3.3. In my view: Digitalisation and innovation policy 91
Box 4.1. In my view: The progressive evolution of innovation policy towards societal challenges.....98
Box 4.2. How are countries orienting their STI funding and policies towards societal challenges and the SDGs?..102
Box 4.3. Better data to enable STI roadmapping: the case of the IEA "Tracking Clean Energy Progress Tool"...105
Box 4.4. In my view: Technology deployment for the SDGs..109
Box 4.5. Independent scientific advice for monitoring implementation of the SDGs 111
Box 5.1. What is AI? 122
Box 5.2. ML and (deep) neural networks: What are they?..124
Box 5.3. Applications of AI in different scientific fields Type 128
Box 5.4. In my view: Moderating expectations: What deep learning can and cannot do yet.............132
Box 5.5. AI and the laws of nature 133
Box 6.1. International policy initiatives to promote sharing of research data 142
Box 6.2. Instruments concerning data access reported in the 2017 EC/OECD STI Policy Survey.....144
Box 6.3. In my view – Trust is the key to unlocking data 147
Box 6.4. In my view: Greater clarity in intellectual property (IP) and data-management policies can contribute to promoting open-data practice 151
Box 6.5. Data skills..156
Box 6.6. Possible future scenarios for access to data for STI 158
Box 7.1. What are countries doing to address gender equity in STI?..165
Box 7.2. Overcoming gender stereotypes 168
Box 7.3. Tracking research careers..172
Box 7.4. Overcoming gender bias in decision-making and performance evaluation in STEM...........175
Box 7.5. Gender-blind STI...176
Box 8.1. How has public funding of R&D evolved in recent years?...187
Box 8.2. Measuring national patterns of research funding allocation..190
Box 8.3. Examples of institutional funding supporting strategic/targeted research.............................197
Box 8.4. In my view : A systems world needs systemic thinking about research funding..................199
Box 10.1. Definition of technology governance 223
Box 10.2. In my view: Professor David Guston on "anticipatory governance" 225
Box 10.3. Definitions of RRI in countries of the European Union..226
Box 10.4. Deliberative agenda-setting: Two examples 229
Box 10.5. Test beds: Testing new governance modes for emerging technologies 231
Box 11.1. Government adoption of design toolkits and standards 244
Box 11.2. Mexico City's Mapaton initiative 246
Box 11.3. Predictiv : Online behavioural experiments platform 248
Box 11.4. Central government support for experimentation 249
Box 11.5. Australian Trade Mark Search 250
Box 11.6. GaaP and the case of eResidency in Estonia...251
Box 11.7. Regulating the sharing economy: The experience from Canada..252
Box 11.8. Systems thinking and the public sector 253
Box 12.1. Examples of DSIP systems 265
Box 12.2. In my view: Are science policy makers ready to embrace the digital revolution?..............267
Box 13.1. Machine learning, econometrics, and economics..286
Box 13.2. Using Crunchbase for economic and managerial research..289

Box 13.3. In my view: When innovation waterworks are clogged, we should remove obstructions downstream 291
Box 14.1. The OECD Blue Sky Forum on STI data and indicators 298
Box 14.2. Defining and measuring innovation 299
Box 14.3. Selected senior policy makers' perspectives from the Blue Sky Forum 2016 301
Box 14.4. Blue Sky messages for future OECD work on STI data and indicators 312

Follow OECD Publications on:

　　　http://twitter.com/OECD_Pubs

　　　http://www.facebook.com/OECDPublications

　　　http://www.linkedin.com/groups/OECD-Publications-4645871

　　　http://www.youtube.com/oecdilibrary

　　　http://www.oecd.org/oecddirect/

This book has...　　　　　　　　　　*StatLinks*
A service that delivers Excel® files from the printed page!

Look for the *StatLinks* at the bottom of the tables or graphs in this book. To download the matching Excel® spreadsheet, just type the link into your Internet browser, starting with the *http://dx.doi.org* prefix, or click on the link from the e-book edition.

Abbreviations and acronyms

AI	Artificial intelligence
CAD	Computer-aided design
CASRAI	Consortia Advancing Standards in Research Administration Information
CERIF	Common European Research Information Format
CIMULACT	Citizen and Multi Actor Consultation on H2020
CRIS	Current research information systems
DOI	Digital Object Identifier
DSA	Data-sharing agreement
DSIP	Digital science and innovation policy
EAD	Ethically aligned design
ERB	Ethics review body
EU	European Union
EUR	Euro
GaaP	Government as a Platform
GBAORD	National government budget appropriations
GBP	British pounds
GDP	Gross domestic product
GERD	Gross domestic expenditure on research and development
GPT	General-purpose technology
GPU	Graphic processing unit
GRID	Global Research Identifier Database
GUF	General university funds
HEI	Higher education institution
H2020	Horizon 2020
HPC	High-performance computing
ICT	Information and communication technology
ID	Identification
IPR	Intellectual property rights
ISNI	International Standard Name Identifier
IoT	Internet of things
LIME	Local interpretable model-agnostic explanations
MAM	Metals-based additive manufacturing
MOOC	Massive open online course
NESTI	OECD Working Party on Science and Technology Indicators
nm	Nanometres
NPM	New public management
NPR	Next Production Revolution
NSO	National statistical organisation
OAI-PMH	Open Archives Initiative Protocol for Metadata Harvesting
OAIS	Open Archival Information System
OECD	Organisation for Economic Co-operation and Development
ORCID	Open Researcher and Contributor ID
PI	Principal investigator
PREF	Public Funding of Research
PRI	Public research institution

QSAR	Quantitative structure activity relationship
REI	Research excellence initiative
R&D	Research and development
SbD	Safety by Design
SCALINGS	Scaling up Co-creation: Avenues and Limits for Integrating Society in Science and Innovation
SDGs	Sustainable Development Goals
SME	Small and medium-sized enterprise
SNA	System of National Accounts
S&T	Science and technology
STEM	Science, technology, engineering and mathematics
STI	Science, technology and innovation
TOP	Transparency and Openness Promotion
UPPIs	Unique, persistent and pervasive identifiers
USD	US dollar
VC	Venture capital
ZB	Zettabytes

Executive summary

Science, technology and innovation (STI) activities face several disruptive drivers of change. These include the ongoing slowdown in productivity growth, despite widespread technological change; rapidly ageing populations; the impacts of climate change, and the resulting need for mitigation and adaptation; and globalisation and the growing role of emerging economies. These drivers create opportunities and challenges for STI. They shape societal and policy expectations regarding the purposes of STI, and they affect the ways STI activities are carried out. Many of these drivers give rise to "grand societal challenges", for example, around healthy ageing, clean energy and food security. Challenges like these are also encapsulated in the Sustainable Development Goals (SDGs), which feature increasingly prominently in STI policy agendas.

If well-managed and used in conjunction with social innovation and policy reforms, scientific and technological advances can alleviate many of these challenges. Gene editing could revolutionise today's medical therapies, nanomaterials and bio-batteries could provide new clean energy solutions, and artificial intelligence (AI) could become an important drug discovery tool over the next decade.

But while new technologies like AI and gene editing present great opportunities, they could also lead to considerable harm, if used inappropriately. Preventing, correcting or mitigating such negative effects has become more important – yet more difficult – as technology has become more complex and widespread. The speed and uncertainty of technological change challenge policymakers to exert sufficient oversight of emerging technologies.

Governments therefore need to become more agile, more responsive, more open to stakeholder participation and better informed. Some governments are already experimenting with new anticipatory and participatory approaches to policy design and delivery, but such practices have yet to be adopted widely in STI policymaking.

Digitalisation is changing innovation and science practices

Digitalisation is transforming innovation processes, lowering production costs, promoting collaborative and open innovation, blurring the boundaries between manufacturing and service innovation, and generally speeding up innovation cycles. Data have become a main input to innovative activities, and many innovations are embodied in software or data. This has implications for policy support to business innovation, which (among other things) needs to ensure broad access to data.

Digitalisation is providing new opportunities to engage stakeholders at different stages of the innovation process. Several open, co-creative and socially responsive practices are emerging. Most countries now feature dedicated sites for inclusive innovation, such as maker spaces, living labs and fab-labs, that support the activities of potential "non-traditional" innovators. Established firms are also engaging in more inclusive innovation. For example, practices such as value-based design and standardisation are beginning to emerge; these could become powerful tools for translating and integrating core social values, safeguards and goals into technology development.

Research is becoming increasingly data-intensive. Enhanced access to data promises many benefits, including new scientific breakthroughs, less duplication and better reproducibility of research results, improved trust in science and more innovation. Governments have a role to play in helping science cope with the challenges of open science: by ensuring transparency and trust across the research community and wider society; by enabling the sharing of data across national and disciplinary boundaries; and by ensuring that recognition and rewards are in place to encourage researchers to share data.

AI and machine learning have the potential to increase the productivity of science, enable novel forms of discovery and enhance reproducibility. AI systems have very different strengths and weaknesses compared to human scientists, and are expected to complement them. However, several challenges hinder the widespread use of AI in science, such as the need to transform and transpose AI methods to operate in challenging and varying conditions; concerns regarding the limited transparency of machine learning-based decision-making; the limited provision of specific education and training courses in AI; and the cost of computational resources for leading-edge AI research.

STI policy and governance are becoming more mission-oriented

In line with the SDGs, governments are seeking to redirect technological change from existing trajectories towards more economically, socially and environmentally beneficial technologies, and to spur private STI investments along these lines. This shift has given impetus to a new era of "mission-oriented" STI policy, with governments looking to work more closely with the business sector and civil society to steer the direction of science and technology towards ambitious, socially relevant goals.

However, current trends in public research and development (R&D) spending may not be commensurate with the corresponding ambition and challenges delineated in mission-oriented policies. Since 2010, government R&D expenditures in the OECD as a whole and in almost all Group of Seven countries have stagnated or decreased, not only in absolute amounts and relative to gross domestic product, but also as a share of total government expenditure. The share of government in total funding of R&D decreased by 4 percentage points (from 31% to 27%) in the OECD area between 2009 and 2016. Although this decrease has been compensated in many countries by an increase in R&D tax credits, governments may still find it difficult to steer research and innovation activities in desired strategic directions.

Significant gender imbalances in science and innovation also remain, at a time when workforce diversity is urgently needed to address the SDGs. Deep-rooted structural factors, including gender stereotypes and research career paths that are inimical to family life, are largely to blame. Most countries have included gender diversity as a key objective in their national STI plans. However, policy initiatives remain fragmented, and a more strategic and systemic long-term approach is necessary.

Governments could benefit from embracing digital technologies in the design, implementation and monitoring of STI policies. Tools such as big data, interoperability standards and natural-language processing can provide governments with more granular and timely data to support policy formulation and design. By linking different datasets, these tools can transform the evidence base for STI policy, and help demonstrate the relationships between science and innovation expenditures and real-world outcomes. Monitoring the contribution of STI to the global and multidimensional SDGs is particularly challenging, and will require new developments in statistics and indicators.

Infographic: Adapting to technological and societal disruption

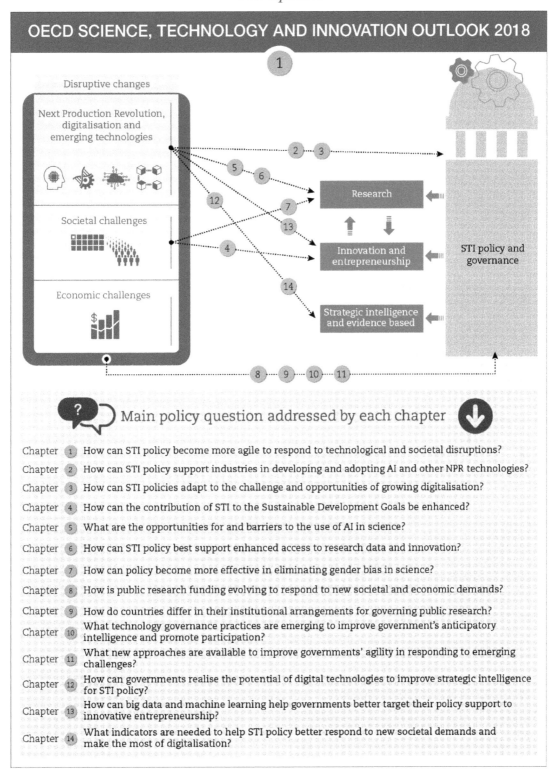

Chapter 1. An introduction to the STI Outlook 2018

This chapter introduces the 2018 edition of the Science, Technology and Innovation (STI) Outlook, distilling the main trends and policy issues from across the chapters into a few key highlights. It is organised into five main sections, starting with the drivers of change disrupting research and innovation and STI policy. Subsequent sections explore their impacts on innovation processes and scientific practices, and raise the question of how STI policy and governance practices can adapt to opportunities and challenges in a fast-changing context.

The statistical data for Israel are supplied by and under the responsibility of the relevant Israeli authorities. The use of such data by the OECD is without prejudice to the status of the Golan Heights, East Jerusalem and Israeli settlements in the West Bank under the terms of international law.

Introduction

Developments in science, technology and innovation (STI) are major drivers of change in modern societies. They are themselves subject to various influences – including a range of societal, economic and technological factors – that shape their activities and outcomes. Public policy is another important influence on STI, because of its funding and regulatory functions. Like its subject matter, STI policy is also subject to multiple influences on its agenda, design and implementation.

Four major trends influencing the direction and design of STI policy stand out. First, support programmes for public research and innovation face growing demand to demonstrate economic and societal relevance and impact. In particular, STI policy increasingly focuses on "challenges", as governments seek to redirect technological change from existing trajectories towards more economically, socially and environmentally beneficial technologies, and to spur complementary private STI investments. This shift has given impetus to a new era of "mission-oriented" STI policy, with governments looking to work more closely with the business sector and civil society to steer the direction of science and technology towards specific goals.

Second, digitalisation is transforming science and innovation processes. Data have become a main input to innovative activities, and many innovations occur in software or data. Aspects of innovation are also accelerating as digital technologies shorten the time needed to perform some tasks. All areas of research are also becoming data-intensive, increasingly relying upon and generating big data. These changes have great potential to improve the productivity of innovation and science, but they require an adaptation of STI policies.

Third, many OECD governments have limited fiscal room for manoeuvre as they seek to reduce their debt burdens. As the latest data available show, current trends in government research and development (R&D) funding in the OECD area may not match the ambition and challenges inherent to mission-oriented policies. Under these conditions, it might be difficult for governments to make the investments in research and innovation activities needed to steer the direction of science and technology.

Fourth, governments can benefit from embracing digital technologies to design, implement and monitor STI policies. Digitalisation is already having a significant impact on the evidence base for STI policy and governance. The growing use of digital tools in research and innovation processes leaves more "digital traces", i.e. digital data that can be used to produce indicators and analysis. Exploiting these traces will provide governments with more granular and timely data, to inform and improve science and innovation policies. Digitalisation can also help meet policymakers' need to demonstrate the relationships between science and innovation expenditures and real-world outcomes.

This chapter introduces the *Science, Technology and Innovation Outlook 2018*, distilling the main trends and policy issues highlighted in the report. As such, it remains largely within the limits set by the chapters' content, and does not aim to survey all of the main trends and issues affecting STI and STI policy today.

What are the economic, societal and technological drivers of STI policy changes?

Combining policy action to address rising economic and societal challenges

The 2016 edition of the STI Outlook described several megatrends that are expected to have a strong impact on research and innovation systems over the next 10-15 years and beyond (Box 1.1). These megatrends are quite slow-moving, which means they remain useful points of reference for thinking about economic, societal and political challenges that STI and STI policy will have to contend with (OECD, 2016). Many are addressed by the Sustainable Development Goals (SDGs) and articulated as "grand societal challenges" that increasingly shape STI policy agendas. At the same time, as these megatrends play out, the variety and degree of uncertainty they generate has unleashed reactionary forces in some countries that challenge much of the post-Second World War economic, political and social consensus.

Box 1.1. Selected megatrends affecting STI

Demography: The world population will continue to grow and is expected to nudge the 10 billion mark by the middle of the 21st century. Africa will account for more than half of this growth, with a significant increase in the number of the continent's young people. In other parts of the world – including in many developing countries – populations will significantly age: the number of people over the age of 80 will account for around 10% of the world's population by 2050, up from 4% in 2010. With a declining share of the population in work, ageing countries will face an uphill battle to maintain their living standards. International migration from countries with younger populations could offset this decline, although it will likely meet resistance. Technologies that enhance physical and cognitive capacities could allow older people to work longer, while growing automation could reduce the demand for labour. Driven by this demographic increase and by the growing number of people living in large cities, the global population will be increasingly urban, with 90% of this growth occurring in Asia and Africa.

Natural resources and energy: The growing population, coupled with economic growth and climate change, will place considerable burdens on natural resources. Severe water stress is likely in many parts of the world, while food insecurity will persist in many predominantly poor regions, exacerbated by climate change. Energy consumption will also rise sharply, contributing to further climate change, absent significant uptake of renewables. Global biodiversity will come under increasing threat, especially in densely populated poorer countries.

Climate change and environment: Mitigating the considerable extent and impacts of climate change will require setting ambitious targets reducing greenhouse gas emissions. The latest report from the Intergovernmental Panel on Climate Change (IPCC, 2018) points out that several critical climate-change impacts could be avoided by limiting global warming to 1.5 degrees Celsius (°C) compared to 2°C, but that this will require rapid, far-reaching and unprecedented changes throughout society. Ambitious targets for waste recycling are also needed, implying the need for a major shift towards a low-carbon "circular economy" by mid-century. This shift will affect all parts of the economy and society, and will be enabled by technological innovation and adoption in both developed and developing economies.

Globalisation: The world economy's centre of gravity will continue to shift eastward and southward, and new players – including governments, certain non-state actors (such as multinational enterprises and non-governmental organisations), and newly emerging megacities – will wield more power. Many of these shifts in power and influence are driven and facilitated by globalisation, which operates through cross-border flows of goods, services, investment, people and ideas, and is enabled by widespread adoption of digital technologies. However, globalisation will inevitably face counter-currents and crosswinds,

> such as geopolitical instability, possible armed conflict and new barriers to trade stemming from increased protectionism.
>
> *Health, inequality and well-being*: The treatment of the infectious diseases that affect the developing world disproportionately is being compromised by growing antibacterial resistance. Non-communicable and neurological diseases are projected to increase sharply, in line with demographic ageing and the global spread of unhealthy lifestyles. Technological advances in DNA sequencing, omics technologies, synthetic biology and gene editing have given researchers new tools to decipher and treat chronic non-communicable diseases. Inequalities and poverty remain a concern in many developed countries, although global poverty continues to decline.
>
> *Source*: Adapted from OECD (2016), OECD Science, Technology and Innovation Outlook 2016, https://doi.org/10.1787/sti_in_outlook-2016-en.

The necessary investments to address these challenges will likely be made in a difficult economic context. According to a recent long-term baseline scenario prepared by the OECD, annual global growth is estimated to slow from 3.5% currently to 2% in 2060 (Guillemette and Turner, 2018). Productivity growth has fallen over the past two decades, especially since the 2008 global financial crisis. This trend, combined with low or declining multi-factor productivity growth in several countries and sectors, has raised concerns about the ability of research and innovation activities to support economic growth and social well-being. Scholars continue to debate the reasons for the slowdown. Some point to slower rates of innovation, which is the root of productivity. Others point to the historical time lag between innovation and its impacts on productivity. They argue that the productivity crisis will end as businesses learn, relevant structural reforms are implemented, complementary investments are made, and recent innovations are broadly adopted and adapted beyond lead innovators. Another potential explanation for some part of the productivity slowdown is mismeasurement of the increasingly digital economy.

Solving increasingly pressing societal challenges at a time when financial resources in several OECD member countries are limited, and the growth engine seems to be stalling, will be difficult. It will require combined actions to solve economic, societal and environmental challenges. With the right policies and incentives in place – notably strong fiscal and structural reform, combined with coherent climate policy – governments can generate growth that will significantly reduce the risks of climate change, while also providing near-term economic, employment and health benefits. For instance, it is estimated that a climate-compatible policy package, including enhanced incentives for innovation, could increase long-run gross domestic product (GDP) by up to 2.8% on average across the Group of Twenty (G20) in 2050 (OECD, 2017a).

Societal and environmental challenges will need to be addressed in a difficult economic context, characterised by low global growth and productivity. Combined policy actions could help solve economic, societal and environmental challenges. Fuelled by innovation, an effective policy to combat climate change could also generate significant growth and new jobs, and enhance well-being.

New emerging technologies hold great potential

New emerging technologies can help address many grand societal challenges. Building on earlier OECD work (OECD, 2017b), Chapter 2 on artificial intelligence (AI) and the technologies of the Next Production Revolution provides many examples of emerging technologies with wide-ranging future applications. For example, gene editing could revolutionise today's medical therapies; nanomaterials and bio-batteries could provide new clean-energy solutions; and AI could become the "primary-drug discovery tool" over the next decade. In the medium term, some technologies now at the demonstration stage could have significant impacts. For instance, new generations of bio-refineries, which transform biomass-waste products into marketable products and energy, have the potential to substantially reduce greenhouse gas emissions.

If well managed and used in conjunction with social innovation and policy reforms, scientific and technological advances have the potential to significantly alleviate many grand societal challenges.

Some of these technologies are already being applied. For example, AI, enabled by ongoing improvements in computer hardware, the widespread availability of large datasets and improved software, has growing applications in various areas of production, from semi-conductors and pharmaceuticals, to more traditional "bricks-and-mortar" industries like mining and construction. However, as Chapter 2 shows, the diffusion of AI and other advanced production technologies is by some measures quite slow, and policies that facilitate diffusion could benefit productivity growth.

Blockchain technology has recently attracted much attention. Together with robotics (e.g. using software robots for process automation) and AI (e.g. for detecting data anomalies and identifying process vulnerabilities), blockchain could significantly change key financial services, from financial transactions to automated contractual agreements. Blockchain technology was first applied in cryptocurrency markets. However, many other applications (e.g. remittances, inter-bank transfers and securities trading) are now emerging in the financial sector (OECD, 2017c) and, as Chapter 2 shows, blockchain is beginning to play roles in production as well. Furthermore, technological convergence – for instance, the combination of technologies such as the Internet of Things (IoT), blockchain, AI and advanced robotics – may open new production frontiers.

How are technological and societal change transforming innovation processes?

The very characteristics of the innovation process are changing as a result of technological opportunities (particularly stemming from the digital transformation of the economy), as well as societal pressures and growing aspirations for more inclusiveness and openness. These changes are unfolding in a more favourable business environment: firms have resumed their R&D investment since the financial crisis, fuelled by restored profitability and the increasingly generous R&D fiscal incentives offered by many governments.

Digitalisation is creating new opportunities for innovation and knowledge exchange

Chapter 3, on innovation policies in the digital age, analyses the impacts of digitalisation on innovation processes. Key phases of the innovation cycle are becoming faster and cheaper. The costs of searching, verifying, manipulating and communicating information and knowledge, as well as the costs of launching innovative goods and services in the market, are falling. Aspects of innovation are also speeding up as competition increases and digital technologies allow some tasks to be performed more quickly, for instance, in design and testing. The growing availability of digital data on customers' needs, and the ability to experiment more easily with data on different customer groups, also helps streamline product and process innovation. As digital technology significantly lowers the cost of versioning, products can be differentiated (and even personalised) much further. As a result, the product cycle can be accelerated, changing the speed of market competition.

Data have become a major input to innovation: basic data on the characteristics of materials or the environment, or on customer demand, can be used to identify a product's optimal features and create "digital twins" of machinery and physical goods, allowing deeper forms of process optimisation. Access to data has become a key parameter in business strategies: companies that control valuable and unique data have a competitive advantage over others. Contrary to physical inputs, data can be re-used and shared, creating new opportunities for collaboration between businesses.

Innovation has also become more collaborative, thanks to both improved conditions on the supply side (data sharing) and stronger demand for collaboration, stemming from increased interdisciplinarity and engagement with a variety of stakeholders. Collaboration can take several forms, such as data sharing, open innovation, digital platforms, and mergers and acquisitions. Interactions along global value chains, which have become increasingly important since the early 2000s, are also affected by digital technologies, with important consequences for the redistribution of high value-added activities among countries (e.g. reshoring of highly automated activities; see De Backer and Flaig, 2017).

The digital transformation has also supported the emergence of new forms of policy support for knowledge transfer. For example, online platforms, networks and communities have emerged as new spaces for knowledge transfer, helping to match supply and demand for technology. They connect firms with global networks of public research centres, individual scientists and freelancers that can help solve specific technological problems. Enhanced options for electronic exchanges have also led to the creation of new models of "off-campus" technology transfer offices (TTOs), such as TTO alliances at the regional, national or sectoral level. These typically result from co-operation between several universities and public research institutes (PRIs), as in Germany (the regional patent agencies) and France (the technology transfer acceleration companies), for example. Pooling specific resources and services (e.g. patent databases and services, marketing and communication activities, and training and experts) often improves efficiency and the quality of services provided by TTOs. Given the broad variety and distribution of these developments, policymakers can play a useful role in promoting integration, co-operation and interoperability between the patchwork of existing and emerging initiatives (OECD, 2018a).

New policy and business practices for inclusive innovation are emerging

Innovation delivers far more than new or improved products and services that provide companies with a competitive edge and contribute to economic growth. Innovation can also be "inclusive", responding to the needs of a broader array of stakeholders. First, innovation can contribute to new or improved products and services for those at a social disadvantage. For example, innovation can provide lower-income groups with greater access to services such as long-distance calling, "e-learning'" and "e-government". The effect of technical progress on prices can also contribute to social inclusion: some information and communication technology (ICT) products, such as laptops and smartphones, have become increasingly affordable and available to a higher number of people. Second, the process of innovation itself can become more inclusive, as previously underrepresented individuals and social groups can now participate in it more easily (OECD, 2017d). Chapter 10 on technology governance presents some emerging business-innovation practices that are more open, co-creative and responsive to social needs. These practices sometimes offer opportunities for individuals and small groups to engage in digital production in dedicated small-scale sites, e.g. maker spaces, living labs and fab-labs. These local workshops are more accessible to potential "non-traditional" innovators – especially young innovators and independent inventors – and are often based on collaboration with universities and local authorities. Innovation can thus become a factor of social inclusion as participating groups develop new skills and broaden their range of opportunities.

Innovation can be inclusive. Several open, co-creative and socially responsive practices are emerging. Dedicated sites, such as maker spaces, living labs and fab-labs, are now found in most countries and support the activities of potential "non-traditional" innovators.

Established firms can also engage in inclusive innovation practices. Chapter 10 on technology governance identifies some more inclusive and open practices (e.g. design ethics) that firms are using at early stages in the innovation cycle. Although such practices are recent and still emerging, they could be powerful tools for translating and integrating core social values, safeguards and goals into technology development. In the field of nanotechnology, for instance, standardisation is seen not only as a means of facilitating commerce through interoperability, but also of promoting health and safety. For example, it can embed knowledge of potentially adverse effects in the design of nanomaterials and nanoproducts. In many initiatives, the value added lies as much in the result – i.e. "ethical" technologies and products – as in the process itself. In addition to their usual technical work, some standardisation committees (like the Institute of Electrical and Electronics Engineers) also operate as fora for public discussion on issues such as AI.

Government support for business R&D is shifting

Business firms have an essential role in developing, diffusing and using the new wave of technologies. This requires major investment in R&D, as well as in other complementary assets and intangibles in a wide variety of fields. The analysis of business R&D expenditures (BERD) shows that firms have taken up this challenge. BERD has picked up in many countries since the financial crisis and is almost back to its pre-crisis growth trend, both in volume and relative to GDP. This increase is driven by growth in aggregate demand and firms' restored profitability (Figure 1.1, Panel a). It is also driven by relatively new actors in the R&D field – mainly large firms in digital industries, which are investing massively in AI and other Next Production Revolution technologies.

Although the bulk of business R&D is financed by companies, public support helps incentivise these activities and focus them on certain public-policy priorities. Global trends in public support for business R&D are difficult to interpret, as policy approaches differ markedly across countries (Figure 1.1, Panel b). However, the share of BERD that is funded by government through direct support (such as grants) has dropped in all countries since the financial crisis, from 14.1% (2009) to 6.8% (2016) in the United States, and from 7.3% (2010) to 6.3% (2015) in the European Union.

Figure 1.1. Trends in business R&D financed by businesses and government

Index, 2000=100

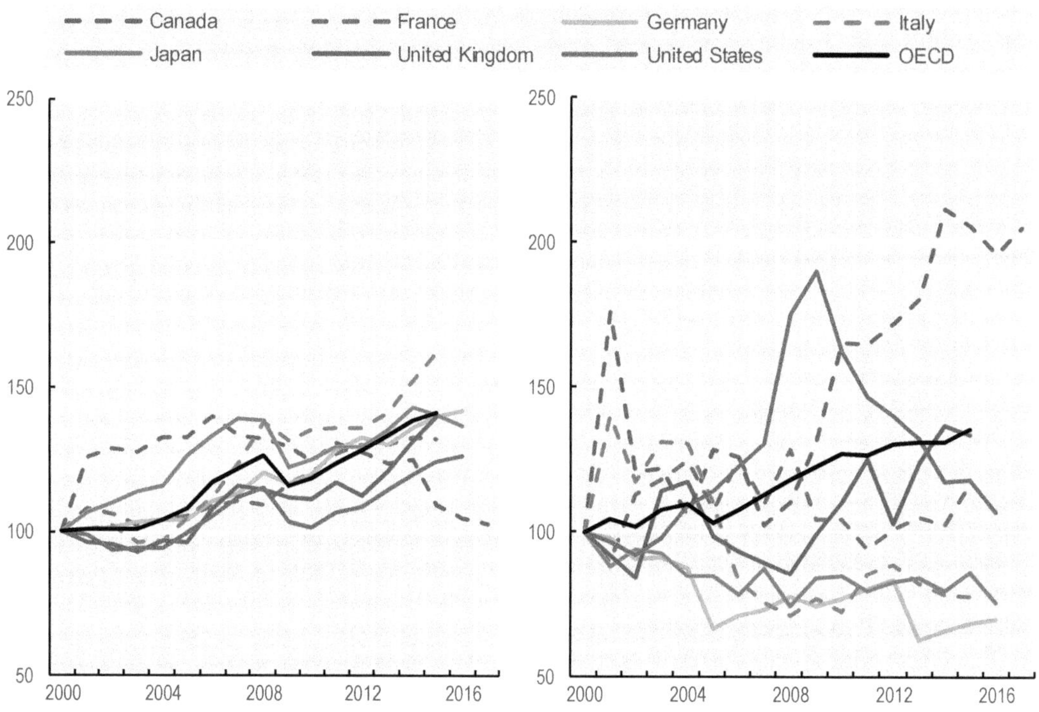

Note: The black line in panel b is BERD financed by government in OECD countries, less the United States.
Source: Calculations based on OECD (2018d), "Research and Development Statistics: Government budget appropriations or outlays for RD (Edition 2017)", OECD Science, Technology and R&D Statistics (database), https://doi.org/10.1787/e724dc33-en (accessed on 26 September 2018)

StatLink https://doi.org/10.1787/888933858069

However, this decrease in direct support for business R&D has been amply compensated by an increase in indirect support through tax incentives over 2006-14 (OECD, 2016, 2017e). When considering total (direct and indirect) government support for business R&D, a majority of countries (i.e. 29 out of the 41 countries for which data are available) increased their support to business R&D, relative to GDP, over 2006-15. This increase is particularly significant in countries where tax incentives account for a large share of total government support (Figure 1.2). It is often related to the reform of indirect support schemes for business R&D, to make them more available, accessible and generous; 12 OECD countries also introduced such schemes over 2000-15 (OECD, 2018b). Thus, the share of tax relief in total government support for business R&D in the OECD area increased on average from 36% to 46% between 2006 and 2015 (OECD, 2018c).

Figure 1.2. Direct government funding and tax support for business R&D, 2015 and 2006

As a percentage of GDP

[Bar chart showing direct government funding of BERD, tax incentive support for BERD, and total government support for BERD in 2006 as a percentage of GDP across countries including Russian Federation, Belgium, France, Ireland, Hungary, Korea, Austria, United States, United Kingdom, Australia, Norway, Slovenia, Canada, Iceland, Netherlands, Portugal, Japan, Czech Republic, Sweden, China, Denmark, Spain, Israel, Brazil, New Zealand, Italy, Greece, Finland, Germany, Turkey, Estonia, Mexico, Poland, Switzerland, South Africa, Lithuania, Slovak Republic, Chile, Latvia.]

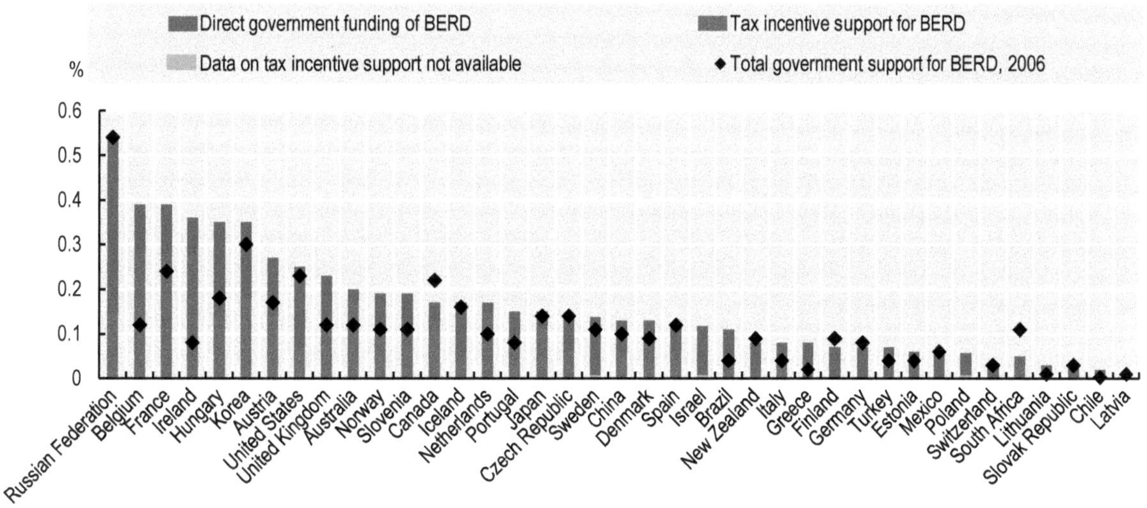

Source: OECD (2017e), OECD Science, Technology and Industry Scoreboard 2017: The digital transformation, https://doi.org/10.1787/9789264268821-en.

StatLink https://doi.org/10.1787/888933858088

Even though R&D tax incentives are considered more cost-efficient and easier to operate than subsidies and grants, they are exclusively allocated to business R&D that reflects market needs. By design, but also owing to specific regulations (e.g. the European Union's Community State Aid rules for R&D and innovation), it is difficult to designate specific fields of research or sectors that would most benefit from these indirect incentives. As a demand-driven policy tool, indirect public funding also offers little margin for governments to influence the amounts allocated, apart from (for example) "capping" the allocated credits. Thus, this shift in the policy mix raises the issue of governments' capacity to influence the direction of private R&D, at a time when achieving societal and environmental goals requires more – and more focused – innovation.

How is science evolving to become more open, automated and gender-friendly?

The 2016 edition of the STI Outlook provided a high-level overview of the main trends and issues set to shape science systems over the next 10-15 years (Box 1.2). While these are still valid, the opportunities and challenges they present continue to unfold, as do the policy responses. Some issues have gained importance in the last two years. These notably include the impact of digitalisation – which this edition of the STI Outlook covers extensively – and the "reproducibility crisis" in science, whereby a growing number of results in scientific publications are difficult or impossible to reproduce by other researchers. The accelerating rollout of open science also places greater emphasis on transparency in science: open-science principles stress open-access publication, open data sharing, and more open and inclusive participation in science itself. Another growing concern is how to support breakthrough research when faced with seemingly decreasing research productivity, and societal challenges of unprecedented scale and scope. It is difficult to demonstrate through data analysis or case studies that new ideas are becoming "harder to find" (Bloom et al., 2017; Jones, 2009). However, several research communities claim that competitive funding mechanisms disadvantage risky, potentially transformative and transdisciplinary research proposals in favour of applied, incremental and mono-disciplinary proposals.

Box 1.2. Key science systems trends and issues

The 2016 edition of the STI Outlook included a chapter on the future of science systems, featuring several key trends and issues that were expected to shape science systems over the next 10-15 years. These trends include:

- fiscal restraint and competing policy demands, placing pressure on government R&D spending

- the growing importance, in some research systems, of non-state funding for public research, including by philanthropists, charities and foundations

- the growing share of public research performed by emerging economies – particularly China, which is now second only to the United States in its overall public expenditure on R&D

- the re-orientation of public science agendas towards "grand societal challenges", with a growing emphasis on the SDGs as a framework for agenda-setting

- the turn towards more challenge-driven public research, placing more emphasis on interdisciplinary research and the interfaces between basic and applied research

- emerging new arrangements for commercialising public R&D, including new TTO-type structures and the use of smarter IP strategies in public research-performing organisations

- growth in citizen science, including "do-it-yourself" science

- greater consideration of the ethical, legal and societal aspects of research, within a framework of "responsible research and innovation"

- emerging new opportunities from the growing digitalisation of science (e.g. regarding automation, big data and more open science), but also significant challenges (e.g. regarding data ownership, conflicting incentives for open science, the costs of maintaining data infrastructures and the availability of skills)

- the growing precariousness of research careers in hyper-competitive research environments, and its negative impacts on certain groups (particularly women)

- shifts in the ways of assessing research performance to reflect the emergence of non-traditional bibliometrics ("altmetrics") and the greater use of public-value criteria to assess the contributions of research to societal challenges

- growing concerns about the "reproducibility crisis" in science

> - the growing gap between scientific evidence and other forms of knowledge and opinion, complicated by the global, multi-dimensional, fast-evolving and complex nature of many grand societal challenges.
>
> *Source*: OECD (2016), OECD Science, Technology and Innovation Outlook 2016, https://doi.org/10.1787/sti_in_outlook-2016-en.

Building upon the wide-ranging assessment of 2016, the 2018 edition of the STI Outlook examines three prominent topics in the current debates on research policy. The first topic is open science and enhanced access to research data, which has several potential benefits, but also faces significant challenges. The second topic is the impact of AI and automation on science, which has the potential to transform science practice over the next decade. The third topic is the long-standing under-representation of women in certain areas of science; while many policy interventions seek to address the issue of gender in science, much remains to be done.

Enhanced access to research data has many benefits

All areas of research are becoming increasingly data-intensive, and big data are no longer the prerogative of experimental physics and astronomy. Chapter 6 highlights the expected benefits of enhanced access to data, i.e. new scientific breakthroughs, less duplication and better reproducibility of research results, improved trust in science and more innovation. These benefits, however, should be balanced against the costs, including the need to protect privacy and security, and prevent malevolent uses. Accordingly, "as open as possible, as closed as necessary" is gradually replacing the "open-by-default" mantra associated with the early days of the open-access movement.

Enhanced access to data poses several outstanding policy challenges. First, governments need to put in place systems and processes to ensure transparency and foster trust across the research community and wider society. For example, while privacy breaches cannot be avoided, the risks should be managed, and the procedures to do this should be clear and transparent. Second, implementation of the FAIR Guiding Principles[1] for policy development and co-operation across communities depends on the development and adoption of a common technical framework (Wilkinson et al., 2016). Policymakers should therefore support bodies (such as the Research Data Alliance) that are building the social and technical infrastructure to enable open data sharing across national and disciplinary borders.

Third, appropriate recognitions and rewards need to be in place to encourage researchers to share data. Data activities should be embedded in evaluation systems to ensure that researchers who provide high-quality research data (including on negative results) are rewarded. Generalising data citation, so that it can be used to incentivise and reward data sharing, also requires new data-citation metrics.

Fourth, the substantial costs of data stewardship and provision entail a long-term financial commitment. Funding them requires understanding not only the business models and value propositions of specific data repositories, but also the research networks in which they are integrated. In some instances, it may make sense to centralise the management of data resources to obtain economies of scale across research systems.

Finally, the additional burden of curating and stewarding data to make them openly available for secondary use is a science-wide human-resource challenge, which will only be met through retraining existing personnel, and providing new education and training opportunities to researchers and professionals in research data-support roles. Data scientists are in high demand in industry, and academic research competes for the best talent – hence the urgent need to develop attractive career paths, to realise the value of enhanced access to public research data.

Government should help science cope with the challenge of open science. This includes ensuring transparency and trust across the research community and wider society; enabling the sharing of data across national and disciplinary boundaries; and ensuring that recognition and rewards are in place to encourage researchers to share data.

Automation could transform scientific practice

AI and machine learning have the potential to increase the productivity of science, enable novel forms of discovery and enhance reproducibility. AI in science has already predicted the behaviour of chaotic systems, tackled complex computational problems in genetics, improved the quality of astronomical imaging, and helped discover rules of chemical synthesis. Since AI systems have very different strengths and weaknesses compared to human scientists, they are expected to augment human abilities in science. Chapter 5 outlines three key technological developments driving the recent rise of AI: improved computer hardware, increased availability of data and improved AI software. Examples are rapidly accumulating of AI being applied across the entire span of scientific enquiry.

Broadening the use of AI in science faces several challenges. First, despite the impressive performance of AI in many areas, the need still exists to further develop AI methods that perform well in constrained and well-structured problem spaces so as to be able to apply these to scientific domains where data are noisy and corrupted and processes are only partially observed. This need exists in climate science, for instance, where the number of variables involved is vast, uncertainty exists on which feedback loops are important, and accurate measurement – although improving – is still a challenge. Creating approaches that work across all data scales – from data-sparse environments to data-rich contexts – will be key. Second, discussions of AI commonly cite the lack of transparency in machine learning-based decision-making as a source of possible concern. As Chapter 5 points out, questions of intelligibility are not confined to machine learning (only a few specialists understand the proofs involved in some leading areas of mathematics, for example). Some existing techniques provide audit trails of machine learning and can help explain its results. But the question of intelligibility is likely to become more salient as AI techniques are used more widely. Third, education and training is a key policy issue. Too few students are trained to understand the fundamental role of logic in AI; bridging this gap will require changes in curricula. Finally, the computational resources for leading-edge AI research are enormous, and can be expensive. The largest computer resources and the largest number of excellent AI researchers are found in the business sector, not in public science.

Several challenges hinder the widespread use of AI in science: the need to transform AI methods to operate in challenging data conditions; concerns regarding transparency in machine learning-based decision-making; the need for more, and more tailored, education and training in AI; and the cost of computational resources for leading-edge AI research.

Removing gender barriers in science requires more joined-up policy

Chapter 7 on gender in the changing context for STI reviews the key issues affecting gender equity in science at different life stages. Gender stereotypes influence educational choices and career expectations even in early childhood. In undergraduate and graduate education, women and men are also unevenly distributed across academic courses, with women significantly less represented in certain science, technology, engineering and mathematics (STEM) fields (particularly engineering, ICT, physics, mathematics and statistics). At the doctoral level, on the other hand, the share of women in certain STEM fields has increased over time, and the "leaky pipeline" between graduate and postgraduate education and training is no longer a major challenge (Miller and Wai, 2015). In research careers, early-stage researchers often hold precarious positions in very competitive environments. Hyper-competition, and its reinforcement of assertive stereotypes, serves as an exclusionary mechanism for those who cannot or will not compete continually. The choice to enter this competition often coincides with "the rush hour of life", i.e. the establishment of partnerships and families, thereby reinforcing gender imbalances.

The changing context for STI increases the need for diversity. While social justice and fairness are important issues in themselves, increasing evidence shows that diversity improves the quality of research outcomes and their relevance for society (Smith-Doerr et al., 2017). Diversity and inclusiveness in STI are a prerequisite for producing the types of knowledge and innovations required to respond effectively to all the SDGs.

Against this backdrop, Chapter 7 lays out a future vision for a more diverse and productive scientific enterprise that recognises and rewards the equivalent and distinct contributions of both men and women. Most countries have adopted this objective, with many national plans identifying gender equity as a strategic priority. The 2017 edition of the EC/OECD STI policy survey shows that this priority has been translated into many specific policy initiatives related to gender in STI. However, the overall policy picture today also points to a fragmented approach, characterised by multiple institutions acting independently and limited co-ordination between education, science and innovation actors. Little systematic evaluation takes place of the effectiveness and long-term impact of the many interventions under way. Engaging in strategic thinking and targeted interventions to create positive feedback loops and strengthen the position of women within STI systems will require co-ordinated actions across actors at multiple levels.

Most countries cite gender diversity as one of the key objectives of their national STI plans. However, policy initiatives remain fragmented. A more strategic and systemic long-term approach is necessary.

How is STI policy responding to societal and technological disruptions?

As disruptive technologies create new challenges and opportunities, the terms of reference for STI policy making are changing. Meeting societal challenges has become a prominent goal, and mission-oriented policies to do this, with defined goals and within defined timeframes, are increasingly popular. However, governments' capacity to engage in such directive policies and significantly affect the major outcomes – particularly since the share of R&D in government spending has declined overall in OECD member countries – is questionable.

Societal challenges: From shaping the STI agenda to influencing specific policy actions

As revealed by a recent survey reported in Chapter 9, on the governance of public research policy, most OECD countries have STI strategies explicitly referencing societal challenges. Out of the 35 countries surveyed, 33 (94%) have a national STI strategy or plan in place. Meeting major societal challenges is an objective in most of these strategies (30 (90%) of 33 strategies). Key priority themes include sustainable growth, health improvements and efficient transportation systems. Strategies often refer to the SDGs, which have become an important political framework globally. However, as shown in Chapter 4 on STI policies for the SDGs, addressing societal challenges is rarely the main rationale for STI policy initiatives, although many competitive-funding schemes include societal impacts as selection criteria. References to STI in the SDGs are often more implicit than explicit. Greater effort outside of STI policy arenas is needed to demonstrate the role of research and innovation in helping to meet the SDGs. This will require a closer alignment of existing STI governance structures (e.g. policy advice, steering and funding, co-ordination, evaluation and monitoring) with the emerging "global governance framework" for the SDGs.

There is a lack of explicit reference to STI in the SDGs, and too little reference to the SDGs in STI. STI governance structures should be more closely aligned with the emerging "global governance framework" for the SDGs.

The magnitude and transnational scope of global challenges and the size of investment needed to address them, demands international co-ordination and co-operation of research efforts. International co-operation in STI provides parties with access to knowledge and expertise and enables cost sharing while avoiding duplication of research efforts. International co-operation among scientists has never been higher nor more diversified as shown by data on co-authored publications and co-patenting. At the same time, however, an erosion of multilateralism in other policy areas threatens international cooperation in STI. The challenge for STI policymakers is to demonstrate the benefits of international co-operation more forcefully in terms of economic, societal and environmental impacts. International STI co-operation for global challenges will also require mechanisms to ensure equitable sharing of the burden of global research efforts as well as the benefits.

Toward a new type of strategic steering to cope with economic and societal challenges

There are growing calls to support economic growth and address societal challenges through strategic steering of STI. As Chapter 4 on STI policies for the SDGs points out, reframing STI policy is not straightforward, and pleas to transform policy frameworks have yet to outline clear pathways for policymakers and propose new levers for policy. At best, they have suggested incremental reformulation of traditional supply and demand-side instruments, by instilling considerations of sustainability and directionality.

Against this backdrop, new mission-oriented programmes have been proposed, for example in the context of the preparatory discussions for the European Union's "Horizon Europe" plan. Mission-oriented programmes are large-scale interventions aiming to achieve a set mission (goal or solution) within a well-defined timeframe, with an important R&D component. Missions are more concrete than broad grand challenges, because they have clear time-bound targets. Compared to previous mission-oriented policies, the new missions focus more clearly on the demand side and the diffusion of innovations, seek to be coherent with other policy fields, and recognise the roles of both incremental and systemic innovations. They are intended as "systemic" public policies that draw on frontier knowledge to attain specific, often very ambitious, goals (Mazzucato, 2018). The terms of reference of these new mission-oriented programmes are still under development; they include melding the entrepreneurial power of bottom-up projects and the "purposive" top-down steering necessary for transformative innovation.

There exist multiple examples of failed mission-oriented policies. Lessons from these experiences warrant caution and attention to the design and evaluation of mission-oriented approaches. While certain examples of bold programmes can be found in history (notably in the space and defence industries), applying their lessons to another context and/or era will call for different policy and governance arrangements. A major difference is that in many previous missions, the government was the main or sole purchaser of the resulting technological developments. Government labs were also often the main performers of R&D. Today, the private sector performs most R&D in many OECD countries. Moreover, undertaking missions dedicated to grand societal challenges will require significant levels of funding and specific co-ordination mechanisms, involving companies and civil society actors. This means that governments need to favour public-private partnerships, where risks and rewards can be shared. Governments are using deliberative processes to better align innovation strategies and societal priorities. Nevertheless, questions remain over governments' capacity to add directionality to STI processes, given their limited fiscal room for manoeuvre and their existing sets of skills and capabilities.

New mission-oriented programmes could mobilise science and innovation to address societal and economic challenges. However, their governance and design arrangements have yet to be developed and tested.

Supporting the development and uptake of emerging technologies requires a mix of old and new types of policy interventions

Developing and using effectively and ethically new technologies involves various changes to STI policy making and governance. Some are technology-specific, while others are more cross-cutting. Several new digital technologies call for new types of intervention that require further experimentation and learning. Various chapters, notably Chapter 2 on AI and the technologies of the Next Production Revolution, Chapter 3 on digital innovation and Chapter 6 on enhanced access to data discuss some new tasks for governments. For example, governments can help support the development and sharing of data as part of open-data initiatives. They can act as catalysts and honest brokers in data partnerships, e.g. by co-ordinating and stewarding data-sharing agreements. Although such efforts are generally undertaken at national level, several international and multilateral initiatives have emerged to foster open access to STI data.

At the same time, since the traditional rationales for public intervention remain valid for government support of emerging technologies, STI policy and governance also need to deliver existing policies more effectively. For example, the technologies of the Next Production Revolution (including microelectronics, synthetic biology, new materials and nanotechnology) result from advances in scientific knowledge and instrumentation. Government support is essential to promote basic research, and to provide incentives and appropriate conditions for effective science-industry relationships. Even AI, research on which is led today by large private companies, rests on decades of public research that provided the foundations for today's developments.

Diffusion-oriented policies are also crucial. For complex systems, such as biorefineries, public-private partnerships around demonstrators can help to resolve technical and economic questions about production before the necessary large investments are made. Governments also have roles to play to help small and medium size enterprises understand and eventually adopt emerging technologies. Further downstream, the certification of technologies, such as 3D printing, will support their diffusion by controlling for possible negative impacts, e.g. related to the risk of environment damage.

Numerous specific challenges may hinder the necessary policy changes. Since the Next Production Revolution has implications for a wide range of fields (including digital infrastructure, skills and intellectual property rights) that were previously not closely coordinated or connected in government, it could accentuate co-ordination problems already apparent in many countries. Governments also often lack knowledge and skills in many areas of complex and fast-evolving new technology. Supporting the transition to Industry 4.0 challenges governments to act with greater foresight and technical understanding across multiple policy domains. Accelerated innovation also raises challenges in providing targeted support, as targets may change so rapidly that traditional instruments could become irrelevant. Governments need to adapt: adopting broader targets, moving targets and flexible management are possible avenues.

Several emerging technologies arise in a wide range of fields not previously closely connected in government, creating co-ordination problems. Many governments also lack knowledge and skills relevant to complex, fast-evolving new technologies.

Reaping the benefits of emerging technologies to ensure economic and social progress requires substantial and effective investment in research and innovation

Although the quality and type of research and innovation are as important as the absolute funding amounts allocated, all the policy initiatives outlined above require financial resources. However, whether current trends in public R&D efforts are commensurate with the current and future challenges needing to be addressed is an open question. Government budget allocations for R&D (GBARD) typically rose before the crisis. A few years after the crisis, however, once the additional spending related to stimulus packages and recovery plans had been exhausted, GBARD decreased or flattened in all Group of Seven countries (G7), except Germany. Given that these countries had the largest R&D budgets, GBARD has declined overall in the OECD area (Figure 1.3).

Figure 1.3. Government budget allocations for civil R&D, 2000-08 and 2008-17

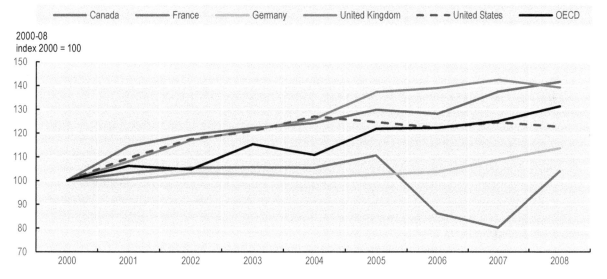

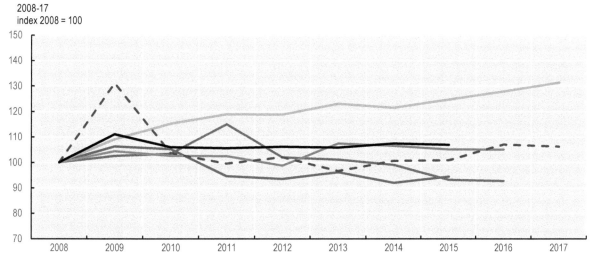

Note: GBARD less defence.
Source: Calculations based on OECD (2018d), "Research and Development Statistics: Government budget appropriations or outlays for RD", OECD Science, Technology and R&D Statistics (database), https://doi.org/10.1787/data-00194-en (accessed on 14 September 2018).

StatLink https://doi.org/10.1787/888933858107

The current trend in public R&D spending may not match the current and future challenges that science and innovation must address. Since 2010, government R&D expenditures in the OECD as a whole and in almost all G7 countries have stagnated or decreased not only in absolute amounts and relative to GDP, but also as a share of total government expenditure.

Comparing the evolution of public budgets for R&D with the budgets for all policy domains combined sheds further light on public funding dynamics (Figure 1.4). A positive correlation exists between the evolution of overall government budgets, and the evolution of the budget for R&D. It is reasonable to assume that the overall budget is one driver of the R&D budget, as governments consider their overall financial position before allocating funds across budget lines. Hence, the slowdown in R&D spending might be partly explained by overall budget restrictions following spending on the recovery packages of 2009 and plans to moderate or reduce public debt. Accordingly, all countries where GBARD increased also experienced an increase in their overall government expenditure (top-right corner). In several countries, public R&D budgets have also increased more rapidly than the overall public budget, thereby increasing the share of R&D in government spending (top right corner, above the diagonal line in the graph). However, since six of the G7 countries experienced an opposite trend, government R&D funding has decreased as a share of total government expenditures for the OECD area as a whole. As discussed in Chapter 8 on new public research-funding approaches and instruments, this might suggest that the policy importance of research and innovation has shifted downwards in many countries. More anecdotally, it echoes some policy officials' frustrations – especially in finance ministries and centres of government – over the absence of sufficiently tangible innovation results stemming from the significant recovery plans implemented in the wake of the financial crisis.

Figure 1.4. Average annual growth of total government budgets and GBARD, 2009-16

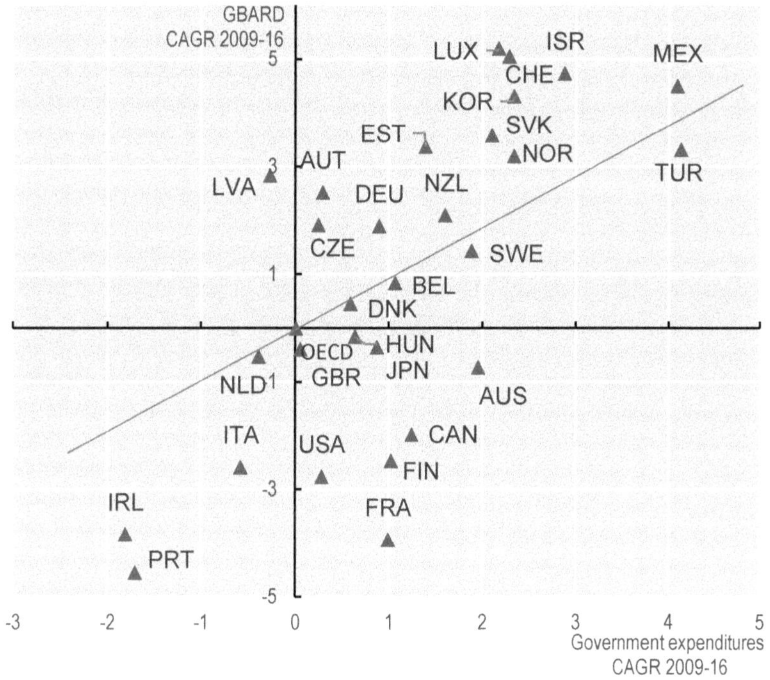

Source: Calculations based on OECD (2018e), "General Government Accounts, SNA 2008 (or SNA 1993): Main aggregates", OECD National Accounts Statistics (database), https://doi.org/10.1787/data-00020-en (accessed on 8 October 2018); and OECD (2018f), "Total GBARD (Government budget allocations for R&D) at current prices and PPP", in Main Science and Technology Indicators, Vol. 2018/2, https://doi.org/10.1787/msti-v2018-2-table57-en.

StatLink ⟶ https://doi.org/10.1787/888933858126

The share of government in total funding of R&D decreased by 4 percentage points (from 31% to 27%) in the OECD area between 2009 and 2016 (Figure 1.5); it only increased in five countries. Hence, the weight of government in total R&D funding has dropped, given growth in business expenditure on R&D has recovered. As previously discussed (Figure 1.2), adding R&D tax credits to public R&D budgets modifies the overall picture, since the tax credits increased significantly during this period. However, tax credits do not enhance government's capacity to influence the direction of R&D, as they are direction-neutral by design. Accordingly, the reduced share of government in R&D funding could lead to less government influence on the overall direction of science and innovation.

> *The shift of the policy mix towards R&D tax incentives decreases governments' capacity to influence the direction of private R&D towards socially desirable goals, at a time when the need for a more strategic orientation of research and innovation is becoming more pressing.*

Business R&D has picked up in several countries in recent years and can therefore compensate somewhat for lower public spending. However, most firms focus on applied research and experimental development. Funding for basic research – without which many of the new developments linked to the digital revolution would not have happened – may be particularly at risk in the coming years.

Figure 1.5. Change in the share of government in the direct funding of gross domestic expenditure on R&D, 2009-16 (or latest year available)

In percentage points

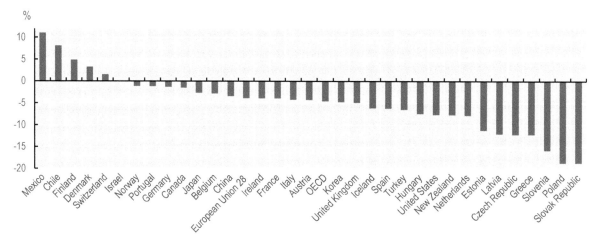

Source: OECD (2018g), "Main Science and Technology Indicators", OECD Science, Technology and R&D Statistics (database), https://doi.org/10.1787/data-00182-en (accessed on 2 October 2018).

StatLink https://doi.org/10.1787/888933858145

Faced with government austerity measures, politicians and senior public-sector leaders in many countries are increasingly demanding hard evidence on the outcomes of research funding; they want to know what works, and what does not. Science and innovation spending is no longer exempt from pressures to provide quantitative evidence of impact. Against this backdrop, policymakers need to shift more attention to supporting monitoring systems, evaluation frameworks and data infrastructures (Chapter 12). As reported in Chapter 9 on the governance of public research policy, 19 of the 34 OECD countries surveyed have independent specialised agencies in charge of evaluating and monitoring the performance of higher education institutions (HEIs) and PRIs.

As shown in Chapter 8 on new public research-funding approaches and instruments, the growing demand for evidence of economic and societal impacts, in addition to scientific excellence, also affects the traditional modes of allocating government funds to public research institutions. Funding instruments have become more complex to respond to the growing number of economic and social objectives to be met by science and innovation. Although the range of options available to policymakers has expanded beyond traditional institutional "block" funding and competitive "project funding", these instruments' growing complexity and diversity creates new challenges (e.g. related to co-ordination and evaluation).

How is STI governance adapting to a fast-changing context?

Confronted with a more rapidly changing and varied research and innovation landscape, governments need to become more agile, more responsive, more open to stakeholder participation and better informed. Governments are experimenting with new approaches to policy design and delivery. They will also benefit from embracing digital technologies when designing, implementing and monitoring STI policies. A new generation of digital tools can produce more granular and timely data to support policy formulation and design. By linking different datasets, these tools can transform the evidence base for STI policy, and help demonstrate the relationships between science and innovation expenditures and real-world outcomes.

New modes of STI governance are emerging

Technological and social changes mean that the way governments work and interact with their policy subjects and partners is shifting. For example, new technologies like AI and gene editing can alter – and even disrupt – society and the economy in unpredictable ways. Preventing, correcting or mitigating the negative effects of technology has also become more important – yet more difficult – as technology itself has become more complex and pervasive. The fast pace of technological change means that policymakers struggle to exert oversight regarding emerging technologies. Traditional "end-of-pipe" regulatory instruments (such as risk assessment) are insufficient under conditions of uncertainty: they often fail to anticipate or address the long-term implications of emerging technologies, and they can be inflexible, inadequate and even stifling for innovation.

The uncertainty and risks created by rapid technological change cannot be borne and directed by the private sector alone: governments must take an active role. Chapter 11 on new approaches in policy design and experimentation argues that governments must evolve to better anticipate, adapt to and mitigate these change processes, as part of their STI policy portfolios and policy-making practices. However, policymakers face the extremely challenging task of balancing the need to maintain stability and confidence in the public system, while rapidly adapting to a new environment and new demands. Yet they must adapt, or governments risk becoming increasingly irrelevant, dysfunctional and disconnected. Despite their potential benefits, many emerging policy approaches – such as design thinking, collective intelligence, behavioural insights, policy experimentation and anticipatory governance – have yet to be widely adopted in STI policy.

Governments face a crucial trade-off between relevance and stability: they must evolve to better anticipate, adapt to and mitigate rapid technological and social changes, while maintaining stability and confidence in the public system.

New forms of technology governance are also required to allow policymakers to respond to technological change in real time. Technology governance is defined as the process of exercising political, economic and administrative authority over the development, diffusion and operation of technology. Chapter 10 argues that technology governance must move "upstream" and become an integral part of the innovation process itself, to steer emerging technologies towards better collective outcomes. This calls for more anticipatory and participatory modes of governance.

Anticipatory approaches can help explore, consult widely on and steer the consequences of innovation at an early stage. They can incorporate public values and concerns, mitigating potential public backlash against technology. They require new capacities within government that blend foresight, engagement and reflexivity to facilitate the acceptance of new technologies, while at the same time assessing, discussing and preparing for new technologies' intended and unintended economic and societal effects. New policy tools, such as normative codes of conduct, test beds, regulatory sandboxes and real-time technology assessments can be useful. Chapter 3 on digital innovation and Chapter 2 on AI and the technologies of the Next Production Revolution also highlight the benefits of environments that facilitate learning, such as test beds and regulatory sandboxes, to help understand the regulatory implications and responses to emerging technologies. Participatory approaches can provide a wide range of stakeholders – including citizens – with effective opportunities to appraise and shape technology pathways.

These practices can help ensure that the goals, values and concerns of society are continuously enforced in emerging technologies as they unfold and that policymakers (and society) will not be taken by surprise. In doing so, they help shape technological designs and trajectories, without unduly constraining innovators.

Towards the next generation of STI data and indicators

Bringing together discussions and perspectives shared at the OECD Blue Sky Forum 2016, Chapter 14 on measurement and analysis in STI presents several key trends affecting the production and use of STI data and statistics. These include the increased connectivity of STI systems across national borders; accelerating digitalisation and its impacts on data availability and integrity; and pressures to demonstrate the impacts of public expenditures on STI in an era of government austerity. These drivers of change create new demands for STI data. They also question traditional forms of statistical definition and classification, which struggle to capture more fluid identities, attitudes and economic pathways. In addition, greater abundance of data place a premium on high quality, trustworthy sources, contributing to redefining the role of STI data experts and providers.

Given the increasingly globalised nature of science and innovation activities, the difficulties of national statistics in capturing the creation and circulation of knowledge and related financial flows across countries are a major concern. International collaboration is therefore required. While innovation systems do not change overnight, in times of fast-paced, disruptive change, timelier and more frequent data also become more critical. Timeliness is also essential for measuring processes that may be short-lived, such as entrepreneurship and business dynamics.

A key issue for STI policymakers is monitoring the link between science and innovation on the one hand, and the whole range of global sustainability concerns on the other, from poverty and hunger eradication, to equality and climate action. Those links are not easily traced, nor are they easily exposed through the sole use of indicators. The multidimensional nature of the SDGs implies that monitoring and measuring the overall role of science and innovation in meeting the SDGs will require accumulating findings from multiple sources. Chapter 4 on STI policies for delivering the SDGs also addresses this issue. It suggests that detailed administrative data and the ability to combine with other data could provide information on the role played by STI "input" commitments to the SDGs. However, obtaining and interpreting such data across countries presents some challenges.

Chapters 4 and 14 coincide in calling for policy frameworks, such as those being developed to monitor SDGs, to consider in their development an appropriate mix of instruments and disciplines for measurement and analysis. This should help address evidence needs and develop solutions that can be globally scaled up to achieve international comparability as well as greater synergies in highly interconnected STI systems. Defining and acting upon such needs requires a more strategic engagement between data producers and policy makers. Evidence from several countries and international initiatives suggest this is a feasible vision.

Monitoring the contribution of science and innovation to the global and multidimensional "grand challenges" remains difficult and will require new statistics and indicators. This effort should be linked to current work to develop indicators to measure overall progress on the SDGs.

The impact of digitalisation on the evidence base for STI policy and governance

Digitalisation is already having a significant impact on the evidence base for STI policy and governance. As more digital tools are used in research and innovation processes, they leave more "digital traces" that can be used for indicators and analysis. At a time when the cost of developing new data sources for responding to specific policy questions can be prohibitive, linking different existing data sources can provide insights that are impossible to obtain by working with the different data components separately.

Chapter 12 outlines the promises of digital science and innovation policy (DSIP), including: (i) streamlining burdensome administrative procedures, to deliver significant efficiency gains within ministries and agencies; (ii) providing more granular and timely data analysis to support STI policy, and improve the allocation of research and innovation funding; (iii) improving the timeliness of performance-monitoring data, to enable more agile short-term policy adjustments; (iv) detecting emerging patterns of change and stability in research, technology and industry, to support short-term forecasting of issues of policy concern; and (v) promoting inclusiveness in STI policy agenda-setting, by opening policy intelligence data to a broader range of stakeholders.

At the same time, policymakers' expectations of DSIP infrastructures should avoid a naïve rationalism that understates the inherent complexity of policy making. DSIP systems can inform policy choices, but they cannot and should not provide a technical fix to what are ultimately political judgements, shaped by competing values and uncertainty. If open by design, DSIP systems could nevertheless be instrumental in embedding various social values in policy making by promoting inclusiveness in science and innovation agenda-setting, making it less technocratic and more democratic.

DSIP systems cannot provide a mere technical solution to policy making, which remains inherently complex and based on political judgements. But they could help embed various social values in policy decisions, by promoting inclusiveness in science and innovation policy agenda-setting.

Irrespective of the policy setting, an embedded, routine use of DSIP will depend for their adoption not just on digital technologies, but also on favourable social and administrative conditions. Organisations and individuals also need to have assurance that data about their funding, activities and results will be handled appropriately and protected when needed.

Chapter 13 on targeting entrepreneurship support on firms with high growth potential presents one possible application of big data and machine learning to a real-world STI policy problem (Box 1.3). It highlights some of the potential benefits, but also the limits of using these techniques in STI policy. Chapter 14 on measuring STI raises other concerns about using big data and machine learning, for example: the possibility that datasets contain defects and biases; difficulties in evaluating big-data techniques and analysis; and complexities in explaining these techniques to decision-makers and the public.

Box 1.3. Targeting entrepreneurship support on high-growth potential firms

Considering that only a tiny minority of new firms contribute to economic growth, some scholars have questioned the effectiveness of untargeted entrepreneurship policy, arguing that public resources should be concentrated on firms with the highest growth potential. This, in turn, poses the related question of whether it may be possible to identify high-potential firms ex ante. One difficulty in identifying successful entrants is the lack of detailed data on the characteristics of firms and entrepreneurs at the moment they create the company. As many firms are very small, limited public information is available from administrative sources. In this challenging context, policymakers can use big data and innovative predictive analytics (e.g. machine learning) to help target successful high-growing entrants.

There are important caveats in using such digital tools in this way. Significant unpredictability will remain about start-up success, as idiosyncratic and unobservable factors will always play an important role in rapidly changing markets. Periods of disruptive changes do not lend themselves well to policies aiming to pick the "best" firms for targeted support. Most innovations in turbulent times emerge through trial and error among various combinations of technological and social innovations. In such contexts, a subset of firms with higher growth potential are not "revealed" to the world; their potential for growth emerges and increases through interactions with their environment, allowing faster learning and greater investment for some. Hence, direct and targeted policy interventions will always have to be complemented with horizontal reforms, to ensure an overall business environment conducive to entrepreneurship and experimentation. In practice, this means striking the right balance between targeting and promoting experimentation. However, evidence generated through big data and machine-learning techniques could influence this balance in the near future, pushing it more towards targeting (Chapter 13).

Private sector companies are increasingly contributing to the evidence base for STI policy, for example, as owners of bibliographic databases and providers of add-on services. The digitalisation of STI policy presents further opportunities for private-sector involvement. Although this presents several benefits, relying on the private sector for DSIP systems and components also creates potential risks for the public sector. For example, reliance on proprietary products and services may lead to discriminatory access to data, even if the data concern research activities funded by the public sector. Moreover, the public sector's adoption of commercial standards for metrics may drive the emergence of private platforms exhibiting network effects that are difficult to challenge.

Conclusion

Taking various angles and approaches, the STI Outlook 2018 focuses on the policy changes needed to respond to the disruptions currently unfolding in technology, the economy, the environment and society. The 13 thematic chapters and this overarching introduction focus on many of the key policy questions. Taken together, they provide insights on the challenges at stake and a range of possible policy responses.

All chapters feature concrete examples of national policy initiatives, with the aim of contributing to the process of international policy learning. Complexity and uncertainty characterise many aspects of the relationship between developments in STI and the economic and social challenges faced by countries at all income levels. Consequently, the need is ever greater for the exchange of information on policies and the factors found to condition their successes and failures.

Note

[1] Findability, accessibility, interoperability and re-use.

References

Bloom, N. et al. (2017), "Are Ideas Getting Harder to Find?", *NBER Working Paper*, No. 23782, National Bureau of Economic Research, Cambridge, MA, http://www.nber.org/papers/w23782.

De Backer, K. and D. Flaig (2017), "The future of global value chains: Business as usual or 'a new normal'?", *OECD Science, Technology and Industry Policy Papers*, No. 41, OECD Publishing, Paris, https://doi.org/10.1787/d8da8760-en.

Guillemette, Y. and D. Turner (2018), "The Long View: Scenarios for the World Economy to 2060", *OECD Economic Policy Papers*, No. 22, OECD Publishing, Paris, https://doi.org/10.1787/b4f4e03e-en.

IPCC (2018), "Global Warming of 1.5 °C, an IPCC special report on the impacts of global warming of 1.5°C above pre-industrial levels and related global greenhouse gas emission pathways, in the context of strengthening the global response to the threat of climate change, sustainable development, and efforts to eradicate poverty", *Intergovernmental Panel on Climate Change*, Geneva, http://www.ipcc.ch/report/sr15.

Jones, B. (2009) "The Burden of Knowledge and the 'Death of the Renaissance Man': Is Innovation Getting Harder?", *The Review of Economic Studies*, Vol. 76/1, pp. 283-317, Oxford University Press, Oxford, https://doi.org/10.1111/j.1467-937X.2008.00531.x.

Mazzucato, M. (2018), *Mission-Oriented Research & Innovation in the European Union – A problem-solving approach to fuel innovation-led growth*, Directorate-General for Research and Innovation, European Commission, Brussels, https://ec.europa.eu/info/sites/info/files/mazzucato_report_2018.pdf.

Miller, D.I. and J. Wai (2015), "The bachelor's to Ph.D. STEM pipeline no longer leaks more women than men: a 30-year analysis", *Frontiers in Psychology*, Vol. 6/37, Frontiers Media, Lausanne, http://dx.doi.org/10.3389/fpsyg.2015.00037.

OECD (2018a), "The policy mix for science-industry knowledge transfer: Towards a mapping of policy instruments and their interactions", *Working Party on Innovation and Technology Policy document*, OECD, Paris, DSTI/STP/TIP(2017)7/REV2.

OECD (2018b), "OECD time-series estimates of government tax relief for business R&D", TAX4INNO Project 674888, Deliverable 2.3: Summary report on tax expenditures, Version 29 May 2018, OECD, Paris, http://www.oecd.org/sti/rd-tax-stats-tax-expenditures.pdf.

OECD (2018c), "OECD review of national R&D tax incentives and estimates of R&D tax subsidy rates", TAX4INNO Project 674888, Deliverable 3.3: Summary report on tax subsidy rates – core countries, Version 18 April 2018, OECD, Paris, http://www.oecd.org/sti/rd-tax-stats-design-subsidy.pdf.

OECD (2018d), "Research and Development Statistics: Government budget appropriations or outlays for RD (Edition 2017)", *OECD Science, Technology and R&D Statistics* (database), https://doi.org/10.1787/e724dc33-en (accessed on 26 September 2018).

OECD (2018e), "General Government Accounts, SNA 2008 (or SNA 1993): Main aggregates", OECD *National Accounts Statistics* (database), https://doi.org/10.1787/data-00020-en (accessed on 08 October 2018).

OECD (2018f), "Total GBARD (Government budget allocations for R&D) at current prices and PPP", in *Main Science and Technology Indicators*, Vol. 2018/2, OECD Publishing, Paris, https://doi.org/10.1787/msti-v2018-1-table57-en.

OECD (2018g), "Main Science and Technology Indicators", *OECD Science, Technology and R&D Statistics* (database), https://doi.org/10.1787/data-00182-en (accessed on 2 October 2018).

OECD (2017a), *Investing in Climate, Investing in Growth*, OECD Publishing, Paris, https://doi.org/10.1787/9789264273528-en.

OECD (2017b), *The Next Production Revolution: Implications for Governments and Business*, OECD Publishing, Paris, https://doi.org/10.1787/9789264271036-en.

OECD (2017c), "Going digital", in *OECD Digital Economy Outlook 2017*, OECD Publishing, Paris, https://doi.org/10.1787/9789264276284-4-en.

OECD (2017d), *Making Innovation Benefit All: Policies for Inclusive Growth*, OECD, Paris, https://www.innovationpolicyplatform.org/system/files/Inclusive%20Growth%20publication%20FULL%20for%20web.pdf.

OECD (2017e), *OECD Science, Technology and Industry Scoreboard 2017: The digital transformation*, OECD Publishing, Paris, https://doi.org/10.1787/9789264268821-en.

OECD (2016), *OECD Science, Technology and Innovation Outlook 2016*, OECD Publishing, Paris, https://doi.org/10.1787/sti_in_outlook-2016-en.

Smith-Doerr, L., S.N. Alegria and T. Sacco (2017), "How Diversity Matters in the US Science and Engineering Workforce: A Critical Review Considering Integration in Teams, Fields, and Organizational Contexts", *Engaging Science, Technology and Society*, Vol. 3 (2017), p. 15, http://dx.doi.org/10.17351/ests2017.142.

Wilkinson, M. et al. (2016), "The FAIR Guiding Principles for scientific data management and stewardship", *Scientific Data 3*, Article No. 160018 (2016), Nature Publishing Group, London, http://dx.doi.org/10.1038/sdata.2016.18 .

Chapter 2. Artificial intelligence and the technologies of the Next Production Revolution

By

Alistair Nolan

Mastering the technologies of the Next Production Revolution requires effective policy in wide-ranging fields, including digital infrastructure, skills and intellectual property rights. This chapter examines a selection of policy initiatives that aim to enable this transformation process and ensure it benefits society. Developing and adopting new production technologies is essential to raising living standards and countering declining labour productivity growth in many OECD countries. Digital technologies can increase productivity in many ways. Artificial intelligence (AI) could spur the development of entirely new industries. And technologies enabled by advances in digital technology, such as biotechnology, 3D printing and new materials, promise important economic and social benefits. This chapter has two parts. The first covers individual technologies, their applications in production and their specific policy implications. These technologies are: AI, blockchain, 3D printing, industrial biotechnology, new materials and nanotechnology. The second part of the chapter addresses two cross-cutting policy issues relevant to future production: access to and awareness of high-performance computing, and public support for research (with a focus on public research for advanced computing and AI).

Introduction

Developing and adopting new production technologies is essential to raising living standards and countering the declining labour productivity growth in many OECD countries over recent decades. Rapid population ageing – the dependency ratio in OECD countries is set to double over the next 35 years – makes raising labour productivity more urgent. Digital technologies can increase productivity in many ways. For example, they can reduce machine downtime, as intelligent systems predict maintenance needs. They can also perform work more quickly, precisely and consistently, as increasingly autonomous, interactive and inexpensive robots are deployed. New production technologies will also benefit the natural environment in several new ways. For example, nanotechnology is helping to develop materials that cool themselves to below ambient temperature without consuming energy.[1]

This chapter examines a selection of policies aiming to enable the Next Production Revolution. With the exceptions of artificial intelligence (AI) and blockchain, it describes only briefly some of the many transformational uses of digital technology in production, as these developments are reviewed in (among other publications) OECD (2017, 2018a). Instead, the chapter emphasises policy initiatives and policy research findings that have arisen recently, or were not addressed in OECD (2017).

This chapter has two parts. The first covers individual technologies and their specific policy implications, namely AI and blockchain in production, 3D printing, industrial biotechnology, new materials and nanotechnology. The second addresses just two of the many cross-cutting policy issues relevant to future production, namely: access to and awareness of high-performance computing (HPC), and public support for research. Particular attention is given to public research related to computing and AI, as well as the institutional mechanisms needed to enhance the impact of public research.

Production technologies: Recent developments and policy implications

AI in production

The *Oxford English Dictionary* defines artificial intelligence as "the theory and development of computer systems able to perform tasks normally requiring human intelligence". Expert systems – a form of AI drawing on pre-programmed expert knowledge – have been used in industrial processes for close to four decades (Zweben and Fox, 1994). However, with the development of deep learning using artificial neural networks[2] – the main source of recent progress in the field – AI can be applied to most industrial activities, from optimising multi-machine systems to enhancing industrial research (Box 2.1). Furthermore, the use of AI in production will be spurred by automated machine learning processes that can help businesses, scientists and other users employ the technology more readily. Currently, with respect to AI that uses deep learning techniques and artificial neural networks, the greatest commercial potential for advanced manufacturing is expected to exist in supply chains, logistics and process optimisation (McKinsey Global Institute, 2018). Some survey evidence also suggests that the transportation and logistics, automotive and technology sectors lead in terms of the share of early AI-adopting firms (Boston Consulting Group, 2018).

Box 2.1. Recent Applications of AI in Production

A sample of recent uses of AI in production illustrates the breadth of the industries and processes involved:

- In pharmaceuticals, AI is set to become the "primary drug-discovery tool" by 2027, according to Leo Barella, Global Head of Enterprise Architecture at AstraZeneca. AI in preclinical stages of drug discovery has many applications, from compound identification, to managing genomic data, analysing drug safety data and enhancing in-silico modelling (AI Intelligent Automation Network, 2018).

- In aerospace, Airbus deployed AI to identify patterns in production problems when building its new A350 aircraft. A worker might encounter a difficulty that has not been seen before, but the AI, analysing a mass of contextual information, might recognise a similar problem from other shifts or processes. Because the AI immediately recommends how to solve production problems, the time required to address disruptions has been cut by one-third (Ransbotham et al, 2017).

- In semiconductors, an AI system can now assemble circuitry for computer chips, atom by atom (Chen, 2018); Landing.ai has developed machine-vision instruments to identify defects in manufactured products – such as electronic components – at scales that are invisible to the unaided eye.

- In the oil industry, General Electric's camera-carrying robots inspect the interior of oil pipelines, looking for microscopic fissures. If laid side by side, this imagery would cover 1 000 square kilometres every year. AI inspects this photographic landscape and alerts human operators when it detects potential faults (Champain, 2018).

- In mining, AI is being used to explore for mineral deposits, optimise the use of explosives at the mine face (taking into consideration the cost of milling larger chunks of unexploded material later on), and operate autonomous drills, ore sorters, loaders and haulage trucks. In July 2017, BHP switched to completely autonomous trucks at a mine in Western Australia (Walker, 2017).

- In construction, generative software uses AI to explore every permutation of a design blueprint, suggesting optimal building shapes and layouts, including the routing of plumbing and electrical wiring, and linking scheduling information to each building component.

- AI is exploring decades of experimental data to radically shorten the time needed to discover new industrial materials, sometimes from years to days (Chen, 2017).

- AI is also enabling robots to take plain-speech instructions from human operators, including commands not foreseen in the robot's original programming (Dorfman, 2018).

> - Finally, AI is making otherwise unmanageable volumes of Internet of things (IoT) data actionable. For example, General Electric operates a virtual factory, permanently connected to data from machines, to simulate and improve even highly optimised production processes. Used for predictive maintenance, AI can process combined audio, video and sensor data, and even text on maintenance history, to greatly surpass the performance of traditional maintenance practices.

Beyond its direct uses in production, the use of AI in logistics is enabling real-time fleet management, while significantly reducing fuel consumption and other costs. AI can also lower energy consumption in data centres (Sverdlik, 2018). In addition, AI can assist digital security: for example, the software firm Pivotal has created an AI system that recognises when text is likely to be part of a password, helping to avoid accidental online dissemination of passwords. Meanwhile, Lex Machina is blending AI and data analytics to radically alter patent litigation (Harbert, 2013). Many social-bot start-ups also automate tasks, such as meeting scheduling (X.ai), business-data and information retrieval (butter.ai), and expense management (Birdly). Finally, AI is being combined with other technologies – such as augmented and virtual reality – to enhance workforce training and cognitive assistance (Box 2.2).

> **Box 2.2. In my view: AI and digitalisation for workforce training and assistance**
>
> Globalisation has increased the demand for customisation, with small product runs requiring agile supply chains. The adaptability demanded of workers is increasing, and established training methods are no longer sufficient. Digitalisation and AI could revolutionise how workers are trained, both on and off the job. Digitalisation itself has drastically lowered the investment in hardware necessary for on-the-job training, as powerful computers allow accurate interactive simulation of complex production processes. For example, human-in-the-loop simulation using virtual-reality headsets has lowered the hardware costs of digital training systems from thousands of dollars to a few hundred. The cost of augmented-reality systems and multimodal interfaces will also continue to decrease, while their performance in factory conditions continues to improve.
>
> The key challenge to reaping the full benefits of digitally delivered training and assistance systems lies in the training material itself. Training courses require specialist knowledge, often from heterogeneous sources, and adaptation to context (worker experience, culture, existing skills, time available, characteristics of the manufacturing operation where training is required, etc.). Today, training material is largely developed manually, which is costly and time-consuming. AI has begun to provide solutions to this challenge. Chatbots and similar systems are now able to interact with workers using natural language, providing answers and context-specific help that often draw on multiple databases.
>
> More significantly still, connected AI is set to tap into collective experience to improve training and cognitive assistance. Shared training databases can contain data on the cumulative experience of many workers undergoing training, as well as their subsequent performance, their responses in unexpected situations and other variables. If training systems are scaled up to serve communities of thousands of users, they will be enormously useful.

Over time, a major effect of AI on production could be the creation of new industries

Beyond such applications, a main effect of AI on future production could be the creation of entirely new industries, based on scientific breakthroughs enabled by AI, much as the discovery of DNA structure in the 1950s led to a revolution in industrial biotechnology and the creation of vast economic value (the global market for recombinant DNA technology has been estimated at around USD 500 billion [US dollars]).[3] Approximately 40 years separated the elucidation of DNA structure and the emergence of a major biotech industry, and around 100 years passed between the scientific revolution in quantum physics and the recent birth of quantum computing (Box 2.5). Such observations underscore the importance of basic research and the importance of long time horizons in some aspects of research policy.

AI: specific policies

Several types of policy affect the development and diffusion of AI. These include: regulations governing data privacy (because of the critical importance of training data for AI systems); liability rules (which particularly affect diffusion); research support (Section 3.2); intellectual property rules; and, systems for skills. Other policies are most relevant to the (still uncertain) consequences of AI. These could include: competition policy; economic and social policies that mitigate inequality; policies for education and training; measures that affect public perceptions of AI; and, policies related to digital security. Well-designed policies for AI are likely to have high returns, because AI can be widely applied and accelerate innovation (Cockburn et al., 2018). Some of the policies concerned – such as those affecting skills – are relevant to any important new technology. This section focuses on policies most specifically affecting AI in production, namely, policies that affect the availability of training data, measures to address hardware constraints, and the design of regulations that do not unnecessarily hinder innovation.

Training data are critical

Wissner-Gross (2016) reviews the timing of the most publicised AI advances over the past 30 years and notes that the average length of time between significant data creation and major AI performance breakthroughs has been much shorter than the average time between algorithmic progress and the same AI breakthroughs. Among many examples, Wissner-Gross cites the performance of Google's GoogLeNet software, which achieved near-human level object classification in 2014, using a variant of an algorithm developed 25 years earlier. But the software was trained on ImageNet, a huge corpus of labelled images and object categories that had become available just four years earlier.[4]

Many tools that firms employ to manage and use AI exist as free software in open source (i.e. their source code is public and modifiable). These include software libraries such as TensorFlow and Keras, and tools that facilitate coding such as GitHub, text editors like Atom and Nano, and development environments like Anaconda and RStudio. Machine learning-as-a-service platforms also exist, such as Michelangelo, Uber's internal system that helps teams build, deploy and operate machine-learning solutions. The challenges in using AI in production relate to its application in specific systems and the creation of high-quality training data.

Without large volumes of training data, many AI models are inaccurate. A deep-learning supervised algorithm may need 5 000 labelled examples per item and up to 10 million labelled examples to match human performance (Goodfellow, Bengio and Courville,

2016). The highest-value uses of AI often combine diverse data types, such as audio, text and video. In many uses, training data must be refreshed monthly or even daily (McKinsey Global Institute, 2018). Consequently, companies with large data resources and internal AI expertise, such as Google and Alibaba, have an advantage in deploying AI. Furthermore, many industrial applications are still somewhat new and bespoke, limiting data availability. By contrast, sectors such as finance and marketing have used AI for a longer time (Faggella, 2018).

In the future, research advances may make AI systems less data-hungry. For instance, AI may learn from fewer examples, or generate robust training data (Simonite, 2016). In December 2017, the computer program AlphaZero famously achieved a world-beating level of performance in chess by playing against itself, using just the rules of the game, without recourse to external data. In rules-based games such as chess and Go, however, high performance can be achieved based on simulated data. For the time being, external training data must be cultivated for real-world applications.

Governments can take steps to help develop and share training data

Many firms hold valuable data which they do not use effectively (whether through lacking in-house skills and knowledge, lack of a corporate data strategy, lack of data infrastructure, or other reasons). This can be the case even in firms with enormous financial resources. For example, by some accounts, less than 1% of the data generated on oil rigs are used (*The Economist*, 2017). However, many AI start-ups, and other businesses using AI, could create value from data they cannot easily access. To help address this mismatch, governments can act as catalysts and honest brokers for data partnerships. Among other measures, they could work with relevant stakeholders to develop voluntary model agreements for trusted data sharing. For example, the US Department of Transportation has prepared the draft "Guiding Principles on Data Exchanges to Accelerate Safe Deployment of Automated Vehicles". The Digital Catapult in the United Kingdom also plans to publish model agreements for start-ups entering into data-sharing agreements (DSAs).

Government agencies can also co-ordinate and steward DSAs for AI purposes

DSAs operate between firms, and between firms and public research institutions. Co-ordination could be helpful in cases where all data holders would benefit from data sharing, but individual data holders are reluctant to share data unilaterally, or are unaware of potential data-sharing opportunities. For example, a total of 359 offshore oil rigs were operational in the North Sea and the Gulf of Mexico as of January 2018. AI-based prediction of potentially costly accidents on oil rigs would be improved if this statistically small number of data holders were to share their data (in fact, the Norwegian Oil and Gas Association has asked all members to have a data-sharing strategy in place by the end of 2018).

The Digital Catapult's Pit Stop open-innovation activity (which complements the Catapult's model DSAs mentioned earlier) is an example of co-ordination aiming to foster DSAs. Pit Stop brings together large businesses, academic researchers and start-ups in collaborative problem-solving challenges around data and digital technologies. Also in the United Kingdom, the Turing Institute operates the Data Study Group, to which major private and public-sector organisations bring data-science problems for analysis: Institute researchers are thereby able to work on real-world problems using industry datasets, while businesses have their problems solved and learn about the value of their data. In a model that promotes data sharing without DSAs, Japan has developed the Industrial Value Chain

Initiative, a collaborative cloud-based platform/repository where member firms share data to help implement digital applications.

Governments can promote open-data initiatives and ensure that public data are disclosed in machine-readable formats for AI purposes

Open-data initiatives exist in many countries, covering diverse public administrative and research data (Chapter 6). To facilitate AI applications, disclosed public data should be machine-readable. A further measure to encourage AI could consist in ensuring that copyright laws allow data and text mining, providing this does not lead to substitution of the original works or unreasonably prejudice legitimate interests of the copyright owners. Governments can also promote the use of digital data exchanges[5] that share public and private data for the public good.

Technology itself may offer novel solutions to use data better for AI purposes

Sharing data can require overcoming a number of institutional barriers. Data holders in large organisations can face considerable internal bureaucracy before receiving permission to release data. Even with a DSA, data holders worry that data might not be used according to the terms of an agreement, or that client data will be shared accidentally. In addition, some datasets may be too big to share in practical ways: for instance, the data in 100 human genomes could consume 30 terabytes (30 million megabytes). Uncertainty over the provenance of counterpart data can also hinder data sharing or purchase. Ocean Protocol,[6] an open-source protocol built by the non-profit Ocean Protocol Foundation, is pioneering a system linking blockchain and AI, to address such concerns and incentivise secure data exchange. By combining blockchain and AI, data holders can obtain the benefits of data collaboration, with full control and verifiable audit. Under one use case, data are not shared or copied. Instead, algorithms go to the data for training purposes, with all work on the data recorded in the distributed ledger. Ocean Protocol is currently building a reference open-source marketplace for data, which users can adapt to their own needs to trade data services securely. Governments should be alert to the possibilities of using such technology in public open-data initiatives.

Governments can also help resolve hardware constraints for AI applications

As AI projects move from concept to commercial application, specialised and expensive cloud-computing and graphic-processing unit (GPU) resources are often needed. Trends in AI experiments show extraordinary growth in the computational power required. According to one estimate, the largest recent experiment, AlphaGo Zero, required 300 000 times the computing power needed for the largest experiment just 6 years before (OpenAI, 2018). Indeed, AlphaGo Zero's achievements in chess and Go involved computing power estimated to exceed that of the world's ten most powerful supercomputers combined (Digital Catapult, 2018).

An AI entrepreneur might have the knowledge and financial resources to develop a proof-of-concept for a business, but lack the necessary hardware-related expertise and hardware resources to build a viable AI company. To help address such issues, Digital Catapult runs the Machine Intelligence Garage programme, which works with industry partners – such as GPU manufacturer NVidia, intelligent processing unit-producer Graphcore, and cloud providers Amazon Web Services and Google Cloud Platform – to give early-stage AI businesses access to computing power and technical expertise.

Care is needed to avoid regulating AI in ways that unnecessarily dampen innovation

Algorithmic transparency, explainability and accountability are among the key concerns in discussions on AI regulation (OECD, 2018b). While this chapter does not examine these questions, a few overarching observations are relevant. First, economy-wide regulation of AI may not be optimal at this time: the technology is still young, and many of its impacts are still unclear (Chapter 10). While international experience on the regulation of AI is still limited, there are grounds for thinking that regulation should specifically cover identified harms arising in particular sectors and applications, and addressed by those agencies already responsible for regulating the relevant sectors. A broad trade-off exists between the accuracy of algorithms and their scrutability. This trade-off highlights the risk of universal regulation of transparency and explainability dampening innovation. New and Castro (2017) argue that an overall approach emphasising algorithmic accountability might best protect society's needs, while also encouraging innovation. The impacts of any adopted regulation, whatever its form, should be closely monitored. Finally, regulatory reviews should be frequent, because AI technology is changing rapidly.[7]

Blockchain in production

Blockchain – a distributed ledger technology – has many potential applications in production (Box 2.3). Blockchain is still an immature technology, and many applications are only at the proof-of-concept stage. The future evolution of blockchain involves various unknowns, for example with respect to standards for interoperability across systems. However, similar to the 'software as a service' model, "blockchain as a service" is already provided by companies such as Microsoft, SAP, Oracle, Hewlett-Packard, Amazon and IBM. Furthermore, consortia such as Hyperledger and the Ethereum Enterprise Alliance are developing open source-distributed ledger technologies in several industries (European Commission, 2018).

Adopting blockchain in production creates several challenges: blockchain involves fundamental changes in business processes, particularly with regard to agreements and engagement among many actors in a supply chain. When many computers are involved, the transaction speeds may also be slower than some alternative processes, at least with current technology (fast protocols operating on top of blockchain are under development). Blockchains are most appropriate when disintermediation, security, proof of source and establishing a chain of custody are priorities (Vujinovic, 2018). A further challenge relates to the fact that much blockchain development remains atomised: the scalability of any single blockchain-based platform – be it in supply chains or financial services – will depend on whether it is interoperable with other platforms (Hardjano et al., 2018).

Blockchain: Possible policies

Regulatory sandboxes are designed to help governments better understand a new technology and its regulatory implications, while at the same time giving industry an opportunity to test new technology and business models in a live environment (Chapter 10). Evaluations of the impacts of regulatory sandboxes are sparse (Financial Conduct Authority (2017) is an exception[8]). Blockchain regulatory sandboxes mostly focus on Fintech, and are being developed in countries as diverse as Australia, Canada, Indonesia, Japan, Malaysia, Switzerland, Thailand and the United Kingdom (European Commission, 2018). Pursuant to proper impact assessment of such schemes, and being sure to design selection processes that avoid benefitting some companies at the expense of others, the scope of sandboxes could be broadened to encompass blockchain applications in industry and other non-financial sectors.

By using blockchain in the public sector, governments could also raise awareness of blockchain's potential, when it improves on existing technologies. Technical issues also need to be resolved, such as how to trust the data placed on the blockchain. Trustworthy data may need to be certified in some way. Blockchain may also raise concerns about competition policy, as some large corporations begin to mobilise through consortia to establish blockchain standards, e.g. for supply-chain management.

Box 2.3. Blockchain : Potential applications in production

By providing a decentralised, consensus-based, immutable record of transactions, blockchain could transform important aspects of production when combined with other technologies. For example:

- A main application of blockchain is tracking and tracing in supply chains. One consequence could be a reduction in counterfeiting: in the motor-vehicle industry alone, firms lose tens of billions of dollars a year to counterfeit parts (Williams, 2013).

- Blockchain could replace elements of enterprise resource-planning systems. The Swedish software company IFS has demonstrated how blockchain can be integrated with enterprise resource-planning systems in the aviation industry. Commercial aircraft have millions of parts. Each part must be tracked, and a record kept of all maintenance work. Blockchain could resolve current failures in such tracking (Mearian, 2017).

- Blockchain is being tested as a medium permitting end-to-end encryption of the entire process of designing, transmitting and printing 3D computer-aided design (CAD) files, with each printed part embodying a unique digital identity and memory (European Commission, 2018). If successful, this technology could incentivise innovation using 3D printing, protect intellectual property and help address counterfeiting.

- By storing the digital identity of every manufactured part, blockchain could provide proof of compliance with warranties, licences and standards in production, installation and maintenance (European Commission, 2018).

- Blockchain could induce more efficient utilisation of industrial assets. For example, a trusted record of the usage history for each machine and piece of equipment would facilitate developing a secondary market for such assets.

- Blockchain could authenticate machine-based data exchanges, implement associated micro-payments and help monetise the IoT. In addition, recording machine-to-machine exchanges of valuable information could lead to "data collateralisation", giving lenders the security to finance supply chains and helping smaller suppliers overcome working-capital shortages (Maerian, 2017) (Chapter 13). By providing verifiably accurate data across the production and distribution processes, blockchain could also enhance predictive analytics.

> Blockchain could further automate supply chains through the digital execution of "smart contracts", which rely on pre-agreed obligations being verified automatically. Maersk, for example, is working with IBM to test a blockchain-based approach for all documents used in bulk shipping. Combined with ongoing developments in the IoT, such smart contracts could eventually lead to full transactional autonomy for many machines (Vujinovic, 2018).

3D printing

3D printing is expanding rapidly, thanks to falling printer and materials prices, higher-quality printed objects and innovation in methods. Recent innovations include 3D printing with novel materials, such as glass, biological cells and even liquids (maintained as structures using nanoparticles); robot-arm printheads that allow printing objects larger than the printer itself (opening the way for automated construction); touchless manipulation of print particles with ultrasound (allowing printing electronic components sensitive to static electricity); and hybrid 3D printers, combining additive manufacturing with computer-controlled machining and milling. Research is also advancing on 3D printing, with materials programmed to change shape after printing.

Most 3D printing is used to make prototypes, models and tools. Currently, 3D printing is not cost-competitive at volume with traditional mass-production technologies, such as plastic injection moulding. Wider use of 3D printing depends on how the technology evolves in terms of the print time, cost, quality, size and choice of materials (OECD, 2017). The costs of switching from traditional mass-production technologies to 3D printing are expected to decline in the coming years as production volumes grow, although it is difficult to predict precisely how fast 3D printing will diffuse. Furthermore, the cost of switching is not the same across all industries and applications.

3D printing: Specific policies

OECD (2017) examined policy options to enhance 3D printing's effects on environmental sustainability. One priority is to encourage low-energy printing processes (e.g. using chemical processes rather than melting material, and automatic switching to low-power states when printers are idle). Another priority is to use and develop low-impact materials with useful end-of-life characteristics (such as compostable biomaterials). Policy mechanisms to achieve these priorities include:

- targeting grants or investments to commercialise research in these directions
- creating a voluntary certification system to label 3D printers with different grades of sustainability across multiple characteristics, which could also be linked to preferential purchasing programmes by governments and other large institutions.

Ensuring legal clarity around intellectual property rights, for 3D printing of spare parts for products that are no longer manufactured, could also be environmentally beneficial. For example, a washing machine that is no longer in production may be thrown away because a single part is broken; a CAD file for the required part could keep the machine in operation. However, most CADs are proprietary. One solution would be to incentivise rights for third parties to print replacement parts for products, with royalties paid to the original product manufacturers as needed.

Government can help develop the knowledge needed for 3D printing at the production frontier

Bonnin-Roca et al. (2016) describe another possible policy area. They observe that metals-based additive manufacturing (MAM) has many potential uses in commercial aviation. However, MAM is a relatively immature technology – the fabrication processes at the technological frontier have not yet been standardised – and aviation requires high safety standards. The aviation sector – and the commercialisation of MAM technology – would benefit if the mechanical properties of printed parts of any shape, using any given feedstock on any given MAM machine, could be accurately and consistently predicted. Government could help develop the necessary knowledge. Specifically, the public sector could support the basic science, particularly by funding and stewarding curated databases on materials' properties, and brokering DSAs across users of MAM technology, government laboratories and academia; support the development of independent manufacturing and testing standards; and help quantify the advantages of adopting the new technology, by creating a platform documenting early users' experiences.

Bonnin-Roca et al. (2016) suggest such policies for the United States, which leads globally in installed industrial 3D manufacturing systems and aerospace production. However, the same ideas could apply to other countries and industries. These ideas also illustrate how policy opportunities can arise from a specific understanding of emerging technologies and their potential uses. Indeed, governments should strive to develop expertise on emerging technologies in relevant public structures, which will also help anticipate hard-to-foresee needs for technology regulation.

Industrial biotechnology and the bioeconomy

As part of the bioeconomy, industrial biotechnology involves the production of goods from renewable biomass –i.e. wood, food crops, non-food crops or even domestic waste – instead of finite fossil-based reserves. Much progress has taken place in the tools and achievements of industrial biotechnology (OECD, 2018c). For example, several decades of research in biology have yielded gene-editing technologies and synthetic biology (which aims to design and engineer biologically based parts, devices and systems, and redesign existing natural biological systems). When combined with other scientific and technological advances – for instance in materials science and robotics – the tools are in place to begin a bio-based production revolution. Bio-based batteries, artificial photosynthesis and micro-organisms that produce biofuels are just some examples of recent advances in biotechnology. Notwithstanding these advances, the largest positive medium-term environmental impacts of industrial biotechnology hinge on the development of advanced biorefineries, which transform sustainable biomass into marketable products (food, animal feed, materials, chemicals) and energy(fuel, power, heat) (OECD, 2017).

Industrial biotechnology and the bioeconomy: Specific policies

Strategies to expand biorefining must address the sustainability of the biomass used. Governments should urgently support efforts to develop standard definitions of sustainability (as regards feedstocks), tools for measuring sustainability, and international agreements on the indicators required to drive data collection and measurement. Furthermore, environmental performance standards are essential: regulators often impose sustainability criteria for bio-based products, most of which are not currently cost-competitive with petrochemicals.

Demonstrator biorefineries operate between pilot and commercial scales, and are critical to answering technical and economic questions about production before costly investments are made at full scale. However, biorefineries and demonstrator facilities are high-risk investments, and some aspects of the technologies are not fully proven. Additional study is also required of the economics of large bio-production facilities. Financing through public-private partnerships is needed to de-risk private investments and demonstrate governments' commitment to long-term, coherent policies on energy and industrial production.

Public initiatives for bio-based fuels have existed for decades, but little policy support has been extended to producing bio-based chemicals, which could substantially reduce greenhouse gas emissions and preserve non-renewable resources OECD, 2018c).

With respect to regulations, governments should focus on boosting the use of instruments – particularly standards – to reduce barriers to trade in bio-based products; addressing regulatory hurdles that hinder investment; and establishing a level playing field between bio-based products and biofuels. Better waste regulation could also boost the bioeconomy. For example, governments could promote less proscriptive and more flexible waste regulations, allowing the use of agricultural and forestry residues and domestic waste in biorefineries.

Governments could also lead in supporting the bioeconomy and industrial biotechnology through public procurement. Bio-based materials are not always amenable to public procurement, as they sometimes form only part of a product (e.g. a bio-based screen on a mobile phone), but public purchasing of biofuels (e.g. for public vehicle fleets) is easier (OECD, 2017).

New materials

Advances in scientific instrumentation, such as atomic-force microscopes, and developments in computational simulations have allowed scientists to study materials in more detail than ever before. Today, materials with entirely novel properties are emerging: solids with densities comparable to the density of air; super-strong lightweight composites; materials that remember their shape, repair themselves or assemble themselves into components; and materials that respond to light and sound, are all now realities (*The Economist*, 2015).

The era of trial and error in material development is also coming to an end. Powerful computer modelling and simulation of materials' structure and properties can indicate how they might be used in products. Desired properties, such as conductivity and corrosion resistance, can be intentionally built into new materials. Better computation is leading to faster development of new and improved materials, more rapid insertion of existing materials into new products, and the ability to improve existing processes and products. In the near future, engineers will not just design products, but will also design the materials from which the products are made (Teresko, 2008). Furthermore, large companies will increasingly compete in terms of materials development. For example, a manufacturer of automotive engines with a superior design could enjoy longer-term competitive advantage if it also owned the material from which the engine is built.

New materials: Specific policies

No single company or organisation will be able to own the entire array of technologies associated with a materials-innovation ecosystem. Accordingly, a public-private investment model is warranted, particularly to build cyber-physical infrastructure and train the future workforce (Chapter 6 in OECD, 2017).

New materials will raise new policy issues and give renewed emphasis to longstanding policy concerns. For example, new digital-security risks could arise because in a medium-term future, a computationally assisted materials "pipeline" based on computer simulations could be hackable. Progress in new materials also requires effective policy in already important areas, often related to the science-industry interface. For example, well-designed policies are needed for open data and open science (e.g. for sharing simulations of materials' structures or sharing experimental data in return for access to modelling tools).

Policy co-ordination is needed across the materials-innovation infrastructure at the national and international levels. Major efforts are under way in professional societies to develop a materials-information infrastructure – such as databases of materials' behaviour, digital representations of materials' microstructures and predicted structure-property relations, and associated data standards – to provide decision support to materials-discovery processes (Robinson and McMahon, 2016). International policy co-ordination is necessary to harmonise and combine elements of cyber-physical infrastructure across a range of European, North American and Asian investments and capabilities, as it is too costly (and unnecessary) to replicate resources that can be accessed through web services. A culture of data sharing – particularly pre-competitive data – is required (Chapter 6 in OECD, 2017).

Nanotechnology

Closely related to new materials, nanotechnology involves the ability to work with phenomena and processes occurring at a scale of 1 to 100 nanometres (a standard sheet of paper is about 100 000 nanometres thick). Control of materials on the nanoscale – working with their smallest functional units – is a general-purpose technology with applications across production (Chapter 4 in OECD, 2017). Advanced nanomaterials are increasingly used in manufacturing high-tech products, e.g. to polish optical components. Recent innovations include nano-enabled artificial tissue, biomimetic solar cells and lab-on-a-chip diagnostics.

Nanotechnology: Specific policies

Sophisticated and expensive tools are needed for research in nanotechnology. State-of-the-art equipment costs several million euros and often requires bespoke buildings. It is almost impossible to gather an all-encompassing nanotechnology research and development (R&D) infrastructure in a single institute, or even a single region. Consequently, nanotechnology requires interinstitutional and/or international collaboration to reach its full potential. Publicly funded R&D programmes should allow involvement of academia and industry from other countries, and enable targeted collaborations between the most suitable partners. The Global Collaboration initiative under the European Union's Horizon 2020 programme is one example of this approach.

Support is also needed for innovation and commercialisation in small companies. Nanotechnology R&D is mostly conducted by larger companies, thanks to their critical mass of R&D and production; their ability to acquire and operate expensive instrumentation; and their ability to access and use external knowledge. Policy makers could improve the access to equipment of small and medium-sized enterprises (SMEs) by: 1) increasing the size of SME research grants; 2) subsidising or waiving service fees; and/or 3) providing SMEs with vouchers for equipment use.

Regulatory uncertainties regarding risk assessment and approval of nanotechnology-enabled products must also be addressed, ideally through international collaboration. These uncertainties severely hamper the commercialisation of nano-technological innovation. Products awaiting market entry are sometimes shelved for years before a regulatory decision is taken. This has sometimes led to promising nanotechnology start-ups failing, and to large companies terminating R&D projects and innovative products. Policies should support the development of transparent and timely guidelines for assessing the risk of nanotechnology-enabled products, while also striving for international harmonisation in guidelines and enforcement. In addition, more needs to be done to properly treat nanotechnology-enabled products in the waste stream.

Selected cross-cutting policy issues

Developing a productive base that masters the technologies of the "next production revolution" involves diverse policy challenges, from implementing the types of technology-specific policies discussed above, in Section 2, to developing cross-cutting policies relevant to all the relevant technologies. Figure 2.1 depicts the types and scope of the policies involved. Cross-cutting policies must address issues as diverse as designing micro-economic framework conditions promoting technology diffusion; building fibre-optic cable networks to carry 5G; increasing trust in cloud computing; and designing education and training systems to respond efficiently to changing needs for skills. OECD (2017a) examines many of these issues in detail. This section covers two cross-cutting policy issues only, namely: improving access to and awareness of High-Performance Computing (HPC); and ensuring public support for R&D. It includes subjects, such as the race to achieve quantum computing and possible public research agendas for AI, that were not addressed in OECD (2017).

Figure 2.1. An overview of policies affecting advanced production

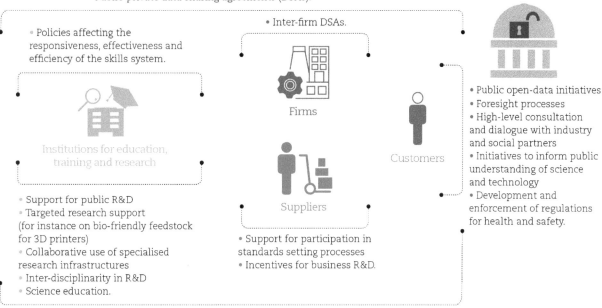

Improve access to HPC

HPC – which involves computing performance far beyond that of general-purpose computers – is increasingly important to firms in industries ranging from construction to pharmaceuticals, the automotive sector and aerospace. Airbus, for instance, owns 3 of the 500 fastest supercomputers in the world. Two-thirds of US-based companies that use HPC say that "increasing performance of computational models is a matter of competitive survival" (US Council on Competitiveness, 2014). The applications of HPC in manufacturing are also expanding beyond design and simulation, to include real-time control of complex production processes. Among European companies, the financial rates of return for HPC use are reportedly extremely high (European Commission, 2016). A 2016 review observed that "[m]aking HPC accessible to all manufacturers in a country can be a tremendous differentiator, and no nation has cracked the puzzle yet" (Ezell and Atkinson, 2016).

As Industry 4.0 becomes more widespread, demand for HPC will rise. But like other digital technologies, the use of HPC in manufacturing falls short of potential. According to one estimate, 8% of US manufacturers with fewer than 100 employees use HPC, yet one-half of manufacturing SMEs could potentially use HPC for prototyping, testing and design

(Ezell and Atkinson, 2016). Public HPC initiatives often focus on the computation needs of "big science". Greater outreach to industry, especially SMEs, is frequently needed. Box 2.4 sets out some possible ways forward, several of which are described in European Commission (2016).

Box 2.4. Getting supercomputing to industry: Possible policy actions

- raise awareness of industrial-use cases, quantifying their costs and benefits
- develop a one-stop source of HPC services and advice for SMEs and other industrial users
- provide low-cost or free experimental use of HPC for SMEs for a limited period, to demonstrate its technical and commercial implications
- establish online software libraries/clearing houses to help disseminate innovative HPC software to a wider industrial base
- incentivise HPC centres with long industrial experience, such as the Hartree Centre in the United Kingdom or Teratec in France, to advise centres with less experience
- modify eligibility criteria for HPC projects, which typically focus on peer reviews of scientific excellence, to include commercial-impact criteria
- engage academia and industry in co-designing new hardware and software, similarly to European projects such as Mont-Blanc9
- include HPC in university science and engineering curricula
- explore opportunities to co-ordinate demand for commercially provided computing capacity.

Public support for R&D

The technologies discussed in this chapter ultimately emerge from science. Microelectronics, synthetic biology, new materials and nanotechnology, among others, have arisen from advances in scientific knowledge and instrumentation. Publicly financed research in universities and public research institutions has often been critical to AI. Furthermore, because the complexity of many emerging production technologies exceeds even the largest firms' research capacities, public-private research partnerships are essential. Hence, the declining public support for research in some major economies is a concern (Chapter 8).

Public R&D and commercialisation efforts have many possible targets, from advancing the use of data analytics and digital technologies in metabolic engineering, to developing bio-friendly feedstocks for 3D printers. One interesting possibility is shaping research agendas to alleviate shortages of economically critical materials (as proposed by the Ames Laboratory's Critical Materials Institute in the United States).

An overarching research challenge relates to computation itself

The processing speeds, memory capacities, sensor density and accuracy of many digital devices are linked to Moore's Law. However, atomic-level phenomena and rising costs constrain further shrinkage of transistors on integrated circuits. Many experts believe a limit to miniaturisation will soon be reached. At the same time (as noted earlier), the computing power needed for the largest AI experiments is doubling every 3.5 months (OpenAI, 2018). By one estimate, this trend can be sustained for at most three-and-a-half to ten years, even assuming public R&D commitments on a scale similar to the Apollo or

Manhattan projects (Carey, 2018). Much, therefore, depends on achieving superior computing performance (including in terms of energy requirements). Many hope that significant advances in computing will stem from research breakthroughs in optical computing (using photons instead of electrons), biological computing (using DNA to store data and calculate) and/or quantum computing (Box 2.5).

Box 2.5. A new computing regime: The race for quantum computing

Quantum computers function by exploiting the laws of subatomic physics. A conventional transistor flips between on and off, representing 1s and 0s. However, a quantum computer uses quantum bits (qubits), which can be in a state of 0, 1 or any probabilistic combination of both 0 and 1 (for instance, 0 with 20% and 1 with 80% probability), while also interacting with other qubits through so-called quantum entanglement (which Einstein termed "spooky action at a distance").

Fully developed quantum computers, featuring many qubits, could revolutionise certain types of computing. Many of the problems best addressed by quantum computers, such as complex optimisation and vast simulation, have major economic implications. For example, at the 2018 CogX Conference, Dr Julie Love, Microsoft's director of quantum computing, described how simulating all the chemical properties of the main molecule involved in fixing nitrogen – nitrogenase – would take today's supercomputers billions of years, yet this simulation could be performed in hours with quantum technology. The results of such a simulation would directly inform the challenge of raising global agricultural productivity and limiting today's reliance on the highly energy-intensive production of nitrogen-based fertiliser. Rigetti Computing has also demonstrated that quantum computers can train machine-learning algorithms to a higher accuracy, using less data than with conventional computing (Zeng, 2018).

Until recently, quantum technology has mostly been a theoretical possibility, but Google, IBM and others are beginning to trial practical applications with a small number of qubits (Gambetta et al., 2017). For example, IBM Quantum Experience10 offers free online quantum computing. In 2017, Biogen worked with Accenture and quantum software company 1QBit on a quantum-enabled application to accelerate drug discovery. In 2017 Volkswagen piloted traffic optimisation experiments using quantum computing (Castellanos, 2017). However, no quantum device currently approaches the performance of conventional computers

---

A need for more – and possibly different – research on AI

Public research funding has been key to progress in AI since the origin of the field. The National Research Council (1999) shows that while the concept of AI originated in the private sector – in close collaboration with academia – its growth largely results from many decades of public investments. Global centres of AI research excellence (e.g. at Stanford, Carnegie Mellon and MIT) arose because of public support, often linked to US Department of Defense funding. However, recent successes in AI have propelled growth in private-sector R&D for AI. For example, earnings reports indicate that Google, Amazon, Apple, Facebook and Microsoft spent a combined USD 60 billion on R&D in 2017, including an important share on AI. By comparison, total US Federal Government R&D for non-defence industrial production and technology amounted to around USD 760 million in 2017.[11]

While many in business, government and among the public believe AI stands at an inflection point, some experts emphasise the scale and difficulties of the outstanding research challenges. Some AI research breakthroughs could be particularly important for society, the economy and public policy. However, corporate and public research goals might not fully align: Jordan (2018) notes that much AI research is not directly relevant to the major challenges of building safe intelligent infrastructures, such as medical or transport systems. He observes that unlike human-imitative AI, such critical systems must have the ability to deal with "distributed repositories of knowledge that are rapidly changing and are likely to be globally incoherent. Such systems must cope with cloud-edge interactions in making timely, distributed decisions and they must deal with long-tail phenomena whereby there is lots of data on some individuals and little data on most individuals. They must address the difficulties of sharing data across administrative and competitive boundaries" (Jordan, 2018).

Other outstanding research challenges relevant to public policy relate to making AI explainable; making AI systems robust (image-recognition systems can easily be misled, for instance); determining how much prior knowledge will be needed for AI to perform difficult tasks (Marcus, 2018); bringing abstract and higher-order reasoning, and "common sense", into AI systems; inferring and representing causality; and developing computationally tractable representations of uncertainty (Jordan, 2018). No reliable basis exists for judging when – or whether – research breakthroughs will occur. Indeed, past predictions of timelines in the development of AI have been extremely inaccurate.

Research and education need to be multi-disciplinary

Interdisciplinary research is essential to advancing production. Materials research involves disciplines such as traditional materials science and engineering, as well as physics, chemistry, chemical engineering, bio-engineering, applied mathematics, computer science and mechanical engineering. Environments supporting interdisciplinary research include institutes (e.g. Interdisciplinary Research Collaborations in the United Kingdom);[12] networks (e.g. the eNNab Excellence Network NanoBio Technology in Germany, which supports biomedical nanotechnology);[13] and individual institutions (e.g. Harvard's Wyss Institute for Biologically Inspired Engineering).[14]

Research and industry can often be linked more effectively

Government-funded research institutions and programmes should have the freedom to assemble the right combinations of partners and facilities to solve scale-up and interdisciplinarity challenges. Investments are often essential in applied research centres and pilot production facilities, to take innovations from the laboratory to production. Demonstration facilities – such as test beds, pilot lines and factory demonstrators – which provide dedicated research environments, with the right mix of enabling technologies and operating technicians, are also necessary. Some manufacturing R&D challenges may need the expertise not only of manufacturing engineers and industrial researchers, but also of designers, equipment suppliers, shop-floor technicians and users (Chapter 10 in OECD, 2017).

Beyond traditional metrics – such as numbers of publications and patents – more effective research institutions and programmes in advanced production may also need new evaluation indicators. These new indicators could assess such criteria as successful pilot-line and test-bed demonstrations; technician and engineer training; membership in consortia; incorporation of SMEs in supply chains; and the role of research in attracting foreign direct investment.

Public-private partnerships can help research commercialisation

Financing business scale-up is a widespread concern. This owes in great part to the fact that many venture-capital firms prefer to invest in software, biotech and media start-ups rather than advanced manufacturing firms, which often work with costlier and riskier technologies (in the United States, only around 5% of venture funding in 2015 targeted the industrial/energy sector) (Singer and Bonvillian, 2017). Partnerships between universities, industry and government can help provide start-ups with the know-how, equipment and initial funding to test and scale new technologies, so that investments are more likely to attract venture funding. Singer and Bonvillian (2017) describe several such collaborations. For example, Cyclotron Road, supported by the US Department of Energy's Lawrence Berkeley Lab, provides energy start-ups with equipment, technology and know-how for advanced prototyping, demonstration, testing and production design. Cooperative Research and Development Agreements – which are struck between a government agency and a private company or university – have also been valuable in providing frameworks for intellectual property rights in such collaborations.

Conclusion

Mastering the technologies of the Next Production Revolution requires effective policy in wide-ranging fields, including digital infrastructure, skills and intellectual property rights. Typically, these diverse policy fields are not closely connected in government structures and processes. Governments must also adopt long-term time horizons, for instance, in pursuing research agendas with possible long-term payoffs. As this chapter has illustrated, public institutions must possess specific understanding of many fast-evolving technologies. One leading authority argues that converging developments in several technologies are about to yield a "Cambrian explosion" in robot diversity and use (Pratt, 2015). Adopting Industry 4.0 poses challenges for firms, particularly small ones. It also challenges governments' ability to act with foresight and technical knowledge across multiple policy domains.

Notes

[1] See Aaswath Raman's 2018 TED talk, "How can we turn the cold of outer space into a renewable resource",
https://www.ted.com/talks/aaswath_raman_how_we_can_turn_the_cold_of_outer_space_into_a_renewable_resource.

[2] Deep learning with artificial neural networks is a technique in the broader field of machine learning that seeks to emulate how human beings acquire certain types of knowledge. The word 'deep' refers to the numerous layers of data processing. The term "artificial neural network" refers to hardware and/or software modelled on the functioning of neurons in a human brain.

[3] AI will of course have many economic and social impacts. In relation to labour markets alone, intense debates exist on AI's possible effects on labour displacement, income distribution, skills

demand and occupational change. However, these and other considerations are not a focus of this chapter.

[4] At its peak, ImageNet reportedly employed close to 50 000 people in 167 countries, who sorted around 14 million images (House of Lords, 2018).

[5] e.g. datacollaboratives.org.

[6] www.oceanprotocol.com.

[7] Microsoft, for instance, is developing a dashboard capable of scrutinising an AI system and automatically identifying signs of potential bias (Knight, 2018).

[8] Even if this assessment covers only the first year of a scheme in the United Kingdom.

[9] http://montblanc-project.eu/.

[10] www.research.ibm.com/quantum.

[11] OECD Main Science and Technology Indicators Database, http://oe.cd/msti.

[12] https://epsrc.ukri.org/funding/applicationprocess/routes/capacity/ircs/.

[13] www.ennab.de.

[14] https://wyss.harvard.edu.

References

AI Intelligent Automation Network (2018), *AI 2020 : The Global State of Intelligent Enterprise*, https://intelligentautomation.iqpc.com/downloads/ai-2020-the-global-state-of-intelligent-enterprise.

Almudever, C.G. et al. (2017), "The engineering challenges in quantum computing", conference paper presented at Design, Automation & Test in Europe Conference & Exhibition (DATE), 27-31 March 2017, Lausanne, pp. 836-845, 10.23919/DATE.2017.7927104.

Azhar, A. (2018), "Exponential View: Dept. of Quantum Computing", *The Exponential View*, 15 July, http://www.exponentialview.co/evarchive/#174.

Bonnin-Roca, J et al. (2016), "Policy Needed for Additive Manufacturing", *Nature Materials*, Vol. 15, pp. 815-818, Nature Publishing Group, United Kingdom, https://doi.org/10.1038/nmat4658.

Boston Consulting Group (2018), "AI in the Factory of the Future : The Ghost in the Machine", The Boston Consulting Group, Boston, MA, https://www.bcg.com/publications/2018/artificial-intelligence-factory-future.aspx.

Carey, R. (2018), "Interpreting AI Compute Trends", 10 July, blog post, *AI Impacts*, https://aiimpacts.org/interpreting-ai-compute-trends/.

Castellanos, S. (2017), "Volkswagen Pilots Quantum Computing Experiments", *The Wall Street Journal*, New York, 8 May 2018, https://blogs.wsj.com/cio/2017/05/08/volkswagen-pilots-quantum-computing-experiments/.

Champain, V. (2018), "Comment l'intelligence artificielle augmentée va changer l'industrie", *La Tribune*, Paris, 27 March, https://www.latribune.fr/opinions/tribunes/comment-l-intelligence-artificielle-augmentee-va-changer-l-industrie-772791.html.

Chen, S. (2018), "Scientists Are Using AI to Painstakingly Assemble Single Atoms", *Science*, American Association for the Advancement of Science, Washington, DC, 23 May, https://www.wired.com/story/scientists-are-using-ai-to-painstakingly-assemble-single-atoms/.

Chen, S. (2017), "The AI Company That Helps Boeing Cook New Metals for Jets", *Science*, American Association for the Advancement of Science, Washington, DC, 12 June, https://www.wired.com/story/the-ai-company-that-helps-boeing-cook-new-metals-for-jets.

Cockburn, I., R.Henderson and S.Stern (2018), "The Impact of Artificial Intelligence on Innovation", NBER Working Paper No.24449, Issued March 18, http://www.nber.org/papers/w24449.

Digital Catapult (2018), "Machines for Machine Intelligence: Providing the tools and expertise to turn potential into reality", Machine Intelligence Garage, Research Report 2018, London, https://www.migarage.ai.

Dorfman, P. (2018), "3 Advances Changing the Future of Artificial Intelligence in Manufacturing", *Autodesk Newsletter*, 3 January 2018, https://www.autodesk.com/redshift/future-of-artificial-intelligence/.

European Commission (2018), "#Blockchain4EU: Blockchain for Industrial Transformations", Publications Office of the European Union, Luxembourg, https://ec.europa.eu/jrc/en/publication/eur-scientific-and-technical-research-reports/blockchain4eu-blockchain-industrial-transformations.

European Commission (2016), "Implementation of the Action Plan for the European High-Performance Computing Strategy", Commission Staff Working Document SWD(2016)106, European Commission, Brussels, https://eur-lex.europa.eu/legal-content/EN/TXT/?uri=CELEX:52016SC0106.

Ezell, S.J. and R.D. Atkinson (2016), "The Vital Importance of High-Performance Computing to US Competitiveness", Information Technology and Innovation Foundation, Washington DC, http://www2.itif.org/2016-high-performance-computing.pdf.

Faggella, D. (2018), "Industrial AI Applications – How Time Series and Sensor Data Improve Processes", Techemergence, San Francisco, 31 May, https://www.techemergence.com/industrial-ai-applications-time-series-sensor-data-improve-processes.

Financial Conduct Authority (2017), "Regulatory sandbox lessons learned report", London, https://www.fca.org.uk/publication/research-and-data/regulatory-sandbox-lessons-learned-report.pdf.

Gambetta, J.M., J.M. Chow and M. Teffen (2017), "Building logical qubits in a superconducting quantum computing system", *npj Quantum Information*, Vol. 3, article No. 2, Nature Publishing Group and University of New South Wales, London and Sydney, https://arxiv.org/pdf/1510.04375.pdf.

Giles, M. (2018a), "Google wants to make programming quantum computers easier", *MIT Technology Review*, Massachusetts Institute of Technology, Cambridge, MA, 18 July, https://www.technologyreview.com/s/611673/google-wants-to-make-programming-quantum-computers-easier/.

Giles, M. (2018b), "The world's first quantum software superstore – or so it hopes – is here", *MIT Technology Review*, Massachusetts Institute of Technology, Cambridge, MA, 17 May, https://www.technologyreview.com/s/611139/the-worlds-first-quantum-software-superstore-or-so-it-hopes-is-here/.

Goodfellow, I., Y. Bengio and A. Courville (2016), *Deep Learning*, MIT Press, Massachusetts Institute of Technology, Cambridge, MA.

Harbert, T. (2013), "Supercharging Patent Lawyers with AI: How Silicon Valley's Lex Machina is blending AI and data analytics to radically alter patent litigation", *IEEE Spectrum*, IEEE, New York, 30 October, https://spectrum.ieee.org/geek-life/profiles/supercharging-patent-lawyers-with-ai.

Hardjano, T., A.Lipton and A.S.Pentland (2018), "Towards a Design Philosophy for Interoperable Blockchain Systems", Massachusetts Institute of Technology, Cambridge, MA, 7 July, https://hardjono.mit.edu/sites/default/files/documents/hardjono-lipton-pentland-p2pfisy-2018.pdf.

House of Lords (2018), "AI in the UK: ready, willing and able?", Select Committee on Artificial Intelligence – Report of Session 2017-19, Authority of the House of Lords, London, https://publications.parliament.uk/pa/ld201719/ldselect/ldai/100/100.pdf.

Jordan, M. (2018), "Artificial Intelligence—The Revolution Hasn't Happened Yet", *Medium*, A Medium Corporation, San Francisco, https://medium.com/@mijordan3/artificial-intelligence-the-revolution-hasnt-happened-yet-5e1d5812e1e7.

Knight, W. (2018), "Microsoft is Creating an Oracle for Catching Biased AI Algorithms", *MIT Technology Review*, Massachusetts Institute of Technology, Cambridge, MA, 25 May, https://www.technologyreview.com/s/611138/microsoft-is-creating-an-oracle-for-catching-biased-ai-algorithms/.

Letzer, R. (2018), "Chinese Researchers Achieve Stunning Quantum Entanglement Record", *Scientific American*, Springer Nature, 17 July, https://www.scientificamerican.com/article/chinese-researchers-achieve-stunning-quantum-entanglement-record/.

Marcus, G. (2018), 'Innateness, AlphaZero, and Artificial Intelligence", arxiv.org, Cornell University, Ithaca, NY, https://arxiv.org/ftp/arxiv/papers/1801/1801.05667.pdf.

McKinsey Global Institute (2018), "Notes from the AI frontier: Insights from hundreds of use cases", discussion paper, McKinsey & Company, New York, April, https://www.mckinsey.com/featured-insights/artificial-intelligence/notes-from-the-ai-frontier-applications-and-value-of-deep-learning.

Mearian, L. (2017), "Blockchain integration turns ERP into a collaboration platform", *Computerworld*, IDG, Framingham, MA, 9 June, https://www.computerworld.com/article/3199977/enterprise-applications/blockchain-integration-turns-erp-into-a-collaboration-platform.html.

National Research Council (1999), *Funding a Revolution: Government Support for Computing Research*, The National Academies Press, Washington, DC, https://doi.org/10.17226/6323.

New, J. and D. Castro (2018), "How Policymakers can Foster Algorithmic Accountability", Information Technology and Innovation Foundation, Washington DC, https://itif.org/publications/2018/05/21/how-policymakers-can-foster-algorithmic-accountability.

OECD (2018a), "Going Digital in a Multilateral World, Interim Report of the OECD Going Digital Project, Meeting of the OECD Council at Ministerial Level", Paris, 30-31 May 2018, OECD, Paris, http://www.oecd.org/going-digital/C-MIN-2018-6-EN.pdf.

OECD (2018b), "AI: Intelligent machines, smart policies: Conference summary", *OECD Digital Economy Papers*, No. 270, OECD Publishing, Paris, https://doi.org/10.1787/f1a650d9-en.

OECD (2018c), *Meeting Policy Challenges for a Sustainable Bioeconomy*, OECD Publishing, Paris, https://doi.org/10.1787/9789264292345-en.

OECD (2017), *The Next Production Revolution: Implications for Governments and Business*, OECD Publishing, Paris, https://doi.org/10.1787/9789264271036-en.

OpenAI (2018), "AI and Compute", OpenAI blog, San Francisco, 16 May, https://blog.openai.com/ai-and-compute/.

Pratt, G.A. (2015), "Is a Cambrian Explosion Coming for Robotics?", *Journal of Economic Perspectives*, Volume 29/3, AEA Publications, Pittsburgh, DOI: 10.1257/jep.29.3.51.

Ransbotham, S et al. (2017), "Reshaping Business with Artificial Intelligence: Closing the Gap Between Ambition and Action", *MIT Sloan Management Review*, Massachusetts Institute of Technology, Cambridge, MA, https://sloanreview.mit.edu/projects/reshaping-business-with-artificial-intelligence/.

Robinson, L. and K. McMahon (2016), "TMS launches materials data infrastructure study," *JOM*, Vol. 68/8, Springer US, New York, https://doi.org/10.1007/s11837-016-2011-1.

Simonite, T. (2016), "Algorithms that learn with less data could expand AI's power", MIT Technology Review, May 24th, Boston, https://www.technologyreview.com/s/601551/algorithms-that-learn-with-less-data-could-expand-ais-power/.

Singer, P.L. and W.B. Bonvillian (2017), "'Innovation Orchards': Helping tech startups scale", Information Technology and Innovation Foundation, Washington DC, http://www2.itif.org/2017-innovation-orchards.pdf?_ga=2.11272691.618351442.1529315338-1396354467.1529315338.

Sverdlik, Y. (2018), "Google is Switching to a Self-Driving Data Center Management System", Data Center Knowledge, August 2nd, https://www.datacenterknowledge.com/google-alphabet/google-switching-self-driving-data-center-management-system.

Teresko, J. (2008), "Designing the next materials revolution", *IndustryWeek*, Informa, Cleveland, 8 October, www.industryweek.com/none/designing-next-materials-revolution.

The Economist (2017), "Oil struggles to enter the digital age", *The Economist*, London, 6 April, https://www.economist.com/business/2017/04/06/oil-struggles-to-enter-the-digital-age.

The Economist (2015), "Material difference", Technology Quarterly, *The Economist*, London, 12 May, www.economist.com/technology-quarterly/2015-12-05/new-materials-for-manufacturing.

U.S. Council on Competitiveness (2014), "The Exascale Effect: the Benefits of Supercomputing for US Industry", U.S. Council on Competitiveness, Washington, DC, https://www.compete.org/storage/images/uploads/File/PDF%20Files/Solve_Report_Final.pdf.

Vujinovic, M. (2018), "Manufacturing and Blockchain: Prime Time Has Yet to Come", *CoinDesk*, New York, 24 May, https://www.coindesk.com/manufacturing-blockchain-prime-time-yet-come/.

Walker, J. (2017), "AI in Mining: Mineral Exploration, Autonomous Drills, and More", *Techemergence*, San Francisco, 3 December, https://www.techemergence.com/ai-in-mining-mineral-exploration-autonomous-drills/.

Williams, M. (2013), "Counterfeit parts are costing the industry billions", *Automotive Logistics*, 1 January, Ultima Media, London, https://automotivelogistics.media/intelligence/16979.

Wissner-Gross, A. (2016), "Datasets Over Algorithms", *Edge.org*, Edge Foundation, Seattle, https://www.edge.org/response-detail/26587.

Zeng, W. (2018), "Forest 1.3: Upgraded developer tools, improved stability, and faster execution", Rigetti Computing blog, Berkeley, https://medium.com/rigetti/forest-1-3-upgraded-developer-tools-improved-stability-and-faster-execution-561b8b44c875.

Zweben, M and M.S. Fox (1994), *Intelligent Scheduling*, Morgan Kaufmann Publishers, San Francisco.

Chapter 3. Perspectives on innovation policies in the digital age

By

Caroline Paunov and Dominique Guellec

Most innovations today are new products and processes made possible by digital technologies or embodied in data and software. This transformation took place first in digital sectors (e.g. software) but has now spread to all sectors, including services (e.g. retail and education) and manufacturing (e.g. automotive). It results in new dynamics, with data as core inputs to research and innovation, more service innovation, the blurring of boundaries between services and manufacturing (servitisation), and greater speed and collaboration in innovation. Innovation policies need to adapt, so as to address data access issues, to become more agile, to promote open science, data sharing and co-operation among innovators, and to review competition and intellectual property policy frameworks. This chapter first assesses the economic mechanisms that characterise digitalisation and reviews the impacts of the digital transformation on innovation in the digital age. It then discusses how these changes affect business dynamics. Based on these insights, it draws lessons for innovation policies and concludes by providing perspectives on the future.

Introduction

Most innovations today are new products and processes, enabled by digital technologies or embodied in data and software. These digital innovations are an outcome and a component of digital technologies, which allow collecting, processing, manipulating, storing and diffusing data automatically, using machines. Progress in electronics (Moore's law) and data science have introduced a new way of using technologies. Advances in artificial intelligence (AI), a set of technologies that emulate certain aspects of human intelligence, promise further progress in the manipulation of digitalised information and knowledge.

These changes are driven by advances in science and innovation, and are themselves drivers of science and innovation. Many dimensions of the digital world differ from the physical, tangible world, and innovation processes and outcomes are being transformed as a consequence. Although this transformation first occurred in the digital sectors, it is now widespread and involves many tangible sectors, such as the agro-food and automotive industries. For example, the Internet of Things (IoT) connects the physical and digital worlds, allowing every object and location in the physical world to become part of the digital world.

With those sweeping transformations under way, it is pertinent to evaluate whether – and in what directions – policy support for innovation should adapt. This chapter assesses the economic mechanisms that characterise digitalisation and reviews the impacts of the digital transformation on innovation in the digital age. The chapter discusses how these changes affect business dynamics. Based on these insights, it draws lessons for innovation policies and provides perspectives on the future.[1]

Changes in innovation characteristics induced by the digital transformation

Digital technologies have lowered information-related production costs and changed the characteristics of innovation (Figure 3.1).

The processes and products embodying or implementing digital technologies are characterised by their "fluidity". Fluidity means that data can circulate, and be reproduced, shared or manipulated instantaneously, at any scale and at no cost. Once available, digitised knowledge (i.e. knowledge that takes the form of data) can be shared instantaneously between any number of actors, notwithstanding geographic distance and other (natural or institutional) barriers, with each actor having full access to it (OECD, forthcoming). This characteristic affects all economic processes, like the commercialisation of new products and the diffusion of knowledge. Fluidity allows scaling up to serve entire markets much more rapidly, i.e. achieving "scale without mass", facilitating both competition by new entrants and "winner-takes-all (or most)" market dynamics. This ease of scaling digital goods contrasts with tangible goods, which are subject to physical production and distribution constraints (e.g. manufacturing and transportation costs).

Figure 3.1. Characteristics of innovation in the digital age

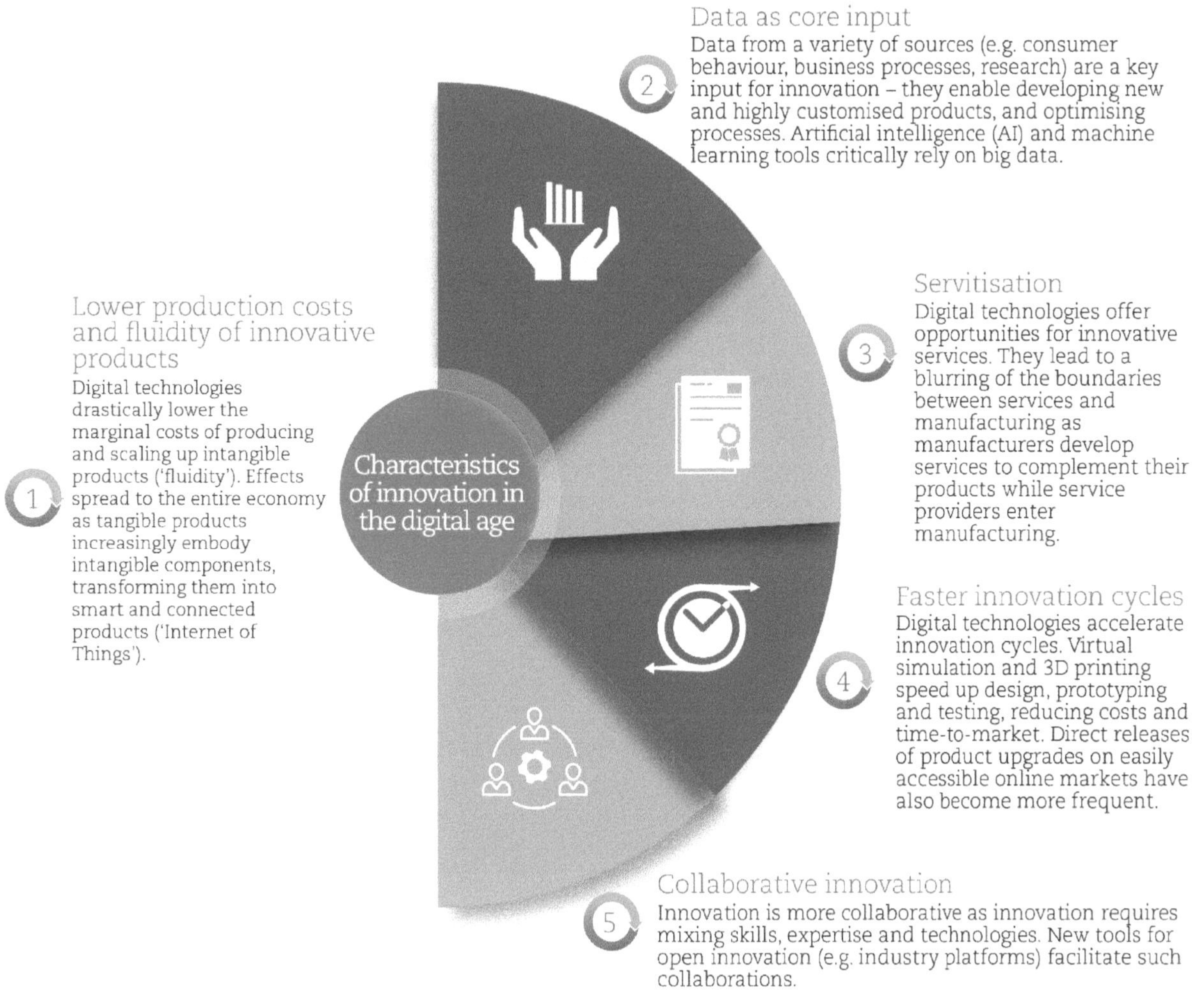

Data as core input
Data from a variety of sources (e.g. consumer behaviour, business processes, research) are a key input for innovation – they enable developing new and highly customised products, and optimising processes. Artificial intelligence (AI) and machine learning tools critically rely on big data.

Lower production costs and fluidity of innovative products
Digital technologies drastically lower the marginal costs of producing and scaling up intangible products ('fluidity'). Effects spread to the entire economy as tangible products increasingly embody intangible components, transforming them into smart and connected products ('Internet of Things').

Servitisation
Digital technologies offer opportunities for innovative services. They lead to a blurring of the boundaries between services and manufacturing as manufacturers develop services to complement their products while service providers enter manufacturing.

Faster innovation cycles
Digital technologies accelerate innovation cycles. Virtual simulation and 3D printing speed up design, prototyping and testing, reducing costs and time-to-market. Direct releases of product upgrades on easily accessible online markets have also become more frequent.

Collaborative innovation
Innovation is more collaborative as innovation requires mixing skills, expertise and technologies. New tools for open innovation (e.g. industry platforms) facilitate such collaborations.

Digital technologies have drastically reduced several types of costs, notably: 1) the marginal costs of producing intangible-intensive goods and services; 2) the costs of searching, verifying, manipulating and communicating information and knowledge; and 3) the costs of launching new goods and services – specifically those with high information and knowledge content – on the market (Haskel and Westlake, 2017). The costs of verifying the reputation and trustworthiness of potential partners through digital technology such as blockchain are lower. This increases the chances a successful search will result in actual matches (between supply and demand of labour, inputs, products, etc.) on the market, thereby reducing production costs and improving product quality (Goldfarb and Tucker, 2017).

Digital technologies are also increasingly embedded in many tangible products. They transform them into smart, connected products (e.g. connected cars and agricultural machinery equipped with sensors) that are able to produce and exchange data about their own status and performance, or the environmental conditions around them (the Internet of Things, IoT). Based on the data they generate, these products are key enablers of a wide range of services and process innovations. For instance, IoT applications can be used to track in real time the trajectory and storage conditions of food throughout the supply chain.

New possibilities for handling data have made them core inputs of innovation in all sectors of the economy (OECD, 2015a). Data feed into innovations in multiple ways; for example, data on consumer behaviour can be used to customise services or to develop entirely new services (such as on-demand mobility services like Uber, which rely on instantaneous information about demand and supply to organise transportation). Data generated in production processes (e.g. managerial and technical data), public-sector data (e.g. transportation and patient files) and research data (e.g. experimental data) are less visible, but equally important. All these types of data are relevant – albeit to different degrees – to innovation.

In this context, the deployment of AI and machine learning further increases the expected value of data. Machine learning requires large numbers of observations before the software is able to perform the expected task, although much research is currently taking place in AI to reduce the amount of data needed to train a program. The development of the IoT also means that data generation is increasing steadily, as more devices and activities are connected.

Because of data's growing importance, many businesses make large investments to access data, whether by setting up data-gathering systems, acquiring data-rich companies (Microsoft notably acquired LinkedIn to take control of its data) or contracting with partners. At the same time, many businesses still need to develop best applications and data analytics infrastructures to bring value from data analytics to their business.

The digital transformation also creates opportunities for innovation in services as digital technologies reduce costs, while allowing greater fluidity in reaching and interacting with consumers, and tracking their behaviour. In particular, innovation opportunities arise for: 1) new services, such as predictive-maintenance services using IoT data, on-demand transportation services and web-based business services; 2) renting as a service or sharing instead of selling equipment; and 3) customising products (i.e. adapting products to each customer's specific needs, thanks to software and data capabilities).

Servitisation is disruptive to business practices, as it removes the boundaries between manufacturing and services, and requires entirely new business models. Many manufacturing firms' strategy and innovation activity now follows the "3 S" model: sensors, software and service. For instance, Bosch has installed software-monitored sensors on many of its car parts, which allow the company to offer its customers better maintenance services. Conversely, service firms like Amazon and Google are also entering the manufacturing industry, producing home appliances, mobile phones, computer chips, etc.

The lower cost of launching new products and processes using the Internet and online platforms facilitates versioning and experimenting products for differentiated customers. Lower costs can also produce more frequent innovation: software can be updated daily (or even more frequently), with new versions downloaded from the Internet. The changes are widespread, extending far beyond the purely digital sectors. In the automotive industry, although the hardware (the car itself) might last for years with little change, the software is frequently updated. In the music industry, the reduced cost of disseminating music through the Internet has generated an increase in creation, to satisfy consumers' highly differentiated and fast-changing tastes.

In addition to the reduced costs of launching and diffusing products, another driver of the digital transformation is the cumulative nature of upgrades, reducing product cannibalisation (i.e. the creative destruction of its own product by a company): when a firm issues an innovation, it may simply add to products that are already on the market and it can be downloaded as an "add-on". Contrary to a new car model, for example, the new digital product will not replace the firm's existing products; rather, it will enhance them.

The acceleration in versioning and innovation is not synonymous with more rapid technological progress and productivity. Many of these improvements are small. Technical change may have become more staged and continuous, but is not necessarily more rapid. Nonetheless, access to these incremental innovations benefits end consumers as they have access to advanced versions. If consumer feedback on versions is integrated effectively in innovation processes, then versioning may also boost innovation.

Where "superstar" effects are in place, a small advantage over competitors might allow a firm to seize all of the market – hence increasing the expected reward in case of a successful (even minor) advance. This also increases the risk for firms, as a setback or a lag – however small – could mean losing all of the market. This creates competitive pressure and, consequently, firms have an interest in updating and launching new versions to gain or maintain lead positions, even at the margin.

Thanks to the reduced costs of (and greater need for) collaboration, innovation has become more collaborative. The reduced costs come from the growing role of data in collaboration, whereas the greater need comes from the evolution of demand (e.g. addressing grand challenges, or designing mobile phones to integrate knowledge from various fields). This enhanced collaboration can take different forms and follow different paths: data sharing, open innovation, innovation ecosystems, platforms (hubs), mergers and acquisitions (often driven by the need to combine various types of competences), and global value chains (which integrate technology in successive stages, along an ordered line).

Successfully harnessing the potential of digital technologies requires combining different technologies used for varying purposes into coherent systems. Actors may also engage in collaborative innovation processes to hedge against the risks from disruptive innovations by competitors; these risks will be higher in the context of general-purpose technologies (GPTs).

New forms of open innovation allow collaborating much more actively than previously with large communities of experts and consumers. External sourcing practices (procurement) involving tournaments, collaborations, open calls and crowdsourcing are new ways for firms to address innovation challenges; some of these practices could become permanent, while others could be one-off only. Examples of corporate initiatives include the BMW Customer Innovation Lab, IBM InnovationJam, Dell IdeaStorm, Procter & Gamble's Connect+Develop and GE Fuse (Board of Innovation, n.d.[5]). These practices are also conducted through intermediary online platforms, such as Innocentive, IdeaConnection, Innoget, Hypios and NineSigma.

Finally, digital technologies can be characterized as relatively young, far-ranging and fast-evolving GPTs, affecting all sectors of the economy. Hence, their current and future development generates much uncertainty. This is particularly true of AI, a set of technologies that can emulate functions normally accomplished by human intelligence, based on pattern recognition and prediction. Not only is AI expected to transform economic activity, it also raises complex societal and ethical issues. However, this transformation could take some time, as the number of possible applications is far greater than the number of current applications (Brynjolfsson, Rock and Syverson, 2017). Although recent research points to decreasing productivity of innovative activities over the past few decades (Bloom et al., 2017), some scholars expect AI to reverse this trend (Cockburn, Henderson and Stern, 2018).

Changes in market structures and dynamics

The transformations in innovation processes and outcomes affect business dynamics and market structure, with consequences on the distribution of performance and rewards among businesses, individuals and regions.

On the one hand, as large volumes of data are fluid and potentially available to everyone at a low marginal cost (notwithstanding obstacles to data access, which can be substantial, but are due to market actors, not to physical costs), the costs of market entry and expansion for new firms requiring such data are lowered. Hence, digitalisation potentially creates a more level playing field in terms of access to data inputs (providing that no regulatory or strategic barriers are in place). Increasingly digitalised information and knowledge become accessible to all, creating more equality of opportunities. This applies not only to many scientific or public-sector databases, but also to certain valuable private-sector data (e.g. scientific publications subject to copyright). For example, the US National Institutes of Health database[2] allows researchers to access information on privately and publicly funded clinical studies from around the world, including study protocols, purposes and results. The database of Genotypes and Phenotypes also provides access to data and results from studies that have investigated the interaction of genotype and phenotype in humans (Sheehan, 2018). Such potentially widespread and free access contrasts with physical goods that do not allow for such widespread access and use.

This increased access to data has spurred dynamic entrepreneurial activity based on digital innovation in several markets. These include the transportation sector (with the emergence of platform-based car-sharing and ride-hailing applications) and retail (with the emergence of start-ups specialised in data analytics, to optimise inventories and personalise sales). Many highly successful start-ups have been created by students using digital technologies and data to illustrate these new dynamics of the intangible economy. Famous examples include Facebook (Mark Zuckerberg), Snapchat (Evan Spiegel), Dropbox (Arash Ferdowski and Drew Houston) and Invite Media (Nat Turner).

Entrepreneurial activity linked to disruptive business models has also helped improve consumer welfare in ways that are not always easy to assess. For example, digital maps, encyclopaedias and social media have massively improved consumer welfare. However, the disruptive business models behind those services mean that routinely used metrics – such as gross domestic product (GDP) – are no longer adequate to capture the improvements, requiring novel approaches to track them (Box 3.1). Work conducted in the context of the OECD-wide Going Digital project documents often unmeasured contributions of the digital economy to well-being.

Box 3.1. In my view: GDP and well-being in the digital economy

Erik Brynjolfsson (Massachusetts Institute of Technology [MIT] and National Bureau of Economic Research) and Avinash Collis (MIT)

One of the fundamental objectives in economics is to assess people's well-being. Economists, policy makers and journalists routinely use changes in GDP and metrics derived from it – such as productivity – as proxies for changes in well-being. However, GDP was never meant to be a measure of welfare. It is a measure of production. In some cases, GDP and welfare are correlated, but in many other situations, this is not the case. In

> fact, in some cases, the change in GDP can even have the opposite sign from the change in welfare.
>
> Treating GDP as a proxy for welfare is particularly problematic for digital goods, such as online encyclopaedias, search engines, social media and digital maps. Most of these are available at zero price to consumers and are therefore largely excluded from GDP. As the production and consumption of such goods grows, GDP does not change, but welfare does increase. A growing number of goods are transitioning from traditional physical goods to free digital goods. While these types of goods were counted in GDP measures, they are excluded from GDP once they transition to free digital goods. The encyclopaedia industry offers an excellent illustration of such a transition. Previously, people bought and paid for physical copies of encyclopaedias, such as Encyclopaedia Britannica, and these transactions contributed to GDP. Over the past 15 years, however, Wikipedia has replaced Encyclopaedia Britannica as the premier reference source. Because it has zero price, Wikipedia is excluded from GDP measures. As a result, the contribution of encyclopaedias to GDP has decreased, because people have shifted from paying for Encyclopaedia Britannica to consuming Wikipedia for free. Nonetheless, consumers are clearly better off.
>
> In theory, consumer surplus is a better measure of consumer welfare than GDP. In practice, it is challenging to measure consumer surplus in a scalable manner, since this requires estimating demand curves. In Brynjolfsson, Eggers and Gannamaneni (2016), we propose a new way of directly measuring consumer welfare, using massive online choice experiments while staying within the neoclassical framework. Our approach takes advantage of the fact that in recent years, it has become much easier to collect data online on a large scale. These advances have been essential to creating alternative measures of the economy, including ours. Our approach can be scaled easily to hundreds of thousands of goods, by running several thousand choice experiments every day. This approach can be implemented more frequently than the consumer price index and can be used to track changes in well-being over time. Moreover, goods – including non-market goods – can be easily added or removed from the basket.
>
> The system of national accounts centred on GDP was one of the greatest inventions of the 20th century. In the 21st century, the proliferation of digital data, combined with an infrastructure that allows surveying millions of people easily, cheaply and quickly, provides an opportunity to develop new measures of welfare. These can be used to supplement and extend existing national accounts.

Digital platforms also enable entrepreneurship by lowering set-up costs for newcomers. For example, e-commerce platforms (e.g. Alibaba, Amazon and eBay) allow new ventures to offer products to the market without paying extra for marketing. Such platforms also gather very accurate information on the activities of the companies that use them (e.g. who their customers are, how their sales are evolving, and what they spend on marketing); this puts them in a favourable position to provide funding to these companies, as the information asymmetry (a usual barrier to funding small and medium-sized enterprises [SMEs]) is minimal. For example, Amazon proposes a range of financial products to businesses trading on its platform (OECD, forthcoming).

But the fluidity of data may contribute to industry concentration thanks to three factors. One factor is the natural advantage of platforms (defined as Internet-based structures that organise the interaction between different actors) in increasing market efficiencies. Important efficiency gains can be derived from combining data to exploit optimally the information and knowledge they contain, providing natural advantages to large aggregators of data. Similarly, providing combined services on a single platform, and bringing together a larger group of users, offers considerable consumer benefits. In other words, several small platforms that provide fewer services, have fewer users each and build on less data would be much less efficient than a single, large and more diversified platform. Such economies of scale are characteristic of a natural monopoly.

The second factor promoting concentration is "scale without mass", a consequence of the increasingly intangible composition of products. The larger the intangible component, the easier it is to expand production to the entire market, at little or no supplementary cost. In the case of software, the cost of producing an additional unit is close to zero, as no further set-up costs are involved. The much smaller number of employees relative to the sales of certain digital companies compared to companies operating in traditional industries illustrates this dynamic. At the same time "scale without mass" allows successful competitors to grow quickly, as fewer overhead costs are incurred even as production is expanded to the full market.

A third factor is the scarcity of certain factors – notably skills – that are complementary to data and are required to exploit data efficiently (OECD (2017a) and Nedelkoska and Quintini (2018)). Such scarcity also tends to favour concentration: up to a certain group size, skilled workers are more efficient when employed jointly (in certain firms or regions) thanks to intra-team knowledge exchanges.

The balance between the factors favouring and hampering concentration varies over time and sectors, and is influenced by policies. Polarised market structures, characterised simultaneously by the dynamics of concentration and massive new entry, are also possible. Such market structures have a few giants, with a long tail of smaller and fast-changing niche producers, and a shrinking space for medium-sized businesses. Using data from a retailer with both online and offline channels, Brynjolfsson, Hu and Smith (2010) show that the variety of products available and purchased online is higher than for those offline, reflecting more opportunities for niche products in the online economy.

Similar distribution and dynamics apply to other economic variables, i.e. the incomes of individuals (with diverse skills, positions and employers) and the wealth of places (with large cities increasing their advantage over rural regions). Skewness is reinforced by the fact that markets are now globally integrated; in the past, national borders shielded places, people and firms from foreign competition, limiting global concentration.

Creating value out of data requires complementary assets – namely, individual skills, collective and organisational competencies (i.e. the right institutional setting to exploit information), and data-assessment tools. In the digital age, data are the main input to many production processes; these data are fluid, contrary to the physical inputs that prevailed previously and limited mobility. The best performers can access and use many of the data available (whereas they could hardly access and use all of the physical resources available), leveraging their advantage more than in the past, where the lowest performers could still secure easier access to certain resources. Any entrepreneur can potentially access a wide range of data and leverage their efficiency advantage, however small (as the whole market becomes integrated). This is true at the individual level, allowing top entrepreneurs to command larger production teams and take decisions with key data (Garicano and Rossi-Hansberg, 2006); at the organisational level, allowing firms with the strongest capacities to leverage data better; and at the geographical level, as the top cities or regions worldwide can access and exploit a wide range of available data to build their prosperity (Kerr and Kominers, 2015). The growing prosperity of cities also reflects the complementarity of non-codified social knowledge with codified, digital knowledge. Gaspar and Glaeser

(1998) suggest that the reduced communications costs may most benefit those that already communicate much, meaning that falling costs would benefit cities most, further driving concentration.

Implications for innovation policies

The new context and features of innovation require changes to the targets, mechanisms, instruments of innovation policies and to the policy mix of innovation. This is because, as discussed in the previous sections, digitalisation is affecting essentially all mechanisms that drive innovation, exactly those mechanisms that innovation policies are targeting. Therefore, all innovation policy instruments are affected (Figure 3.2). Some instruments will adapt their target or content to digital innovation while essentially preserving their processes; that includes for instance policies supporting entrepreneurship, SMEs or generic technologies. Other domains will go through in-depth transformations, including sometimes of their rationale: this includes science policy, with its move towards open science and or policies supporting university-industry linkages, with a move towards co-creation.

Figure 3.2. Policy issues and instruments requiring change to be effective in the digital age

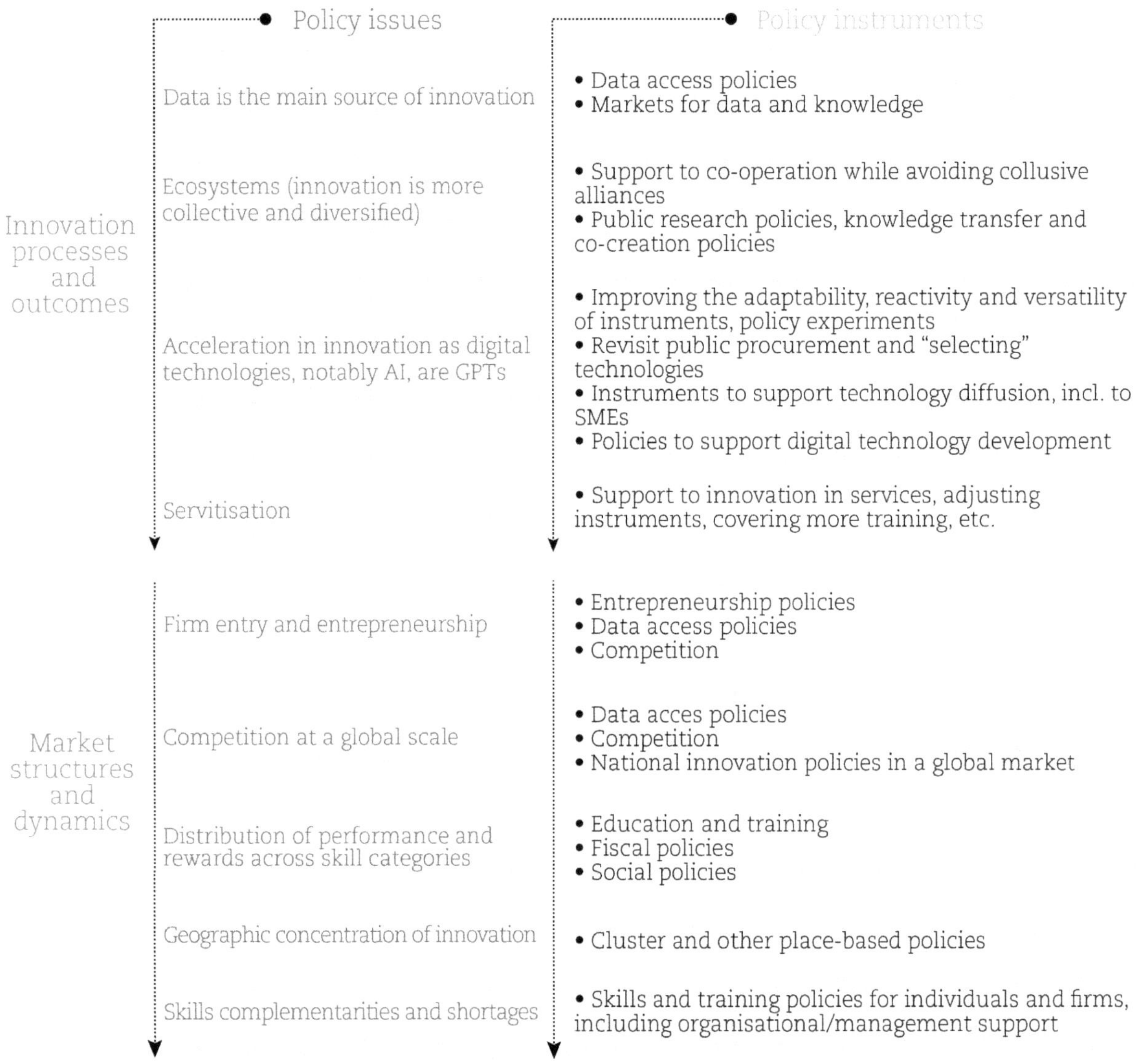

	Policy issues	Policy instruments
Innovation processes and outcomes	Data is the main source of innovation	• Data access policies • Markets for data and knowledge
	Ecosystems (innovation is more collective and diversified)	• Support to co-operation while avoiding collusive alliances • Public research policies, knowledge transfer and co-creation policies
	Acceleration in innovation as digital technologies, notably AI, are GPTs	• Improving the adaptability, reactivity and versatility of instruments, policy experiments • Revisit public procurement and "selecting" technologies • Instruments to support technology diffusion, incl. to SMEs • Policies to support digital technology development
	Servitisation	• Support to innovation in services, adjusting instruments, covering more training, etc.
Market structures and dynamics	Firm entry and entrepreneurship	• Entrepreneurship policies • Data access policies • Competition
	Competition at a global scale	• Data acces policies • Competition • National innovation policies in a global market
	Distribution of performance and rewards across skill categories	• Education and training • Fiscal policies • Social policies
	Geographic concentration of innovation	• Cluster and other place-based policies
	Skills complementarities and shortages	• Skills and training policies for individuals and firms, including organisational/management support

Source: Guellec and Paunov (2018).

This section discusses eight principles for the design of innovation policies in the digital age (Figure 3.3).

First, as **data now constitute new input to innovation**, access to data – and to the tools that gather and help interpret them – will influence who participates in digital innovation, and in what ways. Innovation policy must therefore address data access. The goal should be to ensure the broadest access to those data and knowledge that facilitate competition (e.g. through alternative uses of the same data), re-use (i.e. producing a gain in efficiency) and transparency (e.g. creating the ability to check the validity of results obtained on a given dataset). However, data-access policy has to take into account the diversity of data, as access issues differ across data categories, as well as economic and non-economic constraints. This includes incentives to produce the data in the first place, competition, intellectual property, privacy and ethics.

Figure 3.3. Eight principles for innovation policies in the digital age

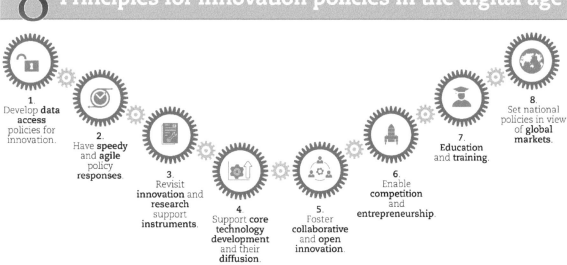

Certain data (e.g. customer data or product-design data) are trade secrets. They cannot be shared without endangering the firm's competitive position, or even its very existence. Opening access to such data might allow firms with the most effective data-processing capabilities to take control of the relevant markets, turning established companies into suppliers and possibly reducing competition, as data-based markets are more prone to "winner-takes-all (or most)" dynamics than other markets.

Government should also create the appropriate conditions to promote the emergence of data markets. The development of knowledge markets, which previously focused on intellectual property (IP) rights and now encompass all data, has been viewed positively by economists (Yanagisawa and Guellec, 2009). Not only does trading data facilitate the exchange of data for innovation purposes, but it also allows putting a price tag on data generation and curation for future use, facilitating the generation of more data.

Second, accelerated innovation cycles owing to digital innovation should be matched by **adequate policy experiments** to support innovation, which means rethinking the types of instruments used and their implementation.

Approaches to ensure rapid and agile policy responsiveness include policy experiments that operate in "start-up mode" where experiments can be deployed, evaluated and modified, and then scaled up or down, or abandoned quickly. Using digital tools to design innovation policy and monitor policy targets is another option to spur faster and more effective decision-making. For instance, some governments use "agent-based modelling" (a form of AI) to anticipate the impact of policy variants on different types of businesses. Another approach is to shift emphasis from instruments that target specific groups of recipients or technologies to ones that are more flexible. Such instruments include tax reliefs, certain regulations, and intellectual property (IP) rights, as well as simplified innovation support schemes (e.g. 'sector-agnostic' and "single-window" grant application processes). Mission-oriented programmes that set a goal, but do not impose the means to reach it, could help. Of course, the specific drawbacks of such instruments (e.g. the lack of selectivity, resulting in a deadweight loss) compared to targeted instruments should be considered and weighted against the advantage of greater flexibility. Another option is to provide the necessary autonomy and agility to choose the proper technological avenues to achieve a stated policy objective. In the United States, the Defense Advanced Research Projects Agency has successfully boosted fundamental defence research, thanks to its organisational flexibility at the administrative level and the significant authority granted to programme directors (Azoulay et al., 2018). Similar programmes have been adopted by other countries, including Canada, to spur game-changing technological breakthroughs.

Third, traditional **support tools for research and innovation should be revisited** to ensure their effectiveness. Service innovation, which receives little support from traditional instruments, is progressing and sector boundaries are increasingly blurred; technological change can take unexpected directions, owing to the novel application of digitalisation to traditional technological fields, which can generate surprising and sudden changes in the technological trajectory. To provide an example of services tools, the Netherlands has implemented an experimental scheme called 'service design vouchers for manufacturing SMEs' to support manufacturing SMEs in developing services that are related to their products.

The functioning of the intellectual property (IP) system is also changing and requires policy attention. To take but one example, AI can create patentable inventions. This raises the question of who should own them: the original AI programmer, the user of the AI software that generated the invention, or the owners of the data to which AI is applied? In addition, patent grants require that the invention be "non-obvious to a person skilled in the art". If an AI system is considered to be such a "person", this might put the bar much higher for patentability in certain domains where AI is now a major research tool (e.g. pharmacy or combinatorial chemistry). However, trademarks may gain new importance as anchors for online search (Bechtold and Tucker, 2014).

Fourth, **policy should support the development of core generic (or multi-purpose) digital technologies** to facilitate downstream innovation and address societal challenges. Businesses are currently investing heavily in these technologies. Initial technological developments were primarily sponsored by governments. This is true not only of the Internet, but also of AI – which was developed almost exclusively through academic research for more than five decades, before businesses got involved in the late 2000s. Hence, governments need to keep investing in core technologies to prepare future waves of innovations. They also need to ensure these multi-purpose digital technologies are developed to serve not only commercial purposes, but also social and environment purposes. Public research is often best placed to do just that. Such investments benefit from collaboration in technology development and around AI's economic, ethical, policy and legal implications. Institutions such as the Digital Catapult Centre in the UK were created to promote the early adoption of advanced digital technologies by innovative firms, for instance, by facilitating access to advanced technology testbeds to experiment and prototype new IoT products and services; and providing the computational power and expertise needed to develop AI solutions.

Aside from development, technology diffusion and adoption also deserve specific policy attention, with differences across firms and sectors requiring the application of suitable diffusion support services. An example of a policy initiative is Germany's SME 4.0 Competence Centres that support SMEs to be aware of, test and adopt new digital technology solutions for their business, each centre focusing on specific technologies or application areas. Another example is the CAP'TRONIC programme in France which aims to help SMEs enhance their competitiveness by integrating digital solutions and embedded software in their products. SMEs participating in the programme can access technical seminars, trainings and workshops, counselling services and expert support to develop their digital innovation projects.

Along the same lines, governments should **apply digital technologies to their own activities, including public research** (e.g. data gathering, analysis, sharing, simulation etc.). This includes the following:

- Increasing data access: data are a core driver of open science, which is widely seen as a way of increasing the quality and reducing the cost of research. Open access allows reusing data, reproducing results, testing a diversity of hypotheses on the same empirical basis, facilitating cross-disciplinary collaboration, etc. (OECD, 2015a; and Dai, Shin and Smith, 2018).

- Offering specific training and capacity-building activities: scientists need to master digital tools (e.g. data curation, simulation and deep learning), so that they can either implement them or collaborate with team members who are using them. For example, enhancing researchers' digital skills is one of the key objectives of Norway's digitalisation strategy for the higher education sector, 2017-20 (Government of Norway, 2018).

- Developing research tools and infrastructures: new instruments (e.g. data-sharing platforms and super-computing facilities for AI) may be critical to research and require new investments. Japan's High Performance Computing Infrastructure programme, for example, requires an annual investment of more than USD 120 million (US dollars) to build a high-performance computing infrastructure that universities and public research centres can use to conduct R&D in various fields.

- Engaging in partnerships: research organisations should partner with industry to leverage industry progress in advanced digital technologies, with a view to applying it to public research.

Fifth, growing interactivity and collaboration in innovation justify policies supporting co-operation and **open innovation** between industry and academia, but also among businesses. The reduced cost of collaboration stemming from digitalisation has not reduced the barriers to collaboration (such as differing regulatory regimes and diverging incentives), but it has made the social cost of not collaborating higher, as more opportunities are lost. Such policies need to consider new forms of collaboration towards innovation. Online platforms, in particular, support small-scale entrepreneurship, by offering opportunities to identify adequate niche markets. Many governments have created platforms where public research and universities can advertise their inventions, knowledge and capacities, and businesses can post their own needs. The two sides can then interact and agree on deals. Other ways to support collaboration include new types of cluster policies, such as Canada's Innovation Superclusters Initiative.

Sixth, support for **competition and entrepreneurship** is needed to find the right balance in the digital age between static efficiency – where scale benefits are important – and dynamic efficiency – which drives innovation. This is a complex area, where the fundamentals of competition policy are called into question by digital innovation in the presence of network effects, standards, etc. (OECD, 2016, 2017b, and 2018a). For instance, it is difficult to determine exactly what constitutes a "dominant position", as market positions are permanently threatened by new entrants. Arguably, digital innovation requires firms to be large, in order to achieve economies of scale; hence, weakening dominant firms (e.g. through aggressive anti-trust action) could weaken innovation. Data concentration may also shape competition dynamics (OECD (2016). On the other hand, several small firms and regulators have complained that large companies engage in certain behaviours (e.g. product tie-ins or preventive takeovers) that may hamper competition and innovation, as they prevent small players from accessing the market. Policies that recognise economies of scale, while ensuring equal access to markets and resources, would help support the long tail of firms (particularly SMEs) and regions (including rural areas with limited innovation capacities) (see the report by Planes-Satorra and Paunov (2017) on inclusive innovation policies).

Seventh, **preparing individuals for the digital transformation** is essential to increase the pool of skilled workers and empower their participation. It is important that innovation authorities collaborate with those in charge of education and labour market policies to ensure the right skills needed for digital innovation are being developed. Innovation authorities have an important role to play in informing other government authorities of new skills demands as businesses engage in digital innovation and that arise with rapid and broad technological change. There are often new mixes of skills for innovation, e.g. innovation in the automotive industry increasingly requires strong capabilities in software engineering and AI, in addition to traditional core competences in mechanical and electronic engineering. Fostering interdisciplinarity (particularly of computer sciences with specific traditional disciplines) is increasingly important, requiring interdisciplinary degrees with an important digital component (see, for example, MIT undergraduate degrees on computer science and biology, and on computer science, economics and data science) (MIT, 2018).

Eighth, data fluidity creates the need to set **national policies targeting global markets**. Digitalisation facilitates the circulation of knowledge, including across national borders, reducing governments' ability to restrict the benefits of policies to their own country. While data sharing clearly generates benefits at a global level, data distribution across countries is not equal. Governments must facilitate data access across borders, while ensuring that ethical and economic standards are respected.

Responding to the new imperatives of the digital transformation, several STI strategies place objectives related to digital transformation at the core of their strategic orientations, often in active consultation with the public (Box 3.2). Developing these strategies also requires engaging with the public to establish a social licence, by demonstrating the beneficial aspects of these technologies and addressing public concerns through better information and appropriate action (e.g. protecting privacy and developing certain applications for the public good). A lack of engagement with society creates the risk of a significant future backlash, with negative impacts on the development and deployment of these technologies.

Box 3.2. STI strategies aiming to achieve digital transformation

- Germany's *New High-Tech Strategy* sets priorities for research and innovation, listing the "digital economy and society" as its first priority. The High-Tech Strategy supports science and industry's implementation of Industry 4.0. It considers the successful development and integration of digital technologies within industrial application sectors as key to the country's future competitiveness. It also supports smart services, big-data applications (particularly focusing on SMEs), cloud computing, digital networks, digital science, digital education and digital-life environments.
- The Estonian *Research and Development and Innovation Strategy 2014-20, "Knowledge-based Estonia"*, aims to increase the economy's knowledge intensity and competitiveness. It identifies information and communication technologies (ICT) (e.g. their use in industry, cybersecurity and software development) as one of three key priority areas for investment in research, development and innovation. The other two priority areas are resource efficiency, and health technologies and services.
- *France Europe 2020: A Strategic Agenda for Research, Technology Transfer and Innovation* places research at the centre of France's policy priorities. It views research (including basic research) as key to addressing the main emerging scientific, technological, economic and social challenges, and promoting competitiveness. France Europe 2020's priorities include strengthening research in breakthrough technologies, and investing in digital training and infrastructures.
- Slovenia's *Smart Specialisation Strategy* includes Industry 4.0 as one of the three key priority areas for action. It highlights the need to optimise and digitalise production processes, and to apply a range of enabling technologies (e.g. robotics, nanotechnologies and modern production technologies for materials) to specific priority areas (e.g. smart buildings, the circular economy and mobility).
- Austria's *Open Innovation Strategy* is the country's response to the challenges of digital transformation and globalisation. Its main objective is to open up, expand and further develop the innovation system in order to boost its efficiency and output orientation, and improve innovation actors' digital literacy. The Open Innovation Strategy formulates 14 measures around 3 action areas: 1) developing a culture of open innovation and teaching open-innovation skills among all age groups; 2) creating heterogeneous open-innovation networks and partnerships across disciplines, industry branches and organisations; and 3) mobilising resources and creating adequate framework conditions for open innovation.
- Japan's Fifth Science and Technology Basic Plan emphasises the importance of achieving "Society 5.0", also defined as a "super-smart society". To that end, it sets the development of cutting-edge ICTs and the IoT as top science and technology policy priorities. The Basic Plan also encourages further developing AI, while minimising risks and limiting automated decision-making.

Source: Planes-Satorra and Paunov (forthcoming).

The future of innovation policies in the digital context

With the rise of AI, digital innovation will continue to expand and even accelerate, involving all fields of technology and all types of innovation. AI allows using large quantities of data more effectively and applying digitalisation to new areas, like driving vehicles. The coupling of AI and the IoT will create virtual twins of most real-world processes, creating a basis for more innovation. Consequently, the trends identified in this chapter will continue: the role of data in the innovation process will continue to grow, innovation will accelerate, and knowledge will become more fluid. Changing innovation policies to reflect these trends will become even more important.

More than a piecemeal approach, a broader strategy is needed, which factors in the profound changes to innovation caused by digitalisation. This strategy would reshape innovation policies and link them more closely with other policy areas. What was formerly a "plus" is now a "must": with digitalisation, most products are new products, and innovation becomes ubiquitous. Consequently, innovation is directly affected by all policy domains, and what happens in other domains could affect innovation. Better linking policy areas to sustain innovation is a major challenge for governments in the digital era. Several governments have become aware of this issue (Box 3.2), but they are still at an early stage in conceptualising and developing integrated responses, and much learning will be needed in the future.

Governments will also benefit from digitalisation, using digital technologies to adapt their policies and improve policy design, implementation, monitoring and evaluation (Chapter 12 on digital science and innovation policy). The availability of larger amounts of data, and the ability to analyse them more rapidly, will strengthen policy processes. Data are available on all aspects of the innovation process, i.e. technologies, firms, innovation projects, innovation funding, business creation, and, crucially, government policies and programmes themselves. By implementing the appropriate analytical tools, governments will be able to improve their diagnosis (e.g. of technological trends and obstacles to innovation across corporate categories), in order to adopt and evaluate the corresponding policies. They could do this rapidly, facilitating policy experimentation. The way forward also requires developing strategies to leverage and interpret different data sources, to achieve informed decision-taking in a fast-changing environment (Brynjolfsson and Mitchell, 2017).

This chapter covered five dimensions of change for innovation in the context of the digital transformation and identified policy implications. Based on this assessment, the chapter outlined several key implications for innovation policy. The "In my view" box, by Luc Soete (Box 3.3), describes other dimensions of change in the digital age and outlines the new challenges posed by digitalisation. It also discusses challenges for innovation policy as it aims to become more agile and contribute to achieving wider societal goals.

Box 3.3. In my view: Digitalisation and innovation policy

Luc Soete, University of Maastricht

How has current digitalisation affected innovation processes and outcomes? Let me just add to the five dimensions listed in this chapter a few of what I would call "low hanging digital-fruit opportunities". First, the ubiquitous use of data as core input presents opportunities going beyond pure consumer behaviour and is now willingly encroaching on other aspects of human behaviour, i.e. social interactions, attention-seeking, interactive entertainment, health diagnostics, political choices and many more. Second, advertising, has a new and central role, now fully transformed from a supply-based attention-seeking activity, to an information-service and participatory activity. Third, opportunities exist to exploit hidden and underutilised sources of capital – both physical capital (trading, sharing and renting out flats; driving services; second-hand goods; and equipment of all sorts) and human capital (activating underutilised talents and skills). We observe this almost daily in our perception of the current digital age.

The coming digital age, however, raises many more new challenges. These can be best described in terms of the further diffusion and development of some key GPT features of digitalisation, such as AI, robotics and machine learning. These GPTs are likely to lead to further optimisation in production, distribution and service provision, and to increased predictability, also allowing full autonomy. The extensive use of – and access to – data as the core, essential input is likely to cover all sectors – not just personal data on social media, customers and transaction data, and patient data, but also data on education and learning, on delivered public-administration services (such as taxes and social security), and all sorts of behavioural data. The sky is the limit. The likely impact will now go way beyond consumers and economics, influencing citizens in everything they do, including their employment, possible deskilling or reskilling and job security.

The chapter considers eight principles of innovation policy that are crucial to the coming digital age. While this list may seem complete, it resembles a mixture of well-known policy challenges, mostly unrelated to "digitalisation" and specific new digital issues – such as the first issue, access to data.

Moreover, some of the objectives may be hard to reach, e.g. there are limits to the "agility" of innovation policy when it comes to developing speedy and agile policies. How experimental can regulatory policy be? The country examples provided when discussing outcome-focused and anticipatory regulation are interesting, but can they be generalised? The "innovation principle" proposed by the European Commission also comes to mind, but it is quite difficult to implement in reality.

In terms of policies, I would propose an alternative approach, focusing more explicitly on the possible conflicts or trade-offs between, on the one hand, the current policy challenges discussed in the chapter (i.e. privacy/protection; public data sharing versus private data ownership; regulatory boundaries when data go beyond consumer or customer data, such as in the case of patient data) and on the other hand, the future digital challenges and opportunities (i.e. how to enable production and distribution optimisation across the board; how to develop machine learning and predictability, including autonomy, and within which sets of rules and responsibilities; what kind of public-private interactions using individual data; and how to address future employment concerns).

> When confronted with such intertemporal potential conflicts or trade-offs in policy making, it would be best to focus on the future digital opportunities in achieving societal goals, such as the "grand challenges" or the Sustainable Development Goals, as guiding principles for the future digital "direction" the coming digital age should take. In a certain sense, the rate and speed of the digital transformation is "out of control". There is very little that governments or policy makers can do, apart from facilitating its further diffusion through increased training and education in relevant areas of data analysis, AI, robotics and machine learning. However, many citizens across the OECD member countries and beyond are asking more fundamental questions, such as the purpose of these new technologies. What is AI good for? What problems will machine learning solve? In my view, the emergence of the new digital age presents policy makers with a unique opportunity to focus innovation policy on the "direction" of technical change. That direction is ultimately a public responsibility, which governments should pursue readily and actively.

Notes

[1] The chapter builds on the digital and open innovation project of the OECD. Guellec and Paunov (2018) provides a more detailed discussion on the implications of the digital transformation on innovation policy. This work builds on and contributes to the OECD-wide Going Digital project.

[2] http://www.clinicaltrials.gov.

References

Azoulay, P. et al. (2018), "Funding Breakthrough Research: Promises and Challenges of the "ARPA Model"", *NBER Working Paper*, No. 24674, National Bureau of Economic Research, Cambridge, MA, http://dx.doi.org/10.3386/w24674.

Bechtold, S. and C. Tucker (2014), "Trademarks, Triggers, and Online Search", *Journal of Empirical Legal Studies*, Vol. Forthcoming, http://dx.doi.org/10.2139/ssrn.2266945.

Bloom, N. et al. (2017), "Are Ideas Getting Harder to Find?", *NBER Working Paper*, No. 23782, National Bureau of Economic Research, Cambridge, MA, http://dx.doi.org/10.3386/w23782.

Board of Innovation (n.d.), *Open Innovation & Crowdsourcing Examples*, https://www.boardofinnovation.com/list-open-innovation-crowdsourcing-examples/ (accessed on 12 April 2018).

Brynjolfsson, E., F. Eggers and A. Gannamaneni (2016), "Using Massive Online Choice Experiments to Measure Changes in Well-being", *NBER Working Paper*, No. 24514, National Bureau of Economic Research, Cambridge, MA, http://dx.doi.org/10.3386/w24514.

Brynjolfsson, E., Y. Hu and M. Smith (2010), "Research Commentary: Long Tails vs. Superstars: The Effect of Information Technology on Product Variety and Sales Concentration Patterns", *Information Systems Research*, Vol. 21/4, pp. 736-747, http://dx.doi.org/10.1287/isre.1100.0325.

Brynjolfsson, E. and T. Mitchell (2017), "What can machine learning do? Workforce implications", *Science*, Vol. 358/6370, pp. 1530-1534, http://dx.doi.org/10.1126/science.aap8062.

Brynjolfsson, E., D. Rock and C. Syverson (2017), "Artificial Intelligence and the Modern Productivity Paradox: A Clash of Expectations and Statistics", *NBER Working Paper*, No. 24001, National Bureau of Economic Research, Cambridge, MA, http://dx.doi.org/10.3386/w24001.

Cockburn, I., R. Henderson and S. Stern (2018), "The Impact of Artificial Intelligence on Innovation", *NBER Working Paper*, No. 24449, National Bureau of Economic Research, Cambridge, MA, http://dx.doi.org/10.3386/w24449.

Dai, Q., E. Shin and C. Smith (2018), "Open and inclusive collaboration in science: A framework", *OECD Science, Technology and Industry Working Papers*, No. 2018/07, OECD Publishing, Paris, http://dx.doi.org/10.1787/2dbff737-en.

Garicano, L. and E. Rossi-Hansberg (2006), "Organization and Inequality in a Knowledge Economy", *The Quarterly Journal of Economics*, Vol. 121/4, pp. 1383-1435, http://dx.doi.org/10.1093/qje/121.4.1383.

Gaspar, J. and E. Glaeser (1998), "Information Technology and the Future of Cities", *Journal of Urban Economics*, Vol. 43/1, pp. 136-156, http://dx.doi.org/10.1006/JUEC.1996.2031.

Goldfarb, A. and C. Tucker (2017), "Digital Economics", *NBER Working Paper*, No. 23684, National Bureau of Economic Research, Cambridge, MA, http://dx.doi.org/10.3386/w23684.

Government of Norway (2018), *Digitalisation strategy for the higher education sector 2017-2021*, Ministry of Education and Research, Oslo, https://www.regjeringen.no/en/dokumenter/digitalisation-strategy-for-the-higher-education-sector-2017-2021/id2571085/sec5 (accessed on 29 May 2018).

Guellec, D. and C. Paunov (2018), "Innovation policies in the digital age", *OECD Science, Technology and Industry Policy Papers*, No. (forthcoming).

Haskel, J. and S. Westlake (2017), *Capitalism without Capital : The Rise of the Intangible Economy*, Princeton University Press, Princeton.

Kerr, W. and S. Kominers (2015), "Agglomerative forces and cluster shapes", *The Review of Economics and Statistics*, Vol. 97/4, pp. 877-899, http://dx.doi.org/10.1162/REST_a_00471.

MIT (2018), *Interdisciplinary Undergraduate Degrees < MIT*, MIT Course Catalog - Bulletin 2017-18, http://catalog.mit.edu/interdisciplinary/undergraduate-programs/degrees/ (accessed on 29 May 2018).

Nedelkoska, L. and G. Quintini (2018), "Automation, skills use and training", *OECD Social, Employment and Migration Working Papers*, No. 202, OECD Publishing, Paris, http://dx.doi.org/10.1787/2e2f4eea-en.

OECD (forthcoming), *Vectors of the Digital Transformation*, OECD Publishing, Paris.

OECD (2018a), *Rethinking Antitrust Tools for Multi-Sided Platforms*, http://www.oecd.org/competition/rethinking-antitrust-tools-for-multi-sided-platforms.htm (accessed on 24 September 2018).

OECD (2018b), *Going Digital in a Multilateral World - Document for the meeting of the Council at the Ministerial Level, 30-31 May 2018*.

OECD (2017a), *Future of work and skills - Paper presented at the 2nd Meeting of the G20 Employment Working Group*, http://www.oecd.org/els/emp/wcms_556984.pdf (accessed on 24 May 2018).

OECD (2017b), *Algorithms and Collusion: Competition Policy in the Digital Age*, http://www.oecd.org/daf/competition/Algorithms-and-colllusion-competition-policy-in-the-digital-age.pdf (accessed on 24 September 2018).

OECD (2016), *Big Data: Bringing Competition Policy to the Digital Era*, http://www.oecd.org/daf/competition/big-data-bringing- (accessed on 24 September 2018).

OECD (2015a), "Making Open Science a Reality", *OECD Science, Technology and Industry Policy Papers*, No. 25, OECD Publishing, Paris, http://dx.doi.org/10.1787/5jrs2f963zs1-en.

OECD (2015b), *Data-Driven Innovation: Big Data for Growth and Well-Being*, OECD Publishing, Paris, http://dx.doi.org/10.1787/9789264229358-en.

Planes-Satorra, S. and C. Paunov (2017), "Inclusive innovation policies: Lessons from international case studies", *OECD Science, Technology and Industry Working Papers*, No. 2017/2, OECD Publishing, Paris, http://dx.doi.org/10.1787/a09a3a5d-en.

Planes-Satorra, S. and C. Paunov (forthcoming), "The digital innovation policy landscape in 2019 (working title)", *OECD Science, Technology and Industry Policy Papers*, No. (forthcoming), OECD Publishing, Paris.

Sheehan, J. (2018), *Digital Health Innovation:Policy and Standards*, Presentation to the OECD Digital Health Innovations workshop, 12 April 2018, the Hague, https://www.innovationpolicyplatform.org/system/files/Panel%201.5.%20Sheehan.pdf (accessed on 01 June 2018).

Yanagisawa, T. and D. Guellec (2009), "The Emerging Patent Marketplace", *OECD Science, Technology and Industry Working Papers*, No. 2009/9, OECD Publishing, Paris, http://dx.doi.org/10.1787/218413152254.

Chapter 4. STI policies for delivering on the Sustainable Development Goals

By

Mario Cervantes and Soon Jeong Hong

Science, technology and innovation (STI) policies play an important role in helping countries achieve the Sustainable Development Goals (SDGs). However, STI policies and frameworks must embed the SDGs to address them effectively. This chapter identifies and successively discusses in five sections the priority areas for action to embed the SDGs more fully within STI policy frameworks. This includes (1) support for "mission-oriented" R&D partnerships between public research, business and other stakeholders relating to specific challenges; (2) stronger support for interdisciplinary research that is inclusive of gender and citizens; (3) international STI co-operation on "global public goods", such as climate, biodiversity and global public health; (4) closer alignment of national-level STI governance structures with the emerging "global governance framework" for the SDGs; and (5) seizing the opportunities of digital technologies to address the SDGs. Finally, the chapter stresses the need to embrace digital technologies, including the necessary data infrastructures and policies, to help address the SDGs.

Introduction

The age-old adage that "necessity is the mother of invention" is a reminder that since ancient times, humans have invented tools and technologies to satisfy basic human needs, such as shelter, food, water and energy – four of the 17 Sustainable Development Goals (SDGs). The SDGs aim to achieve socially inclusive economic development within the ecological boundaries of the earth's capacity to sustain human activity. However, the challenges they present, and more generally, the "sustainability agenda" itself, bring into question the dominant focus on economic growth and the rate of innovation inherent in most countries' science, technology and innovation (STI) policy frameworks. Of course, economic growth and societal challenges are not mutually exclusive. Some countries have chosen to invest in SDG-enhancing innovation that can be introduced to the market, thereby contributing to their own economic growth.

Figure 4.1. The SDGs

Source: Global Reporting Initiative (n.d.), "Sustainability Disclosure Database", http://database.globalreporting.org.

The SDGs also represent a challenge from the standpoint of STI policy because of their interdependencies. Solutions to achieve the Goals cannot be solely technological: they must also involve social innovation and collaboration with stakeholders, beyond the traditional government-science-industry interface. At the same time, the SDGs themselves only reference STI implicitly, rather than explicitly. For example, innovation features explicitly in only one of the Goals, SDG 9: "to build resilient infrastructure, promote inclusive and sustainable industrialisation and foster innovation" (Figure 4.1). The term "science" is absent in the description of the Goals. Among the 169 targets, 14 targets explicitly refer to "technology", and another 34 relate to goals in technological terms (United Nations, 2015, 2016). The remaining 121 targets include certain technological dimensions, but technology is only one of many means to implement them.

This chapter identifies and discusses the priority areas for action to embed the SDGs more fully within STI policy frameworks. This includes redirecting resources towards specific challenges through "mission-oriented" R&D partnerships between public research, business and other stakeholders. Initiating ambitious international co-operation will need to be initiated to protect, produce and preserve "global public goods" (e.g. climate, biodiversity and global public health). This contrasts with the present situation, where national competitiveness is still the main driver of STI activities. Better interlinkages between development aid and STI policies for SDGs could help leverage limited public resources, especially in developing countries, where societal challenges are especially acute. At a more holistic level, a closer alignment of STI governance structures and functions (e.g. policy advice, steering and funding, co-ordination, and evaluation and monitoring) with the emerging "global governance framework" for the SDGs will be key to co-ordinating these two policy domains. Finally, the chapter stresses the need to embrace digital technologies, including the necessary data infrastructures and policies, to help address the SDGs.

The need to reset overarching STI policy frameworks

STI policy frameworks will need to evolve to pinpoint the challenges raised by the SDGs. Policymakers, scientists, analysts and laypersons are calling for reframing innovation policy to consider not just the changing nature of innovation (i.e. globalised, technological and non-technological, open and digital), but also its responsiveness to societal demands for inclusiveness and other societal challenges, such as epitomised by the SDGs (OECD, 2017; Weber, 2017). This push for a more pro-active and responsive innovation policy is illustrated in the recent calls for "directionality" and "mission-oriented" innovation strategies to tackle grand challenges. Such calls also apply to traditional science policy, reflecting concerns about responsible innovation and research – especially in fields (e.g. artificial intelligence [AI], gene editing and neurosciences) where science and technology move faster than legal and ethical rules. The transition towards open science and open data also challenges purely "national" and "scientific peer-based" science-governance models, rendering science not only more permeable, but also more transparent and accountable to society (Dai, Shin and Smith, 2018).

Reframing STI policy is not straightforward. Pleas for "transforming" innovation-policy frameworks have not outlined clear pathways for policymakers, nor have they proposed new levers for government policy. At best, they have proposed incremental reformulation of traditional supply and demand-side instruments (such as R&D funding, human-capital development, networking and clustering policies, and regulatory and demand-led approaches), by instilling sustainability and directionality considerations (Box 4.1).

Box 4.1. In my view: The progressive evolution of innovation policy towards societal challenges

Ian Hughes, Senior Research Fellow, MaREI Centre, Environmental Research Institute, University College Cork, Ireland

For many years, policymakers have developed innovation models and policy instruments to target investments in science and technology in order to maximise their economic impacts. More recently, the focus of innovation policy has broadened significantly not only to include innovation for economic growth, but also to address the formidable twin challenges of environmental sustainability and sustainable development. This expanded scope means that policymakers increasingly need to use multiple policy framings to achieve the diverse outcomes many governments are now demanding from their investments in innovation.

Innovation for economic growth

For decades, the National Innovation System (NIS) framework, aimed primarily at fostering economic growth, has dominated innovation policy. Innovation policies within the NIS framework aim to stimulate firms to increase their innovation activities in order to spur job creation, boost competitiveness and increase gross domestic product (GDP) growth. The policy instruments under the NIS model include support for basic research in universities; favourable tax treatment and direct subsidies for R&D in firms; and support for creating linkages between the various actors in the system to build their innovative capacities. Such policies include cluster policies, to stimulate collaboration between firms; research centres, to increase links between firms and higher education institutions; education policies, to support firms' absorptive capacities; support for high-growth innovative firms; and support for the commercialisation of public research. The NIS framework remains the central framing used by innovation policymakers today. Its continued importance is reiterated in the OECD Innovation Strategy 2015, which stresses that innovation must continue to provide the foundation for new businesses, new jobs and productivity growth, and is an important driver of economic growth and development.

Innovation for environmental sustainability

The emergence of acute environmental challenges – including climate change, resource depletion and pollution – has led to the recent development of System Innovation (SI), a second framework for innovation policy. SI is a horizontal policy approach combining technologies and social innovations to tackle systemic problems, such as sustainable housing, mobility and health care. It involves many actors outside of government (as well as different levels of government) and takes a longer-term view. While the NIS framework aims to strengthen and enhance the productivity of an existing innovation system, the challenge of attaining environmental sustainability has shown that many current sociotechnical systems are no longer environmentally sustainable. An SI approach, designed to bring about fundamental change in the systems that provide us with energy, food, health and transport (among others), is necessary. Recent OECD work on SI shows that policies aimed at transitioning sociotechnical systems to more environmentally sustainable configurations differ significantly from policies aimed at increasing the economic performance of existing systems (OECD, 2015). Among the challenges facing policymakers in the context of SI is the need to develop a vision of what future sustainable systems will look like, including which technologies are likely to play important roles in the future system; what infrastructures will be needed; and how business models and behavioural patterns will need to change. To facilitate the transition, policymakers will need to lengthen planning and investment horizons; co-ordinate across government ministries and levels; establish and maintain long-term collaborative partnerships; place increased emphasis on diffusing knowledge and existing technology, as well as inventing technology; and manage and overcome resistance to sociotechnical change. As countries respond to the pressing challenge of environmental sustainability, OECD countries are increasingly adopting SI as a supplemental framework to the NIS for guiding innovation-related investment decisions and setting policy objectives.

Innovation for sustainable development and human well-being

With the signing of the UN 2030 Agenda for Sustainable Development, a third challenge for innovation policymakers has emerged, namely innovation for sustainable development. Agenda 2030 aims to deliver a more sustainable, prosperous and peaceful global future, and sets a framework for achieving this objective by 2030. This framework comprises 17 SDGs, which cover the social, economic and environmental requirements for a sustainable future. Innovation will play a key role in achieving the targets across all of the SDGs, most notably concerning good health and well-being; affordable and clean energy; clean water and sanitation; decent work and economic growth; industry, innovation and infrastructure; sustainable cities and communities; responsible consumption and production; and climate action.

A range of emerging disruptive technologies, including AI, robotics, terotechnology, gene editing and biotechnology, have the potential to address many of the challenges in Agenda 2030 and the SDGs. More rapid and equitable diffusion of these technologies will be needed if sustainable development is to be achieved in practice and within the timeframe set down. At the same time, emerging technologies are also raising major ethical, legal, economic, policy and social issues. Anticipating and addressing the wider societal implications of disruptive technologies in both developed and developing countries will be important, not only for protecting the public good, but also for realising the full social and economic potential of technological development.

Both the NIS and the SI frameworks are well articulated, and are increasingly being used by innovation policymakers across the OECD to meet the goals of environmental sustainability and economic growth. A policy framework granting equal weight to sustainable development in decision-making, and placing justice and inclusion at its core, still needs to be developed.

The strategic orientation of research towards the SDGs

The urgency of many global challenges, such as climate change, has revived a long-standing debate on how to apply "mission innovation", defined as "large-scale interventions aimed at achieving a clearly defined mission (goal, solution) within a well-defined timeframe with an important R&D component" (European Commission, 2018). Missions were initially associated with US defence R&D and space programmes, as well as with government-sponsored R&D procurement in areas of national security or independence (such as energy).

Societal needs in areas such as agriculture, health and energy have been recognised in the formulation of modern science policy since the second half of the 20th century, leading to the creation of specialised agencies (e.g. the US National Institutes of Health in the 1940s), research councils and public research laboratories in many OECD countries. However, research-policy agendas have shifted towards environmental and societal challenges in OECD countries since the 2000s (Figure 4.2). The data on national government budget appropriations (GBARD) show an increase in environment and health-related R&D and, to a lesser extent, in earth and space-related R&D. By contrast, growth in the R&D budgets for defence and agriculture has been less strong. Publicly funded energy R&D has also not kept up.

Figure 4.2. Growing societal concerns are changing balances in public R&D budgets

GBARD, OECD index 1981=100, 1981-2015

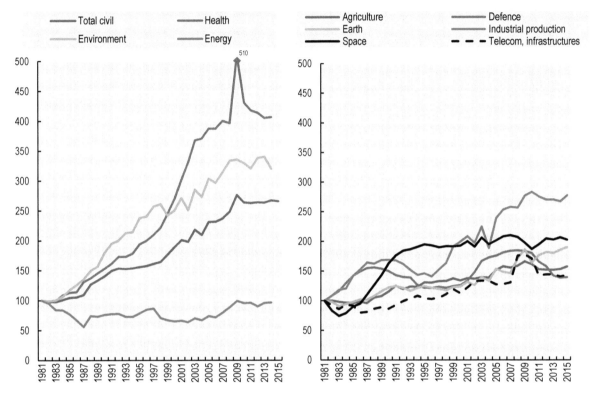

Note: Environment-related R&D budgets include research on controlling pollution and developing monitoring facilities to measure, eliminate and prevent pollution. Energy-related R&D budgets include research on the production, storage, transport, distribution and rational use of all forms of energy, but exclude prospecting and propulsion R&D. Health-related R&D budgets may underestimate total government funding. Efforts to account for the funding of medical sciences through non-oriented research and general university funds help provide a more complete picture.
Source: OECD (2018a), Research and Development Statistics, *www.oecd.org/sti/rds* (accessed on 19 July 2018, IPP.Stat, *https://www.innovationpolicyplatform.org/content/statistics-ipp* (accessed on 19 July 2018).

StatLink https://doi.org/10.1787/888933858164

In 2016, the estimated total public energy research, development and demonstration (RD&D) budget for International Energy Agency (IEA) member countries reached close to USD 16.6 billion (US dollars), just below the 2015 levels. The total public energy RD&D budget of these countries continues to decrease year-on-year, from its recent peak of USD 19.4 billion in 2012. New public and private initiatives – such as the Mission Innovation pledge taken in November 2015 by a group of 20 countries at COP21 in Paris – are attempting to increase investment in renewable-energy R&D and innovation (IEA,

2017). The Breakthrough Energy Coalition is a global group of 28 high net-worth investors from ten countries committed to funding clean-energy companies emerging from Mission Innovation initiatives (Breakthrough Energy Coalition, 2018).

New mission-oriented approaches are also being proposed in the context of the European Union's upcoming Horizon Europe research and innovation programme, which will succeed Horizon 2020. Horizon Europe aims to tackle some of the biggest challenges facing society today, from climate change to inequality, driving collaboration across different industries and bodies in both the private and public sectors (European Commission, 2018). Missions are more concrete than broad "grand challenges", in that they have clear time-bound targets. In a mission-oriented approach, the ambition would not be (for example) "to tackle climate change", but to cut carbon dioxide emissions by a given amount, in a given place over a specified time period. Missions require a "market-shaping" framework, rather than the more traditional and passive "market-fixing" framework focused on correcting market failures (Mazzucato, 2018). Compared to the traditional mission orientation, the new missions focus more clearly on the demand side and the diffusion of innovations; seek coherence with other policy fields; and accept both incremental and systemic innovations.

Lessons from government interventions suggest that although governments have succeeded in some missions (e.g. the Apollo "Man on the Moon" mission), they have also failed in others. These lessons warrant caution, and attention to the design and evaluation of mission-oriented approaches. Some essential interrelated questions arise when analysing the new mission orientation and its potential for addressing global challenges. The technological challenges and measures required to cope with climate change differ radically from those characterising defence and space-related mission R&D programmes, where the main supplier and buyer was the government. Today, the private sector performs most R&D in many OECD countries. Moreover, the outputs of defence and space-related R&D programmes were used by the US Government agencies financing the R&D; hence, transferring the results of R&D from new mission-oriented programmes will not be as straightforward. Without large procurement allowing easy scaling of new technologies, new mission-oriented innovations will probably encounter many of the traditional barriers to technology diffusion and scale.

In many OECD countries, the national governance structures do not appear to favour a "challenge" approach. Such an approach requires strong vertical co-ordination, with significant horizontal alignment. This is especially challenging in countries where ministries have devolved the implementation of strategic research programmes to agencies. To succeed, new mission-oriented approaches will not only need to be linked to the SDGs, but will also require significant levels of funding, as well as specific co-ordination mechanisms involving companies and civil-society actors.

Even before the SDGs emerged as a global agenda for sustainable development, many countries had mobilised STI to address social and environmental challenges, especially at the national level. They relied on a variety of policy instruments, such as supporting public funding programmes in specific sectors, promoting public-private partnerships, introducing regulatory reforms and strengthening governance arrangements. Box 4.2 provides an overview of the frequency of use of expected societal impact when selecting research-project proposals in competitive grant schemes, as declared by policymakers in more than 50 countries in the 2017 EC/OECD STI policy survey (European Commission/OECD, 2017). It also analyses the main societal challenges targeted by STI initiatives designed to address such challenges.

Box 4.2. How are countries orienting their STI funding and policies towards societal challenges and the SDGs?

Information on the criteria for public funding to research was collected on 568 public competitive research grants in more than 50 countries through the 2017 EC/OECD STI policy survey, which gathers quantitative and qualitative data on STI policy (European Commission/OECD, 2017). Figure 4.3 shows that expected societal impact is one of the main criteria used to select projects, ahead of possible commercial applications or even alignment with national goals.

Figure 4.3. Main criteria for funding – competitive research grants

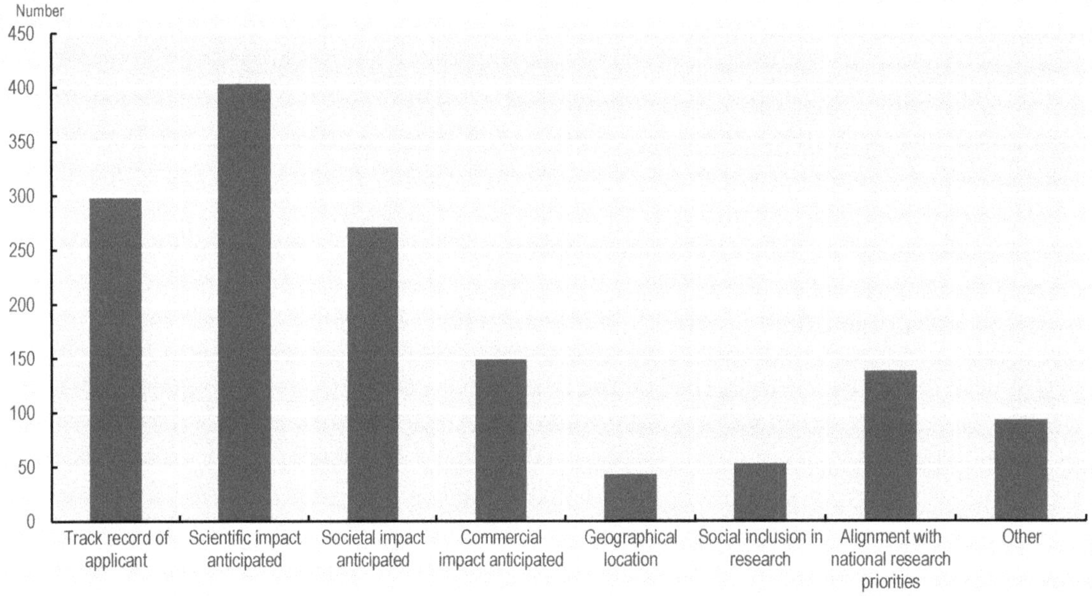

Source: EC/OECD (2017), STIP Compass: International Science, Technology and Innovation Policy (STIP) Database, edition 2017, https://stip.oecd.org.

StatLink https://doi.org/10.1787/888933858183

Figure 4.4 provides a snapshot of 200 STI policy initiatives reported by 17 countries and the EU as targeting societal challenges. These are responses to questions on "research and innovation for society strategy", "research and innovation for health and healthcare", "research and innovation for sustainable development", and "research and innovation for developing countries". Of the 200 policy initiatives, environmental sustainability (SDGs 6, 13, 14, 15) is reported most often as an objective, followed by health and well-being (SDG 3). Energy innovation (SDGs 7) and social development (SDGs 1, 2, 4, 5, 10, and 11) are reported less often.

Figure 4.4. Breakdown of STI initiatives by targeted societal challenges, 2018

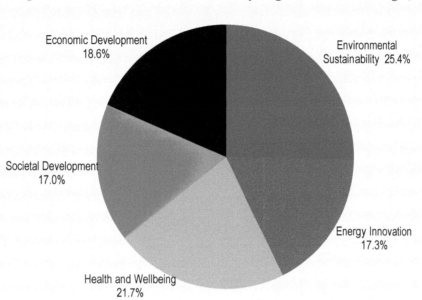

Note: Target sector classification criteria: Environmental sustainability refers to areas related to environmental preservation, global warming and natural ecosystem including clean air, water, land, ocean, greenhouse gas reduction, natural resources management, biodiversity and so on. This target category is relevant to the following SDGs; 6 (clean water and sanitation), 13 (climate action), 14 (life below water), 15 (life on land). Energy innovation refers to areas related to energy efficiency, renewable energy and energy transformation (including electric vehicles) which correspond to SDG 7 (affordable and clean energy). Health and wellbeing refers to areas related to healthcare, disease prevention, vaccination, aging, health promotion and wellbeing which correspond to SDG 3 (good health and wellbeing). Societal development : refers to areas related to make societies and communities more safe, equitable and sustainable which include preventing poverty, quality education, reducing inequalities, demographic change, cities and social infrastructure(including smart city), and so on. This sector is mainly related to the SDG 1 (no poverty), 2 (zero hunger), 4 (quality education), 5 (gender equality), 10 (reduced inequalities), 11 (sustainable cities and communities) and 16 (peace, justice and strong institutions). Economic development refers to areas related to innovation, industry and business development, economic growths which correspond to SDG 8 (decent work and economic growth), 9 (industry, innovation and infrastructure) and 11 (responsible consumption and production)
Source: EC/OECD (2017), STIP Compass: International Science, Technology and Innovation Policy (STIP) Database, edition 2017, *https://stip.oecd.org*.

StatLink https://doi.org/10.1787/888933858202

Of the 200 initiatives in our sample dedicated to societal challenges, 27% provide project grants for public research, 24% for national strategies, agendas and plans, and 12% for grants for business R&D and innovation (Figure 4.5).

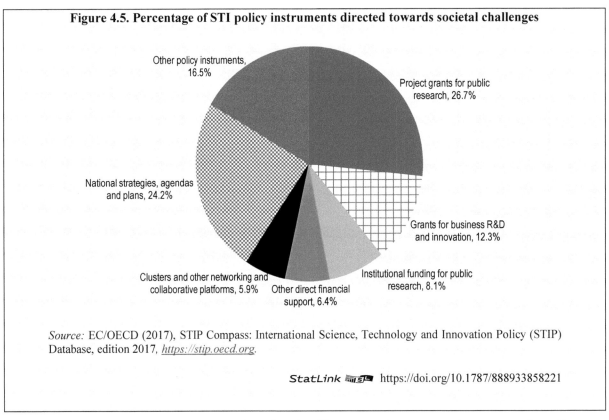

Figure 4.5. Percentage of STI policy instruments directed towards societal challenges

Source: EC/OECD (2017), STIP Compass: International Science, Technology and Innovation Policy (STIP) Database, edition 2017, https://stip.oecd.org.

StatLink https://doi.org/10.1787/888933858221

STI roadmapping is a key strategic policy-intelligence tool to support better targeting of research and innovation activities. Recognising this, the United Nations has called for Member States to develop STI roadmaps for each of the SDGs.[1] Roadmapping was developed by industry to connect short-term (technological) capabilities with long-term strategic goals. Policymakers have adopted and increasingly applied it in the context of large-scale technological or industrial transitions. The shift from industry-led to government-led roadmapping changes the scope of the roadmapping exercise. Rather than focus solely on technical developments, it now includes broader social, political and technological issues. Considering these "sociotechnical" dynamics, a systemic approach – also integrating stakeholder engagement in STI policy design, adaptation and application – may be needed to enhance the effectiveness of STI roadmaps. When framed around "functional needs", STI roadmaps can better inform decision makers to address regulatory, institutional, infrastructural and behavioural changes. Finally, integrating STI roadmapping with other tools (such as patent analysis) can provide insights on the possible contribution of emerging technologies, improve priority-setting and help target demonstration projects. The IEA has developed a new tool to track progress on clean energy R&D investment by technology area and economic sector which could be used to improve STI roadmapping in the energy space (Box 4.3).

> **Box 4.3. Better data to enable STI roadmapping: the case of the IEA "Tracking Clean Energy Progress Tool"**
>
> The new Innovation Tracking Framework of the IEA identifies key long-term "technology innovation gaps" that need to be filled in order to meet long-term clean-energy transition goals. The Framework, which will be progressively expanded and updated, builds on the Agency's leading in-house knowledge and data on technology innovation and investment; its rich history in technology roadmapping and extensive energy-technology trend analysis; and its unique Technology Collaboration Programmes, which bring together expertise from over 6 000 global scientists and engineers in about 40 technology areas.
>
> The Framework has identified around 100 innovation gaps across 35 key technologies and sectors. Innovation gaps within each technology area highlight where R&D investment or general innovation activity needs improvement. To track developments across key innovation gaps over the past year, the IEA has developed a methodology that looks at the following key innovation aspects: investment patterns; key initiatives from the private or public sector; and general technology improvement, using key metrics.
>
> *Source*: IEA (2018), IEA website, *https://www.iea.org/tcep*.

Interdisciplinarity and greater inclusivity

Beyond changes to innovation policy, changes in the performance of scientific research are also necessary. First, interdisciplinary research and transdisciplinary research – which goes beyond research between disciplines to create new disciplines, such as sustainability science – will both be needed to identify positive complimentary interactions in the SDGs, as well as trade-offs that can constrain or cancel progress on other SDGs (International Council for Science [ICSU], 2015).

Second, science policy must also address the issue of gender participation (Chapter 7). Gender equality is one of the 17 SDGs (SDG 5). However, because women participate more in the social sciences than in the natural sciences, they contribute less to the provision of scientific evidence and advice in areas such as climate research and energy research.[2] Moreover, owing to their roles in society, women may suffer the consequences of climate change or poverty more acutely, especially in developing countries. Science policy can play an important role in achieving gender equality: not only should science include women in research education and careers, research designs should also control for gender differences.

Third, science policy will need to recognise and embrace more fully the contributions of citizens in the research priority-setting process, as well as the research enterprise (e.g. citizen science). Citizen participation can be contributory (through the collection and provision of evidence) or collaborative (through mentoring and volunteer activities). Citizen-science activities can also help raise awareness of SDG challenges in local communities and facilitate the behavioural changes necessary to implement social or technological innovations.

These three dimensions of a more inclusive science policy are already having important impacts on the way research priorities are set, funded, evaluated and diffused. Inclusivity might also point towards building scientific capacity within developing countries to help them better harness knowledge production to achieve local goals.

The international STI co-operation imperative

While every country needs STI to meet its own national SDG goals, STI capabilities are unevenly distributed across the globe. Some countries are resource-rich but knowledge-poor, whereas other countries have knowledge that is insufficiently connected to the industrial sector or actual societal needs. International co-operation offers a way for research and innovation actors to come together. It also creates spillovers from technology transfer between companies, research institutions and countries.

Public support for international co-operation in research and innovation is predominantly predicated on enhancing national research excellence, competitiveness and the anticipated returns in terms of national productivity, exports and growth. The "national" perspective in STI policy has served OECD countries very well in the pursuit of economic growth. International co-operation in science, as it emerged in the post-war period, aimed to reinforce national capacities by sharing costs among countries, notably through the creation of international research infrastructures. Meanwhile, the underlying model of "competitive-co-operation" that characterised scientists' interactions helped countries advance their national goals and targets.

Today, this national growth-grounded perspective appears at odds with the need to protect, produce and preserve global public goods, such as a stable climate and biodiversity. The challenge for countries is how to balance their national priorities and goals (e.g. competitiveness and research excellence) and engage in co-ordinated and concerted action at the international level to solve global public-good problems.

Recent OECD analysis based on sample data from ÜberResearch's Dimensions for Funders database, which gathers data from national funding councils, showed that research projects that could relate to one of the 17 SDGs represented only about 11% of the total number of projects funded in 2015. International co-operation occurred in about 2% of these projects, meaning that international co-operation for SDGs represents about 0.2% of all STI projects (OECD, 2017). There exists a lack of dedicated funding for large-scale and longer-term co-operation. Fragmented funding, as well as divergent rules and procedures for research funding, are also a problem. Changing this situation could imply major changes both in the formulation of science and technology national policies and instruments, and the distribution of roles between different actors.

OECD work on international STI co-operation has identified several factors holding back international co-operation, as follows:

- national research focus
- global public-good problems, with individual countries unwilling to pay the costs of action ("tragedy of the commons")
- lack of knowledge of partners' capabilities, especially in developing countries
- lack of trust and legal regimes
- weak intellectual property rights (IPR) protection, especially in less-developed economies
- low government and business capacity in partner countries, including insufficient skills and lack of necessary research infrastructure to enable international co-operation

- national STI governance frameworks that hinder international co-operation if they are not well aligned
- fragmented bottom-up and non-state initiatives (e.g. universities, non-governmental organisations, foundations).

The OECD is currently considering whether to revise its existing principle-based Recommendations on International STI co-operation. These proved useful to countries, by drawing political and funding attention to issues such as IPR enforcement in academic collaboration, and removing barriers to scientist and researcher mobility. The current Recommendations predate the Internet, and were devised at a time when science was less data-driven and intensive. As science becomes increasingly data-driven, international STI co-operation policies and initiatives will need to integrate the data-science infrastructure dimension to ensure relevant data can be accessed and shared among international partners and other stakeholders. Thus, while some of the Recommendations' principles are still valid, they do not offer guidance on how to mobilise contemporary STI for societal challenges, such as the SDGs. Adding new considerations – e.g. incentives for researchers to share their data, while respecting privacy and IPR regulations – would make the Recommendations more relevant.

Moving from a national to an international perspective also means shifting the emphasis from competition to co-operation, including with non-state stakeholders. This may require hybrid funding systems, new types of research bodies and new public-private partnerships that effectively make international STI co-operation for the SDGs and other grand challenges a national priority.

The European Commission's framework programmes are increasingly open to global participation from non-EU countries. The European Union (EU) has increased the number of science and technology agreements with third countries in recent years. EU mobility programmes, like the Marie Curie fellowships, now support researchers from more than 80 countries. The latest communications from the European Commission on Horizon Europe indicate even closer alignment between the EU societal challenges and the SDGs. Indeed, the EU is carrying out a "mapping and gap analysis" of its policies against the SDGs, to determine how STI tools could support actions to fill the gaps or improve policy coherence (European Commission, 2016).

Linking development aid and STI policies

The flow of private-sector capital into developing countries has a major impact on growing new industries, building infrastructure and financing the human-capital development that is essential to STI. Most financial flows from OECD countries to developing countries come from private sources, i.e. investments, migrant remittances and foundations. Financing for STI activities in the context of development assistance remains marginal in absolute terms: according to OECD Development Assistance Committee (DAC) statistics, OECD donor countries only devote around 5% of development assistance to STI activities (OECD, 2017).[3] OECD data also show that philanthropy funding for development, supporting research activities or activities channelled through universities, think tanks, research institutes, etc., amounted to around USD 6 billion over 2013-15 (25% of the three-year total) (OECD, 2017).

There exists growing recognition among donor and recipient countries alike that STI-related official development assistance (ODA) financing could be used to leverage total investment in research and innovation. Donor-country aid agencies and charities, such as the Wellcome Trust, the Bill & Melinda Gates Foundation or Canada's Grand Challenge programme (Figure 4.6), have integrated research and innovation (including social innovation) in their efforts to help developing countries build the necessary government and business capacities to achieve the SDGs.

In response to the 2030 Agenda, the OECD DAC has revised its peer-review methodology. It also agreed in October 2017 on a set of Blended Finance Principles for Unlocking Commercial Finance to the SDGs, which will provide donors with a coherent framework for blending finance activities (OECD, 2018b).

Figure 4.6. Promoting social and technological innovation in developing countries: The approach of Grand Challenges Canada

Define Grand Challenges - significant barriers that, if overcome, will lead to transformational impact.

Launch innovation competitions to find best ideas in low- and middle income countries and Canada.

Invest selectively to catalyze scale and sustainability of the most promising tested innovations.

Model and measure impact to guide resource allocation.

Accelerate impact through provision of individual and collective support to innovator.

Iterate and evolve to improve our value for money.

Source: Grand Challenges Canada, 2018

Many of the vehicles used by aid agencies and charities involve partnerships with firms and community groups to bring new technologies into developing countries. In 2015 alone, USAID was involved in 360 partnerships with the private sector, generating USD 4.9 billion in cash and in-kind contributions. For example, USAID has a long-standing partnership with Merck, which provides doses of the anti-parasite medication Ivermectin to Africa and Latin America, to fight onchocerciasis and elephantiasis. The programme now reaches 250 million people annually, delivering a total of 2 billion doses since its inception in 1987 (National Academies, 2017). The UK Government's aid strategy, Tackling Global Challenges in the National Interest, recognises the importance of research as part of its contribution to aid. The strategy allocates significant new resources to research programmes/initiatives (e.g. the Global Challenges Research Fund and the Newton Fund) to enhance the contribution of science to overcoming key global development challenges.

The entry of research-funding councils and research ministries into ODA programmes has led to some tensions: should collaborative research with developing countries focus only on excellence or should it instead focus on providing technological solutions to development problems? Some argue that there is no trade-off between excellence in science and development sponsored research and indeed there is anecdotal evidence that development funded research is as equally cited as academic research. Another tension is the focus on more applied research and solutions that can be commercialised immediately, as opposed to longer-term basic research projects – yet longer-term basic research is needed to develop institutional learning capacity. Indeed, the mobilisation of development related STI investments will have to confront the challenge of how to translate and scale up solutions so that they address a given challenge and at the same time foster broad-based economic development. Too often STI initiatives in developing countries fail to scale up or to become embedded in a developing country because of lack of entrepreneurship and finance or business environment that is constrained by outdated state regulations or even corruption, which acts as a "tax" on economic activities (Box 4.4).

Box 4.4. In my view: Technology deployment for the SDGs

Alfred Watkins, Chairman, Global Solutions Summit

The global development community has devoted substantial time, attention and resources to encouraging scientists and engineers to find innovative solutions for the SDGs. As a result, we now have proven, effective and affordable solutions to many pressing development problems, including off-grid renewable energy; potable water; off-grid solar-powered irrigation; high-quality community health clinics; and off-grid food storage, refrigeration and processing. These new solutions should (in principle) make it even more affordable and feasible to hit many SDG targets – especially in the least-developed countries, where enormous progress should be possible simply by deploying proven solutions that are already widespread elsewhere. But if this is correct, why are we not on track to achieve the SDGs?

In almost all cases, the binding constraint is not a lack of scientific expertise, technological know-how or proven, cost-effective solutions. The binding constraint is that we have not yet figured out how to address the less glamorous and more mundane organisational, entrepreneurial, financial and business-development issues associated with getting these solutions into the hands of tens – if not hundreds – of millions of people in emerging markets. Tackling this deployment challenge will require progress along a wide range of fronts, almost none of which require scientific expertise. Consider, for example, just a few of the tasks required to supply potable water to the millions of individuals who lack daily access to safe drinking water:

- An innovator or equipment supplier may have developed a cost-effective, efficient and affordable nano-filtration mechanism. But a nano-filter cannot produce potable water without pumps, hoses and cisterns; a power supply (grid, solar, bicycle, diesel), water-quality monitoring equipment; a retail-distribution system; and a payment-collection mechanism. Someone has to organise this supply chain in thousands of communities.
- Those same innovators and equipment suppliers may already be selling purification systems to buyers in the United States or the European Union. However, they do not necessarily have sales contacts in Africa, Asia and Central America, nor do they have the personnel, financial resources and inclination to search for potential customers in numerous far-flung countries. Somebody needs to link the supply of technology with the people who need that technology.

- Somebody has to take responsibility for managing local procurement; organising construction; maintaining and repairing the equipment; obtaining the necessary permits; registering and operating the business; and handling all the other mundane, but essential tasks associated with providing potable water in a single community. In other words, someone – presumably an entrepreneur – has to figure out how to incorporate this game-changing technology into a financially sustainable, efficient and game-changing organisation. The scientist who invented the nano-filter may be an expert in new materials, but may not have the expertise, business acumen, organisational skills and personal inclination to handle these other tasks.
- Last but not least, the households and communities themselves may know in broad terms what they need, but they don't necessarily know where to find it; how to look for it; how to evaluate competing technological solutions; how to organise so many dispersed actors and mundane tasks; how to organise a village enterprise or coop; and how to negotiate terms and conditions with potential partners who are vastly more experienced and sophisticated.

To date, the development community has treated these deployment challenges as an afterthought, on the grounds that – as Ralph Waldo Emerson claimed in 1882 – if we "build a better mousetrap…the world will beat a path to your door." The Global Solutions Summit, convened at UN Headquarters in June 2018, was organised on the premise that technology deployment is not as simple and automatic as Emerson suggested. If that is true, then we can no longer afford to relegate technology deployment to an afterthought in the STI/SDG dialogue. It is an indispensable piece of the puzzle and requires at least as much attention as the quest for new discoveries.

Three important conclusions emerged from the Global Solutions Summit:

Transferring scientific insights from the lab to the last mile should be thought of as a supply chain, with scientists occupying the most upstream position, engineers and inventors in the next spot, and deployment officials filling out the remainder of the supply chain. If STI is going to impact the SDGs, we need mechanisms for passing the baton from scientists and engineers to the diverse groups of non-scientists who are best-suited to implement the essential deployment processes.

Technology deployment requires an effective and efficient deployment ecosystem – one that empowers all the actors in the deployment process to find each other and join forces, and then to transfer the lessons of successful experience from country to country. We need to devote more time and attention to these ecosystem issues.

Bilateral and multilateral development agencies, along with the United Nations, the OECD and others, will not be the ones to deploy these new technologies and development solutions in dozens of countries. They need to figure out how best to empower others – e.g. foundations, NGOs, local entrepreneurs, local universities and technical training institutes – to handle these tasks.

Changing STI governance for sustainability transitions

The contribution of STI to achieving the SDGs will depend on leadership and effective governance arrangements for economic policy making in general and STI systems in particular. At the national level, evidence based on the OECD Country Reviews of Innovation shows that countries' innovation performance depends in part on the quality of STI governance. This quality rests on the set of publicly defined institutional arrangements, incentive structures, etc., that determine how the various public and private actors engaged in socio-economic development interact when allocating and managing resources for STI.

However, national STI governance institutions and structures are not static. Technological and scientific progress, and the global expansion of innovation, have increased the number of actors investing in and setting the agenda for science and technology. Large private firms (such as Alphabet) are investing in basic research in AI. Small entrepreneurial firms are using digital technologies to provide solutions to SDG challenges in developing countries, without any government support. Large charities increasingly shape global agendas for health research, forcing government ministries to re-assess their own priorities. Participatory approaches to STI agenda and priority-setting and evaluation are increasingly common (Chapter 10), as illustrated by the monitoring of the SDGs by independent scientists (Box 4.5).

Box 4.5. Independent scientific advice for monitoring implementation of the SDGs

Before leaving office, former Secretary-General Ban Ki-moon appointed 15 eminent scientists and experts to monitor the implementation of the SDGs and draft the quadrennial Global Sustainable Development Report. The report will be presented to all heads of state at the General Assembly in 2019, without previous government negotiations and agreement. This innovation in UN procedures gives independent scientists an independent say. One of the experts' main tasks will be not to look at the SDGs in isolation, but to study their synergies and possible contradictions. They will also need to consider SDG priorities from the perspective of science and policy.

Source: United Nations (2018), "STI Forum 2018 – Multi-stakeholder forum on science, technology and innovation for the Sustainable Development Goals", 5-6 June 2018, *https://sustainabledevelopment.un.org/TFM/STIForum2018*.

The connection between responsible research and innovation, and the SDGs (e.g. ending poverty; zero hunger; health and well-being; clean water and sanitation; reduced inequalities; climate action; life on land; and peace and justice) is manifest. It reflects the growing scrutiny and accountability underlying the funding of both public and private R&D (Chapter 10). However, STI policy in many OECD countries (and beyond) is driven by an economic rationale: it is a means to correct for market and system failures. STI governance frameworks have not systematically considered sustainability or the knock-on effects of technological progress. In most countries, governments and businesses only deal with the negative or unexpected effects of technological innovations (e.g. neurotoxic pesticides and toxic vaccine adjuvants) once they have emerged.

SDGs bring many challenges to STI governance arrangements and processes. On the one hand, meeting the SDGs and the underlying 169 targets requires greater "directionality" in national research and innovation agendas. On the other hand, interdependence among the various SDG goals means that achieving progress in one goal can leverage progress in another goal, but may also offset progress in yet another goal. Some seemingly effective technologies for solving certain challenges may also generate negative effects on other challenges – for example, solar energy is a zero-carbon renewable source of energy, but solar panels can generate pollution if toxic components are improperly released into the environment. There also exists a risk of conflicting objectives or budgetary arbitrage in the context of limited research funding. The question of how STI is part of the institutional frameworks in countries' national governance systems, and influences public decision-making, is important when designing efficient and acceptable policy tools.

In many countries, the governance of STI policies is still removed from strategic priority-setting, planning and reporting processes for the SDGs. STI data collection has also not caught up with demands for SDG reporting. This is particularly true in developing countries, where STI institutions and co-ordination mechanisms are weak or absent. Until now, STI has not featured prominently in Voluntary National Reviews, which countries undertake voluntarily to report progress on the SDGs at the United Nations High-Level Political Forum, held each July (United Nations, 2018). The UN request for Member States to produce STI roadmaps for the SDGs may lead to closer co-ordination, policy alignment and even integration between the parts of government co-ordinating SDG reporting and those responsible for national innovation strategies.

Policy co-ordination is essential: only a comprehensive and wide-ranging strategy to enhance innovation can help address social and environmental goals, while building a lasting foundation for future economic growth and competitiveness. Current national STI governance approaches are inward-looking and fragmented, while international institutions to drive technological innovation for sustainable development remain relatively weak or are absent altogether.

Several countries, like France, Finland, Brazil and Japan, are attempting to align national STI agendas with the SDGs. The Japanese Government established the SDGs Promotion Headquarters, a new cabinet body comprising all government ministers and headed by the Prime Minister. The purpose of the SDGs Promotion Headquarters is to foster close co-operation among relevant ministries and government agencies, in order to lead the comprehensive and effective implementation of SDG-related measures. The interministerial council adopted the SDG Implementation Guiding Principles in 2016, which represent Japan's national strategy for addressing the major challenges to implementing the 2030 Agenda.

How science will inform the decision-making process in SDG governance systems will depend on the legitimacy, credibility and salience of the contributions of both national and international scientific institutions to the various UN structures (e.g. the High-Level Political Forum and the Global Sustainable Development Report) charged with providing STI input (Box 4.5) (van der Hel and Biermann, 2017). The Technology Facilitation Mechanism supports this process. Its objective is to enhance the effective use of STI for the SDGs, based on multi-stakeholder collaboration between UN Member States, UN entities, civil society, the private sector, the scientific community and other stakeholders (Figure 4.7). If STI is to contribute to the SDGs, its role must be communicated to the public at large: shifting public STI resources from national economy and labour market-related objectives will be difficult without jeopardising acceptance and ownership by the general public (Stramm, 2016). The task of science and technology communities, together with other stakeholders, will be to provide evidence and examples of the various roles STI can play in defining and articulating problems related to the SDGs, and implementing solutions.

Figure 4.7. STI inputs to the SDG process

Schematic illustration

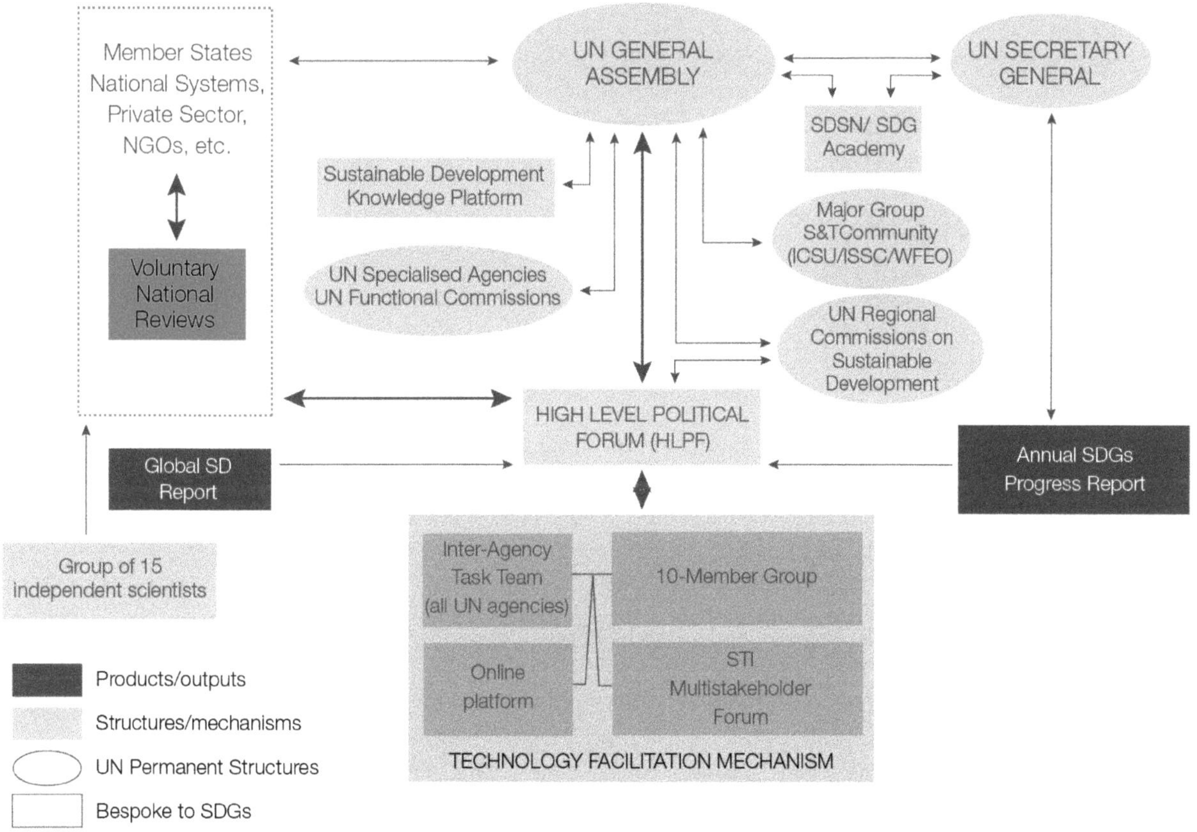

These shematic organograms are illustrative. Arrows reflect institutional links and inputs into different parts of the system, but are not scalar or proportionate.

Source: Adapted from InterAcademy Partnership (InterAcademy, 2017).

One key dimension of the STI governance system is monitoring and measuring the contribution of STI to implementing the SDGs. Monitoring progress on the societal and environmental dimensions of the SDGs will need new indicators. For example, analysis based on detailed budgeting data may provide information on STI "input" commitments to the SDGs, e.g. those relating to poverty or clean water. Intermediate output indicators – such as patents – provide some data and could be used for STI roadmapping exercises.

It is also necessary to explore the contribution of STI through data at the subnational level. New initiatives have developed at the subnational government level: for example, the City of New York's OneNYC[4] has developed indicators based on local data to monitor progress on the SDGs. Non-governmental actors and community groups also help monitor progress: still in the United States, SDG USA[5] conducts research on the measurement and status of US SDGs across the 50 states, highlighting the best state practices and policy options to achieve them.

The drive for improved STI indicators should also capture the multidimensionality and interdependencies inherent in the SDGs. Multidisciplinary research is one example where measurement needs to be improved. The OECD is developing a conceptual approach to measuring transboundary effects (i.e. the impacts of one country's actions on other countries and the contributions to global public goods) within the 2030 Agenda. This approach will begin with a mapping of transboundary effects (which are both explicit and implicit in the SDGs), and a proposal for selecting and assessing relevant indicators (OECD, 2018b).

In parallel, frameworks that measure overall progress on the SDGs (such as the SDG Index and Dashboards, developed by the Sustainable Development Solutions Network (SDSN) and the Bertelsmann Foundation) might do well to support new STI indicator development, e.g. through reciprocal involvement in international statistical bodies (such as the OECD National Experts Group on Science and Technology Indicators, and Eurostat).

The promise of digitalisation

Enabling and converging technologies, notably information and communications technology (ICT), have been a central feature of technological progress. Digital technologies, such as AI, blockchain and 3D printing, hold promise to help accelerate economic development and progress towards the SDGs.

Digitalisation can help existing business solutions scale and disseminate faster. Emerging business models are allowing technologies to diffuse to developing countries, generating positive impact on the SDGs (Table 4.1). Digital solutions can reach people globally, regardless of their income group. Mobile phones and digital payment systems are just two examples of how digitalisation can bring basic banking services to people in developing countries, enabling entrepreneurship and economic activities everywhere.

However, many barriers hinder the deployment of digital technologies, from the need to finance the underlying ICT infrastructure (such as broadband and cloud services) to insufficiently skilled workers who could help firms exploit these technologies. Insufficient, poor or outdated regulation in the ICT sector regarding market access, data privacy and security, and IPR are hampering the deployment of digital technologies, especially in less-developed countries. These impediments to digitalisation are also preventing convergence between ICT and other enabling technologies, including biotechnology (e.g. synthetic biology) and new materials (e.g. graphene), which could help address problems related to human health and agriculture, and reduce carbon dioxide emissions.

Table 4.1. **How the digital transformation can help achieve the SDGs: some examples**

SDG focus areas and targets that benefit most from digital solutions	Possible digital solutions	Digitalisation's potential impact, with illustrative data points
Goal 1: No poverty		
• Scientific education • Data science to support targeted poverty alleviation • Eradicate extreme poverty • Reduce poverty in all its dimensions • Ensure equal rights to economic resources and basic services	• Mobile access to telephony and the Internet, includes need for a device • E-learning • Digital payment systems	• Increases access to opportunities to break free of poverty and improve economic participation • One-third fewer people living on less than USD 1.25 per day thanks to extended Internet coverage
Goal 8: Decent work and economic growth		
• Sustain per-capita economic growth and at least 7 % GDP growth in least-developed countries • Improve global resource efficiency and decouple economic growth from environmental degradation • Achieve full and productive employment and decent work • Reduce youth unemployment • Strengthen capacity of domestic financial institutions and expand access to banking	• Connectivity • E-work, e.g. augmented-reality, cloud-based platforms ("platform as a service"), telecommuting, virtual business • Digital solutions that transform production and consumption patterns	• Boosts growth and helps decouple it from resource consumption • Up to 1.38% GDP growth from 10% increase in broadband penetration • 70% cut in oil consumption in 2030 compared to today from all digital solutions examined
Goal 9: Industry, innovation and infrastructure		
• Infrastructure development • Increase access to ICT and provide universal access to Internet • Develop quality, reliable, sustainable and resilient infrastructure • Promote inclusive and sustainable industrialisation • Upgrade infrastructure and retrofit industries with clean technology • Enhance scientific research and upgrade technological capabilities of industrial sectors, including by increasing the number of R&D workers	• Smart manufacturing, e.g. industrial IoT, machine-to-machine, 3D printing and cyber-physical systems • Data analytics and cloud computing, drones and robotics, embedded system production technology • Smart logistics, e.g. IoT/connected vehicles, load units, products and machines; augmented-reality and wearable technologies; commercial unmanned aerial vehicles; digital warehouses • Optimised fleet and route management • Connectivity, e.g. fixed and/or mobile access to telephony and the Internet; includes need for a device	• Boosts efficient and innovative supply, production and delivery of goods • USD 982 billion in economic benefits to industries from smart manufacturing and smart logistics

Source: GESI (2015), System Transformation: How Digital Solutions will drive progress towards the Sustainable Goals, *http://systemtransformation-sdg.gesi.org/160608_GeSI_SystemTransformation.pdf*.

Data, and related hard and soft digital infrastructures, are important to digitalisation. Much is made of the potential for satellite data to contribute to the SDGs, notably clean water scarcity and sustainable farming. However, access to data – and the computing power and human skills necessary to process and analyse them – is unevenly distributed. Many developing countries lack good-quality government data, as well as basic scientific data on climate, water systems, soil and human health – hence the importance of embedding open-data capabilities in developing countries. The International Science Council, and its Committee on Data for Science and Technology, are working with UN agencies, governments, institutions and other international partners to create regional open-science platforms in Africa, Latin America and the Caribbean (Science International, 2015).

Future outlook

The SDGs aim to achieve economic development that is both socially inclusive and within the ecological boundaries of the earth's capacity to sustain human activity. The main conclusion of this chapter is that for STI to contribute to the three dimensions of the SDGs – i.e. environment, economy and society – the SDGs will need to be more fully embedded within STI policy frameworks. Some avenues for policy action include:

- Instilling greater "directionality" in technology and innovation policies, to focus on the technological and innovation-related targets of the SDGs: this may take the form of challenge or mission-oriented approaches, which must include the demand side and involve stakeholders in policy design and implementation.

- Better use of roadmapping STI for the SDGs, which is a potentially useful tool for identifying technology and technology market gaps: roadmapping should also help address system interlinkages between the various SDGs.

- Stronger support for interdisciplinary research: research should be inclusive of gender and citizens, in order to address the interdependencies inherent in the SDGs.

- Reorienting government-initiated international co-operation in STI towards investments in public-goods problems: co-operation should also foster multi-stakeholder partnerships – including with developing countries – involving business, venture capital and community groups (among others).

- Improved interlinkages between official development assistance and STI policies, including in funding and governance arrangements.

- Better alignment of STI governance arrangements at the national and international levels, with the SDGs at all levels of decision-making, e.g. by linking research agendas and innovation strategies to the SDGs: to meet key sustainable-development challenges, STI actors and institutions must integrate demand and user/citizen/consumer/prosumer perspectives. STI must also play a role in the global governance structures and institutions emerging from the implementation and monitoring of the SDGs at the national, regional and global levels.

- Increased investment in the digital transformation, including in infrastructure and skills will be needed, as well as the removal of outdated regulations that impede technology convergence and the emergence of new business models.

Notes

[1] In the Addis Ababa Action Agenda, UN Member States vowed to "adopt science, technology and innovation strategies as integral elements of our national sustainable development strategies" (para. 119). In the 2017 UN STI Forum, participants highlighted that "the STI roadmaps and action plans are needed at the subnational, national and global levels, and should include measures for tracking progress. These roadmaps incorporate processes that require feedback loops, evaluate what is working and not working, and produce continual revisions that create a real learning environment (IATT, 2018).

[2] The Intergovernmental Panel on Climate Change remains dominated by the contributions of male scientists (80%). This is an improvement over the 1990s, when men performed more than 95% of climate science (Gay-Antaki and Liverman, 2018).

[3] As defined by sector-purpose codes, plus keyword searches in descriptive fields.

[4] https://onenyc.cityofnewyork.us.

[5] https://www.sdgusa.org.

References

Breakthrough Energy Coalition (2018), "Our Commitment", webpage, Breakthrough Energy Coalition, http://www.b-t.energy/coalition/ (accessed 10 September 2018.

Dai, Q., E. Shin and C. Smith (2018), "Open and inclusive collaboration in science: A framework", *OECD Science, Technology and Industry Working Papers*, No. 2018/07, OECD Publishing, Paris, https://doi.org/10.1787/2dbff737-en.

European Commission (2018), *Mission-oriented research and innovation: Inventory and Characterisation of Initiatives*, Study prepared for the European Commission by the Joint Institute for Innovation Policy (JIIP), Joanneum Research, Tecnalia, TNO, VTT, and the Danish Technological Institute (DTI), Publications Office the European Union, Luxembourg, http://dx.doi.org/10.2777/697082.

European Commission/OECD (2017), *STIP Compass: International Science, Technology and Innovation Policy (STIP)* (Database), edition 2017, https://stip.oecd.org.

European Commission (2016), *The role of science, technology and innovation policies to foster the implementation of the sustainable development goals (SDGs)*, Report of the expert group "Follow-up to Rio+20, notably the SDGs" – Study, Publications Office of the European Union, Luxembourg, http://dx.doi.org/10.2777/615177.

Gay-Antaki, M. and D. Liverman (2018), "Climate for women in climate science: Women scientists and the Intergovernmental Panel on Climate Change", PNAS, Vol. 115/9, pp. 2060-2065, https://doi.org/10.1073/pnas.1710271115.

GESI (2015), *System Transformation: How Digital Solutions will drive progress towards the Sustainable Development Goals*, Global e-Sustainability Initiative, Brussels, http://systemtransformation-sdg.gesi.org/160608_GeSI_SystemTransformation.pdf.

Global Reporting Initiative (n.d.), "Sustainability Disclosure Database", http://database.globalreporting.org/.

IATT (2018), "IATT Background Paper: STI for SDG Roadmaps", United Nations Interagency Task Team on STI for the SDGs, https://sustainabledevelopment.un.org/content/documents/19009STI_Roadmap_Background_Paper_pre_STI_Forum_Final_Draft.pdf.

ICSU (2015), *Review of Targets for the Sustainable Development Goals: The Science Perspective, International Council for Science*, Paris, https://council.science/cms/2017/05/SDG-Report.pdf.

IEA (2018), IEA website, https://www.iea.org/tcep/.

IEA (2017), *Energy Technology R&D Budgets: Overview*, International Energy Agency, Paris, https://www.iea.org/publications/freepublications/publication/EnergyTechnologyRDD2017Overview.pdf.

Inter-Academy Partnership (2017) Supporting the Sustainable Development Goals: Merit Based Academies. http://www.interacademies.org/IAP_SDG_Guide.aspx.

ITU (2017), *Fast-forward progress: Leveraging tech to achieve the global goals*, International Transport Union, Paris, https://www.itu.int/en/sustainable-world/Documents/Fast-forward_progress_report_414709%20FINAL.pdf .

Mazzucato, M. (2018), *Mission-Oriented Research and Innovation in the European Union: A problem-solving approach to fuel innovation-led growth*, European Commission, Publications Office of the European Union, Luxembourg, https://doi.org/10.2777/36546.

National Academies (2017), *The Role of Science, Technology and Partnerships in the Future of USAID*, National Academies Press, Washington, DC.

OECD (2018a), *Research and Development Statistics* (database), www.oecd.org/sti/rds (accessed on 19 July 2018, IPP.Stat, https://www.innovationpolicyplatform.org/content/statistics-ipp (data extracted on 19 July 2018).

OECD (2018b), *Better Policies for 2030: An OECD Action Plan on the Sustainable Development Goals*. OECD Publishing. https://www.oecd.org/dac/Better%20Policies%20for%202030.pdf.

OECD (2017), "International co-operation in STI for the grand challenges – insights from a mapping exercise and survey, unpublished paper for official use", OECD, Paris, DSTI/STP(2017)13.

OECD (2016), "The future of science systems", in *OECD Science, Technology and Innovation Outlook 2016*, OECD Publishing, Paris, https://doi.org/10.1787/sti_in_outlook-2016-6-en.

OECD (2015), *System innovation: synthesis report*, OECD Publishing, Paris, https://www.innovationpolicyplatform.org/sites/default/files/general/SYSTEMINNOVATION_FINALREPORT.pdf.

OECD (2010), *OECD Innovation Strategy: Getting a Head Start on Tomorrow*, OECD Publishing, Paris, https://doi.org/10.1787/9789264083479-en.

OECD/IDB (2016), *Broadband Policies for Latin America and the Caribbean: A Digital Economy Toolkit*, OECD Publishing, Paris, http://dx.doi.org/10.1787/9789264251823-en.

Science International (2015), *Open Data in a Big Data World,* Paris: International Council for Science (ICSU), International Social Science Council (ISSC), The World Academy of Sciences (TWAS), InterAcademy Partnership (IAP).

Stamm, A. and A. Figueroa (2016), "Mission Impossible? International Co-operation in STI for Grand Challenges", German Development Institute (DIE), Bonn, https://www.die-gdi.de/en/the-current-column/article/mission-impossible-international-cooperation-in-science-technology-and-innovation-for-grand-challenges-1/.

TWI2050 (2018), "Transformations to Achieve the Sustainable Development Goals", First report prepared by The World in 2050 initiative, International Institute for Applied Systems Analysis (IIASA), Laxenburg, Austria.

United Nations (2018), "STI Forum 2018 – Multi-stakeholder forum on science, technology and innovation for the Sustainable Development Goals, 5+6 June 2018", United Nations, New York, https://sustainabledevelopment.un.org/TFM/STIForum2018.

United Nations (2016), "Perspectives of Scientists on the SDGs", in *UN Global Sustainable Development Report 2017*, Chapter 3, United Nations, New York, https://sustainabledevelopment.un.org/content/documents/10789Chapter3_GSDR2016.pdf.

United Nations (2015), "Transforming our world: the 2030 Agenda for Sustainable Development", United Nations General Assembly, New York, https://sustainabledevelopment.un.org/content/documents/7891Transforming%20Our%20.

van der Hel, S. and F. Biermann (2017), "The authority of science in sustainability governance: A structured comparison of six science institutions engaged with the Sustainable Development Goals", *Environmental Science and Policy*, Vol. 77, pp. 211-220, Elsevier, Amsterdam, https://doi.org/10.1016/j.envsci.2017.03.008.

Weber, K. (2017), "Moving Innovation Systems Research to the Next Level: Towards an Integrative Agenda", *Oxford Review of Economic Policy*, Vol. 33/1, pp. 101-121, Oxford University Press, Oxford, http://dx.doi.org/10.1093/oxrep/grx002.

Chapter 5. Artificial intelligence and machine learning in science

By

Ross D. King and Stephen Roberts

Finding solutions to many of the world's major challenges requires increasing scientific knowledge. Artificial intelligence (AI) has the potential to increase the productivity of science, at a time when some evidence suggests that research productivity may be falling. This chapter first outlines the three key technological developments driving the recent rise in AI: vastly improved computer hardware, vastly increased availability of data and vastly improved AI software. It then describes the promises of AI in science, illustrating its current uses across a range of scientific disciplines. Later sections raise the question of explainability of AI and the implications for science, highlighting gaps in education and training programmes that slow down the rollout of AI in science. The chapter finishes by envisioning a future in which increasingly intelligent AI systems, working with human scientists, help address society's most pressing problems, while expanding scientific knowledge.

Introduction

The world faces many global challenges, from climate change to antibiotic bacterial resistance. Solutions to many – if not all – of these challenges require augmented scientific knowledge. Until quite recently, the role of artificial intelligence (AI) in science received little attention. In the words of Glymour (2004), "despite a lack of public fanfare, there is mounting evidence that we are in the midst of... a revolution – premised on the automation of scientific discovery". Today, AI is regularly the subject of published reports in the most prestigious scientific journals, such as *Science* and *Nature*.

Nevertheless, the scientific community has a poor general understanding of AI. As with many new technologies, opinions polarise towards extremes, from "AI will revolutionise everything" to "AI will have no real impact". The truth, of course, is somewhere in the middle; what is unclear is how close it is to either of the poles. Answering this question is made more complex by the complicated history of AI (Boden, 2006): since its inception in the 1950s, AI has gone through several cycles of enthusiasm and disillusionment.

What differentiates the current situation from previous AI "hype-cycles" is that the underlying computer technology has improved, there are vastly more data, AI is better understood, and – perhaps most importantly as a point of historical difference – the amount of corporate money being invested has increased, and large profits are being made from using AI. Some of the largest companies in the world (e.g. Google, Amazon, Facebook, Tencent, Baidu and Alibaba) have focused their businesses on AI. Taken together, these developments mean that AI will very likely have a huge and growing impact on the world.

As described in Box 5.1, AI has the potential to increase the productivity of science, at a time where evidence suggests research productivity may be falling and new ideas are harder to find (Bloom et al. 2017; Jones, 2005). The use of AI in science could also enable novel forms of discovery, enhance reproducibility and even wield philosophical implications on the scientific process. Three key technological developments are driving the recent rise of AI: vastly improved computer hardware, vastly increased data availability and vastly improved AI software. Several additional factors are also enabling AI in science: AI is well funded, at least in the commercial sector; scientific data are increasingly abundant; high-performance computing is improving; and scientists now have access to open-source AI code. Multiple examples show AI being used across the entire span of scientific enquiry. Furthermore, AI is being applied to all phases of the scientific process, including optimising experimental design.

Box 5.1. What is AI?

AI is the discipline of creating algorithms (computer software) that can learn and reason about tasks that would be considered "intelligent" if performed by a human or animal. "Narrow" AI is the development of solutions to specific tasks that require intelligence, e.g. beating the world's chess or Go champion, driving a car or making a medical diagnosis. "Full" – or general – AI is the development of a system that has equal or greater intelligence to an adult human. It is generally believed that full AI is decades away; hence, this chapter focuses on narrow AI. As AI algorithms focus on the generic ability to learn, rather than solve any particular problem, they are very widely applicable.

At least one current obstacle to achieving the full potential of AI in science is economic. Computational resources, which are essential to leading-edge research in AI, can be extremely expensive. The largest computing resources – and the longest employee lists of excellent AI researchers – are frequently found not in universities or the public sector, but in the private sector. Private-sector work mainly focuses on generating profits, rather than solving outstanding scientific questions. A key policy issue concerns education and training in AI and machine learning (ML). Too few students are trained to understand the fundamental role of logic in AI; most data analysis taught to non-specialists in universities is still based on the classical statistics developed in the early 20th century.

This chapter outlines the technologies driving the recent rise in AI. It describes the promises of AI in science, illustrating its current uses across a range of scientific disciplines. Later sections raise the question of explainability of AI and the implications for science, highlighting gaps in education and training programmes that slow down the rollout of AI in science. The chapter finishes with a vision of AI and the future of science.

Technological drivers are behind the recent rise of AI

Three technological drivers are behind the recent rise of AI:

- *Faster computers*: the modern computer age has been shaped by the exponential increase in computer speeds, in line with "Moore's Law". This means that the supercomputing power needed to beat the world champion (Gary Kasparov) at chess for the first time in 1996 can now fit in a standard mobile phone. To keep up with demand for ever-greater computing power, manufacturers have created a wealth of innovations over the past decades, from multithreading multicore central processing units to large-scale graphics processing units. AI partly owes its recent achievements to the pace of computing advances, allowing AI algorithms to explore complex solutions to large-scale problems. Indeed, some of the most publicised achievements of modern AI, such as playing the game of Go better than any human expert, would not have been possible without vast high-speed computing resources.

- *The scale of data*: with the advent of cheaper sensors, telemetry equipment, ultra-fast computing and cheap data storage at scale, science has undergone a paradigm shift. In a collection of essays published as *The Fourth Paradigm*, Hey et al. (2009) argue that experimental science has undergone a fundamental change. The era of direct experimentation is gone, replaced by the era of data collection. Rather than perform science directly, experiments are designed to record and archive data at an unprecedented scale. Science, namely the evidence-based audit trail of the reasoning of discovery, then takes place within the data. In this sense, much of traditional science has become data science. For most of human history, scientists have observed the universe and the natural world, postulating laws or principles to help generalise the complexity of observations into simpler concepts. Deriving such generalisations from data is akin to finding a hidden structure that is highly explanatory and as such, amenable to intelligent automation.

- *Improved AI software*: significant advances in AI software have taken place in recent years, especially in ML, and more particularly the branch of ML known as deep learning (DL) (Box 5.2).

> **Box 5.2. ML and (deep) neural networks: What are they?**
>
> ML normally refers to the branch of AI focused on developing systems that learn from data. Rather than being explicitly told how to solve a problem, ML algorithms can create solutions by learning from examples (referred to as "training" the ML algorithm).
>
> Often, the terms ML and AI are used interchangeably, and their meaning has certainly changed over the last two decades. From a more recent perspective, ML has grown to encompass data-driven approaches, including traditional computational statistics models, e.g. polynomial regression and logistic classification. In modern parlance, the term AI is used to describe "deeper" models, which have the ability to learn (almost) arbitrarily complex mappings from input to outcome. Such models include deep neural networks and Gaussian processes. Strictly speaking, AI is an extension of ML, augmenting models that learn from example with approaches such as expert systems, logical and statistical inference methods, and planning.
>
> **(Deep) neural networks**
>
> DL and deep neural networks are a type of ML. Recently, DL has transformed the way in which algorithms achieve (or exceed) human-level performance in areas such as game playing and computer vision. DL owes its success to the easy availability of vast amounts of data and vastly more powerful computers, as well as new algorithmic insights. In common with other "non-parametric" methods (such as Bayesian non-parametric models), DL does not specify the functional form of solutions. Instead, it has enough flexible complexity to learn arbitrary mappings, from input to outcome, from many training examples.
>
> Neural networks began in the 1950s, making significant progress in the 1980s and 1990s. Deep models have added complexity, with several "hidden layers" of non-linear functions cascading between input and output. Despite initial investigations of deep neural networks back in the 1990s, high-performance computing of the time did not allow training over large datasets in realistic time periods for well over a decade. It is only more recently that we have seen the truly impressive ability of DL to solve certain classes of problem.

Why AI in science matters

AI systems are now capable of superhuman reasoning. They can accurately remember vast numbers of facts, execute flawless logical reasoning and near-optimal probabilistic reasoning, learn more rationally than humans from small amounts of data and learn from large amounts of data no human could deal with. These abilities give AI the potential to transform science by augmenting human scientific reasoning (Kitano, 2016). ML and AI have the potential to contribute to science in several key ways: finding unusual and interesting patterns in vast datasets; discovering scientific principles, invariance and laws from data; augmenting human science; and combining with robotic systems to yield "robot scientists". The following paragraphs describe key contributions in more detail.

AI might enable novel types of discovery

One motivation for investing in AI for science is that AI systems "think differently". Human scientists – at least all modern ones – are educated and trained in basically the same way; this is likely to impose unrecognised cognitive biases in how they approach scientific problems. AI systems have very different strengths and weaknesses than human scientists. The expectation is that combining both ways of thinking will provide synergies. Indeed, the evidence from human-software symbiosis has shown that the fusion of automated and human exploration of complex systems can yield efficient and effective solution discovery (Kasparov, 2017).

AI in science may become essential in a context where the volume of scientific papers is vast and growing, and scientists may have reached "peak reading"

AI systems and human scientists have complementary reading skills. Human scientists can understand papers in detail (although such understanding is limited by the ambiguities inherent in natural languages), but can only read and remember a limited number of papers. By contrast, AI systems can extract information from millions of scientific papers, but the amount of detail that can be abstracted is severely limited (Manning and Schütze, 1999).

Applying AI in science has philosophical implications, e.g. in terms of better understanding the scientific process

Automating science also has major philosophical implications. If an AI-based mechanism can be built that is judged to have discovered some novel scientific knowledge, then this will shed light on the nature of science (King et al., 2018). To quote Richard Feynman "What I cannot create, I do not understand" (written on his blackboard at the time of his death). Building robot scientists, for example, entails the need to make concrete engineering decisions related to several important problems in the philosophy of science. For instance, is it more effective to reason only with observed quantities, or to also involve unobserved theoretical concepts? This engineering-based approach to understanding science – shedding light on the discovery process by attempting to replicate it through machine processes – is analogous to the AI approach to understanding the human mind through the creation of artefacts (such as machine learning systems using artificial neural networks) that can be empirically shown to possess some of its attributes. Making machines that physically implement different philosophies of science enables empirical comparison of these philosophies. Currently, philosophers of science are generally limited to historical analysis.

AI can combine with robotic systems to execute closed-loop scientific research

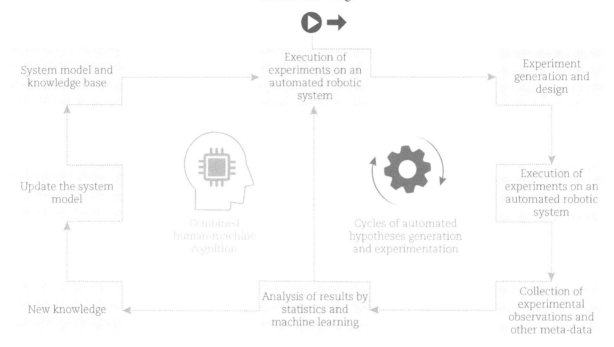

Figure 5.1. Hypothesis-driven closed-loop learning

How iterative cycles of hypothesis-driven experimentation allow for the autonomous generation of new scientific knowledge

The convergence of AI and robotics has many potential benefits for science. It is possible to physically implement a laboratory-automation system that exploits techniques from the AI field to execute cycles of scientific experimentation. The execution of cycles of scientific research is a general approach applicable in many fields of science. Fully automating science has several potential advantages:

- *Faster scientific discovery.* Automated systems can generate and test thousands of hypotheses in parallel, utilising experiments that test multiple hypotheses. Human beings' cognitive limitations mean they can only consider a few hypotheses at a time (King et al., 2004; King et al., 2009).

- *Cheaper experimentation.* AI systems can select experiments utilising greater economic rationality (Williams et al., 2015). The power of AI offers very efficient exploration and exploitation of unknown experimental landscapes, and leads the development of novel drugs (Griffiths and Hernandez-Lobato, 2017; Segler et al., 2018), materials (Frazier and Wang, 2015; Butler et al., 2018) and devices (Kim et al., 2017).

- *Easier training.* Including initial education, a human scientist requires over 20 years and huge resources to be fully trained. Humans can only absorb knowledge slowly through teaching and experience. Robots, by contrast, can directly absorb knowledge from each other.

- *Increased and more productive work.* Robots can work longer and harder than humans, and do not require rest or holidays.

- *Improved knowledge/data sharing and scientific reproducibility.* One of the most important current issues in biology – and other scientific fields – is reproducibility. A 2016 edition of *Nature* observed that: "There is growing alarm about results that cannot be reproduced. Explanations include increased levels of scrutiny, complexity of experiments and statistics, and pressures on researchers" (Alexander et al., 2018). Robots have the superhuman ability to record experimental actions and results. These results, along with the associated metadata and employed procedures, are automatically recorded in full and in accordance with accepted standards, at no additional cost. By contrast, recording data, metadata and procedures adds up to 15% to the total costs of experimentation by humans. Moreover, despite the widespread recording of experimental data, it is still uncommon to fully document the procedures used, the errors made and all the metadata.

Laboratory automation is now essential to most areas of science and technology, but is expensive and difficult to use. The high expense stems from the low number of units sold and the market's immaturity. Consequently, laboratory automation is currently used most economically in large central sites, and companies and universities are increasingly concentrating their laboratory automation. The most advanced example of this trend is cloud automation, where a very large amount of equipment is gathered in a single site, where biologists send their samples and use an application programming interface to design their experiments.

Human-AI interaction

Little research has been done on working scientists' attitude to AI, or the sociological and anthropological issues involved in human scientists and AI systems working together in the future. Compared to humans, AI systems possess a mixture of super- and sub-human abilities. Computers and laboratory robots have traditionally been used to automate low-level repetitive tasks, because they have the super-human capacity to work near flawlessly on extremely repetitive tasks for days at a time. In comparison, humans perform badly at repetitive tasks, especially during extended periods. However, AI systems are sub-human in their adaptability and understanding, and human scientists are still unequalled in conditions that require flexibility and dealing with unexpected situations; they are especially endowed with intuitive functions that might otherwise have been considered low level (King et al., 2018). Given AI systems' mixture of super- and sub-human abilities, investigating how human scientists co-operate with their AI counterparts can be informative. These relationships occur at many levels, from the most profound (deciding on what to investigate, structuring a problem for computational analysis, interpreting unusual experimental results, etc.) to the most mundane (cleaning, replacing consumables, etc.). The growing use of AI systems in science is also expected to profoundly change some sociological aspects of science, such as knowledge transmission, crediting systems for scientific discoveries and perhaps even the peer-review system.[1] Most of the current methods for establishing scientific authority (peer-review, conference plenaries, etc.) are inherently social and designed for human scientists. If AI systems become common in science, such established knowledge-making institutions might have to change to ensure continued academic credibility (King, 2018).

AI across scientific domains

In many scientific disciplines, the ability to record data cheaply, efficiently and rapidly allows the experiments themselves to become sophisticated data-acquisition exercises. Science – the construction of deep understanding from observations of the surrounding world – can then be performed within the data. For many years, this has meant that teams of scientists, augmented by computers, have been able to extract meaning from data, building an intimate bridge between science and data science. More recently, the sheer size, dimensionality and rate of production of scientific data have become so vast that reliance on automation and intelligent systems has become prevalent. Algorithms can scour data at scales beyond human capacity, finding interesting new phenomena and contributing to the discovery process. Box 5.3 shows examples of AI applications in several research fields.

Box 5.3. Applications of AI in different scientific fields Type

AI is increasingly being applied across the span of science, as shown in the examples below.

Physical sciences

Recent work at the forefront of large-scale intelligent data analysis has had massive impact in the physical sciences, particularly in the particle and astrophysics communities, in which event discovery within the data is essential. Such approaches lie, for example, at the core of the detection of pulsars (van Heerden et al., 2016), exoplanets (Rajpaul et al., 2015), gravitational waves (George and Huerta, 2018) and particle physics (Alexander et al., 2018). ML (typically Bayesian) approaches have been widely adopted, not only for purposes of detection, but also to ascertain and remove underlying (and unknown) systematic corruptions and artefacts from large physical-science datasets (Aigrain et al., 2017). They have also been applied to more mainstream regression and classification methods, such as the photometric redshift estimation requirements of the European Space Agency's Euclid mission2 (Almosallam et al., 2015). Furthermore, a significant body of literature considers whether techniques such as deep neural networks can be as valuable to the physical sciences as they have proven in such areas as speech and language understanding. Although complex DL systems play less of a role at present, they will most likely increase their part in extracting insight from data in the coming years.

These illustrations highlight a deep connection between the physical sciences and the field known today as data science, which draws heavily on statistics, mathematics and computer science. A symbiotic relationship exists between data and the physical sciences, with each field offering both theoretical developments and practical applications that can benefit the other, typically evolving through an interactive feedback loop. With the forthcoming emergence of larger and more complex datasets in the physical sciences, this symbiotic relationship is set to grow considerably in the near future.

Chemistry

One of the most prominent applications of AI to chemistry is the planning of organic synthesis pathways. Significant progress has recently been made in this field, both by using the traditional approach of encoding expert chemist knowledge into rules (Klucznik et al., 2018) and by using ML (Segler et al., 2018).

Another active application is drug design (Schneider, 2017). A key step in drug design is learning about quantitative-structure activity relationships (QSARs). The standard QSAR learning problem is: "given a target (usually a protein) and a set of chemical compounds (small molecules) with associated bioactivities (e.g. inhibition of the target), learn a predictive mapping from molecular representation to activity". Almost every type of ML method has been applied to QSAR learning (although no single method has been found superior).

AI is increasingly being integrated with laboratory robotics in drug design to fully automate cycles of research. In 2018, the United Kingdom announced a new facility at the Rosalind Franklin Institute, aiming to transform the UK pharmaceutical industry by pioneering fully automated molecular discovery to produce new drugs up to ten times faster. Similar initiatives are under way in industry, for example at AstraZeneca's new facility in Cambridge, England.

Biomedicine

Probably the most famous AI company in the world is the London-based DeepMind, thanks to its development of AlphaGo, which now beats the best humans at the game of Go, and AlphaGo Zero. DeepMind is actively seeking to deploy its ML technology (DL, reinforcement learning) to medical problems for the UK National Health Service, mostly focusing on image analysis. However, privacy concerns have arisen over the use of health-related data by DeepMind, which is part of the Google suite of companies (Wakefield, 2017).

Related to DeepMind's image processing is the impressive DL method of diagnosing skin cancer using mobile-phone photos (Esteva et al., 2017). Despite the demonstrated success of applying AI to diagnoses, based on image analysis, such applications barely scratch the surface of the potential of AI in cancer diagnosis and treatment.

Many examples of vast-scale algorithmic science projects exist in the physical sciences. The Square Kilometre Array, a radio telescope network currently under construction in Australia and South Africa, will generate more data than the entire global Internet traffic per day when it goes on line. Indeed, the project is already streaming data at almost one terabyte per second. The Large Hadron Collider at CERN, the European Organization for Nuclear Research, discovered the elusive Higgs boson in data streams produced at a rate of gigabytes per second. Meteorologists and seismologists routinely work with global sensor networks that are heterogeneous with regard to their spatial distribution, as well as the type, quantity and quality of data produced. In such settings, problems are not confined to the volumes of data now produced. The signal-to-noise ratio also matters: signals may only provide biased estimates of desired quantities; furthermore, incomplete data complicate or hinder the extraction of automated meaning from data. Data rate alone is hence not the core problem. Data cleaning and curation are of equal importance.

Using AI to select experiments

Addressing the issue of which data and algorithm to employ leads to the issue of intelligently selecting experiments, both to acquire new data and to shed new light on old data. Both these processes can be – and often are – automated. The concept of optimal experimental design may be old, but modern equivalents bring smart statistical models to enable each data run and algorithm choice to maximise the informativeness gained.

Moreover, this optimisation process can consider the costs associated with data recording and computation, enabling efficient and optimal experimentation within a given budget.

In standard ML, the learning algorithm is given all the examples at the start. Active learning is the branch of ML where the learning algorithm is designed to select examples from which to learn; this is a more efficient form of learning. There exists a close analogy between active learning and the process scientists use to select experiments. Active learning proceeds by using existing knowledge to propose where most knowledge will be obtained from a future measurement; the measurement is then taken at this location. Scientific experimental design follows a similar process, with future experiments selected to plug gaps in existing knowledge or test existing theories. Experimental results then help form a better understanding, and so the process repeats. Indeed, scientists do not typically wait patiently and form theories from what they observe; rather, they actively conduct experiments to test hypotheses. Work in active learning (King et al., 2004; Williams et al., 2015) offers an efficient method for balancing the cost of experimentation with the rewards of discovery.

Active learning is a special case of a more generic methodology, Bayesian optimisation and optimal experimental design (Lindley, 1956), which provides an elegant framework for optimally balancing exploration and exploitation in the presence of uncertainty. Bayesian optimisation is at the core of modern approaches. The incorporation of probability theory into experimental design allows algorithms not just to decide where knowledge might be maximised, but also to reduce the uncertainty associated with regions of "experiment space" that are sparsely populated with results. This enables Bayesian experimental approaches not just to "exploit" areas of valuable results, but also to explore hitherto un-investigated experiments.

Explainability: What does it imply in the context of science?

Inscrutability in ML decision-making is commonly cited in discussions of AI as a source of possible concern. The Defense Advanced Research Projects Agency, in the United States, is funding 13 different research groups, working on a range of approaches to make AI more explainable. However, a problem of inscrutability exists in some areas of science – particularly mathematics – independently of the role of machines. Andrew Wiles' proof of Fermat's last theorem ran to over 100 pages and took many mathematicians many years to verify. Will this problem of inscrutability become more salient in science as AI becomes more widespread?

One of the core goals of science is to increase knowledge of the natural world through the performance of experiments. This knowledge should be expressed in *formal logical languages*. Formal languages promote semantic clarity, which in turn supports the free exchange of scientific knowledge and simplifies scientific reasoning. The use of AI systems allows formalising in logic all aspects of a scientific investigation.

AI can, in fact, be used to help formalise scientific argumentation involving many research units (segments of experimental research) and research steps. Making experimental structures explicit renders scientific research more comprehensible, reproducible and reusable.

A major motivation for formalising experimental knowledge is that it can be reused more easily to answer other scientific questions. Many modern AI and ML models can be used to infer the importance of observations, measurements and data features. This insight is often more valuable to scientists than the outcome variables from the models. Techniques such as local interpretable model-agnostic explanations (LIME), for example, offer a good way of explaining the predictions of ML classifiers. LIME can examine "what matters" in the data, by selectively perturbing input data and seeing how the predictions change. Even with the use of DL techniques, if a scientist needs complete audit trails then excellent approaches exist, for example based upon boosted decision trees (a method using multiple decision trees that are additive, rather than averaged).

A key policy concern: Gaps in education and training

A key policy issue concerns education and training. Modifications of the education system often take place at a much slower pace than many other societal changes. Many subjects still taught to children seem more appropriate to the 19th century than the 21st. Three main traditional subjects underlie an understanding of AI: logic, data analysis (statistics) and computer science. Despite being fundamental to reasoning and having a 2 400-year history, logic is currently not taught in schools in most countries, and is almost not taught at all in universities, outside of specialised courses in computer science and philosophy. This means that few students are trained to understand the fundamental role of logic in AI.[3]

The analysis of data is as fundamental a subject as logic, but is also little taught in schools. Most data analysis currently taught to non-specialists in universities is still based on the classical statistics developed in the early 20th century. It deals with such topics as hypothesis testing, confidence intervals and simple optimisation methods – the forms of data analysis also most often reported in scientific papers. However, this type of data analysis presents philosophical and technical problems (Jaynes, 2003).

An even greater problem is that data analysis is taught in a way that resembles more cooking than science: in the presence of data in a form that looks like X, then a t-test should be applied at a 5% one-tail confidence level; if the data are in form Y, then an F-test should be applied at a 1% two-tail confidence level, etc. Unfortunately, such courses convey little understanding of fundamental concepts, meaning that few students understand the fundamentals of data analysis needed for ML. Students should learn about Bayesian statistics and computational intensive methods based on resampling to better understand the reliability of conclusions.

Computer science education has not kept pace with the importance of AI to society. Computer science has also been conflated with "information technology skills" (Royal Society, 2017). Another problem is that in Western countries (as opposed to many developing countries), female students are far outnumbered by male students. It would be very worrisome if this low share were to transfer to the applications of AI in science (Chapter 7).

A general skill shortage also exists in AI. This creates a need for master's conversion courses to transform graduates in other disciplines into scientists qualified to work at the AI/science interface, as well as more PhD positions at that interface. The independent report "Growing the AI Industry in the UK" (Hall and Pesenti, 2017) articulated how the UK Government and industry can work together to build skills and infrastructure, and implement a long-term strategy for AI, and recommended funding to reach these goals.

A vision of AI and the future of science

Despite the impressive performance of AI in many areas, the need still exists to transfer methods that perform well in constrained, well-structured problem spaces (such as game playing, image analysis, text and language modelling) to noisy, corrupted and partially observed scientific problem domains. The problems DL approaches encounter with small (and noisy) datasets compound this issue. Creating a realistic approach that works across all data scales, from data-sparse environments to data-rich environments, requires yet more innovation (Box 5.4). Probabilistic models do offer such capacities, although Bayesian DL is still in its infancy.

Box 5.4. In my view: Moderating expectations: What deep learning can and cannot do yet

Gary Marcus, New York University

Deep learning currently dominates AI research and its applications, and has generated considerable excitement – perhaps somewhat more than is actually warranted. Although deep learning has made considerable progress in areas such as speech recognition and game playing, and contributed to the use of AI in science, as described in this chapter, it is far from a universal solvent, and by itself is unlikely to yield general intelligence.

To understand its scope and limits, it helps to understand what deep learning does; fundamentally, as it is most often used, it approximates complex relationships by learning to classify input examples into output examples, through a form of successive approximation that uses large quantities of training data. It then tries to extend the classifications it has learned to other sets of input "test" data pertaining to the same problem domains. However, unlike human reasoning, deep learning lacks a mechanism for learning abstractions through explicit, verbal definition. Current systems driven purely by deep learning face a number of limitations:

- Since deep learning requires large sets of training data, it works less well in problem areas where data are limited.
- Deep learning techniques can fail if test data differ significantly from training data, as often happens outside a controlled environment. Recent experiments show that deep learning performs poorly when confronted with scenarios that differ in minor ways from those on which the system was trained.
- Deep learning techniques do not perform well when dealing with data with complex hierarchical structures. Deep learning learns correlations between sets of features that are themselves "flat" or non-hierarchical, as in a simple, unstructured list, but much human and linguistic knowledge is more structured.
- Current deep learning techniques cannot accurately draw open-ended inferences based on real-world knowledge. When applied to reading, for example, deep learning works well when the answer to a given question is explicitly contained within a text. It works less well in tasks requiring inference beyond what is explicit in a text.
- The lack of transparency of deep learning makes this technology a potential liability when applied to support decisions in areas such as medical diagnosis in which human users like to understand how a given system made a given decision. The millions or even billions of parameters used by deep learning to solve a problem do not easily allow its results to be reverse-engineered.
- Thus far, existing deep learning approaches have struggled to integrate prior knowledge, such as the laws of physics. Yet dealing with problems that have less to do with categorisation and more to do with scientific reasoning will require such integration.

> - Relatively little work within the deep learning tradition has attempted to distinguish causation from correlation.
>
> Deep learning should not be abandoned, but general intelligence will require complementary tools – possibly of an entirely different nature that is closer to classical symbolic artificial intelligence – to supplement current techniques.

Although they offer impressive performance, many AI approaches provide little in the way of transparency regarding their function. Auditing the reasoning behind decision-making is required in many application domains. For practical systems, where AI makes decisions about people (for example), such an audit trail is essential. Furthermore, few AI algorithms can offer formal guarantees regarding their performance. In safety-critical environments, the ability to provide such bounds and verify failure modes when faced with unusual data is a prerequisite. Some research in this area is already under way, though not commonplace.

It is to be hoped that the collaboration between human scientists and AI systems will produce better science than can be performed alone. For example, human/computer teams still play better chess than either does alone. Understanding how best to synergise the strengths and weaknesses of human scientists and AI systems requires a better understanding of the issues (not just technical, but also economic, sociological and anthropological) involved in human/machine collaboration.

Arguably, advances in technology and the understanding of science will drive the development of ever-smarter AI systems for science. Hiroaki Kitano, President and CEO of Sony Computer Science Laboratories, has called for new Grand Challenge for AI: to develop an AI system that can make major scientific discoveries in biomedical sciences worthy of a Nobel Prize (Kitano, 2016). This may sound fantastical, but the physics Nobel laureate Frank Wilczek (2006) is on record as saying that in 100 years' time, the best physicist will be a machine. If this vision of the future comes to pass, this will not only transform technology, but humans' understanding of the universe (Box 5.5).

> **Box 5.5. AI and the laws of nature**
>
> How can algorithms infer an understanding of the laws of nature? AI algorithms learn solutions from examples. A critical part of these solutions consists in forming a function that generalises, i.e. performs well when presented with data that did not form part of the training examples. This critical generalisation requirement requires AI algorithms to "discover" a problem's systematic trends and properties that are common across all the examples. This ability to find underlying commonality in complex data also allows models to find simple representations, rules and patterns in scientific data. The "laws" of science are such representations. Examples include the blocked adaptive computationally efficient outlier nominators (BACON) algorithm, which "discovered" Kepler's laws of planetary motion (Langley et al., 1987).

Conclusion

The laws of science are compressed, elegant representations offering insight into the functioning of the universe. They are ultimately developed through logical (mathematical) formulation and empirical observation. Both avenues have seen revolutions in the application of ML and AI in recent years. AI systems can formulate axiomatic extensions to existing laws. The wealth of data available from experiments allows science to take place in the data. Science is rapidly approaching the point where AI systems can infer such things as conservation laws and laws of motion based on data only, and can propose experiments to gather maximal knowledge from new data. Coupled with these developments, the ability of AI to reason logically and operate at scales well beyond the human scale creates a recipe for a genuine automated scientist.

Notes

[1] One of the co-authors of this chapter, Ross King, has himself had the experience of wishing to give a robot scientist – Adam – credit as a co-author of a scientific paper, but encountered legal problems, as the lead author needed to sign a declaration stating that all the authors had agreed to the submission. A counter-argument is that not giving machines credit constitutes plagiarism.

[2] http://sci.esa.int/euclid.

[3] The central role of logic is set out in leading AI textbooks, such as Russell and Norvig (2016).

References

Aigrain, S. et al. (2017), "Robust, open-source removal of systematics in Kepler data", *Monthly Notices of the Royal Astronomical Society*, Vol. 471/1, pp. 759-769, Oxford Academic Press, Oxford, https://doi.org/10.1093/mnras/stx1422.

Alexander, R et al. (2018), "Machine learning at the energy and intensity frontiers of particle physics", *Nature*, Vol. 560, pp. 41-48, Springer Nature, https://www.nature.com/articles/s41586-018-0361-2.

Almosallam, I. et al. (2015), "A Sparse Gaussian Process Framework for Photometric Redshift Estimation", *Monthly Notices of the Royal Astronomical Society*, Vol. 455/3, pp. 2387-2401, Oxford Academic Press, https://doi.org/10.1093/mnras/stv2425.

Bloom, N. et al. (2017), "Are Ideas Getting Harder to Find?", *NBER Working Paper*, No. 23782, September 2017, National Bureau of Economic Research, Cambridge, MA.

Boden, M.A. (2006), *Mind as Machine: A History of Cognitive Science*, Oxford University Press, Oxford.

Butler, K.T. et al. (2018), "Machine learning for molecular and materials science", *Nature*, Vol. 559, pp. 547-555, Springer Nature.

Esteva, A., B.Kuprel, R.A.Novoa, J.Ko, S.M.Swetter and H.M.Blau (2017), "Dermatologist-level classification of skin cancer with deep neural networks", *Nature* volume 542, pages 115–118 (02 February 2017), http://www.nature.com/articles/nature21056.

Frazier P.I. and J. Wang (2016), "Bayesian Optimization for Materials Design", in Lookman T., F. Alexander and K. Rajan (eds.), *Information Science for Materials Discovery and Design*, Springer Series in Materials Science, Vol. 225, Springer Nature Switzerland.

George, D. and E.A. Huerta (2018), "Deep Learning for real-time gravitational wave detection and parameter estimation: Results with Advanced LIGO data", *Physics Letters B*, Vol. 778, pp. 64-70, https://doi.org/10.1016/j.physletb.2017.12.053.

Glymour, C. (2004), "The Automation of Discovery", *Daedalus*, Vol. 133/1, pp. 66-77, MIT Press Journals, Cambridge, MA, https://doi.org/10.1162/001152604772746710.

Griffiths, R-R. and J.M. Hernández-Lobato (2017), "Constrained Bayesian Optimization for Automatic Chemical Design", arxiv.org, Cornell University, Ithaca, NY.

Hall, W. and J. Pesenti (2017), "Growing the AI Industry in the UK", independent report, Government of the United Kingdom, London, https://assets.publishing.service.gov.uk/government/uploads/system/uploads/attachment_data/file/652097/Growing_the_artificial_intelligence_industry_in_the_UK.pdf.

Hey, T., S. Tansley and K. Tolle (2009), *The Fourth Paradigm: Data Intensive Scientific Discovery*, Microsoft Research, Redmond, WA.

Jaynes, E.T. (2003), *Probability Theory: The Logic of Science*, Cambridge University Press, Cambridge.

Jones, B.F. (2005), "The Burden of Knowledge and the 'Death of the Renaissance Man': Is Innovation Getting Harder?", *NBER Working Paper*, No. 11360, Cambridge, MA.

Kasparov, G. (2017), *Deep Thinking: Where Machine Intelligence Ends and Human Creativity Begins*, John Murray, London.

Kim, M. et al., (2017), "Human-in-the-loop Bayesian optimization of wearable device parameters", *PLoS ONE*, Vol. 12/9, PLOS, San Francisco, http://doi.org/10.1371/journal.pone.0184054.

King, R.D. (2018), "Tackling AI Impact on Drug Patenting", *Nature*, Vol. 560, correspondence, 16 August, p. 307, Springer Nature.

King, R.D. et al. (2018), "Automating science: philosophical and social dimensions", *IEEE Technology and Society Magazine*, Vol. 37/1, pp. 40-46, IEEE Society on Social Implications of Technology, New York.

King, R.D. et al. (2009), "The Automation of Science", *Science*, Vol. 324/5923, pp. 85-89, Elsevier, NY, http://doi.org/10.1126/science.1165620.

King, R.D. et al. (2004), "Functional genomic hypothesis generation and experimentation by a robot scientist", *Nature*, Vol. 427, pp. 247-252, Springer Nature, http://doi.org/10.1038/nature02236.

Kitano, H. (2016), "Artificial Intelligence to Win the Nobel Prize and Beyond: Creating the Engine for Scientific Discovery", *AI Magazine*, Vol. 37/1, Spring 2016, pp. 39-49, Association for the Advancement of Artificial Intelligence, Palo Alto, CA, https://doi.org/10.1609/aimag.v37i1.2642.

Klucznik, T. et al. (2018), "Efficient Syntheses of Diverse, Medicinally Relevant Targets Planned by Computer and Executed in the Laboratory", *Chem*, Vol. 4/3, pp. 522-532, Elsevier, Amsterdam, https://doi.org/10.1016/j.chempr.2018.02.002.

Langley, P. et al. (1987), *Scientific Discovery: Computational Explorations of the Creative Process*, MIT Press, Cambridge, MA.

Lindley, D.V. (1956), "On a measure of information provided by an experiment", *Annals of Mathematical Statistics*, Vol. 27/4, pp. 986-1005, Institute of Mathematical Statistics, Beachwood, OH.

Manning, C. and H. Schütze (1999), *Foundations of Statistical Natural Language Processing*, MIT Press, Cambridge, MA.

Marcus, G. (2018), "Deep Learning: A Critical Appraisal", preprint available at https://arxiv.org/ftp/arxiv/papers/1801/1801.00631.pdf, submitted 2 January 2018, accessed 13 October 2018.

Rajpaul, V. et al. (2015), "A Gaussian process framework for modelling stellar activity signals in radial velocity data", *Monthly Notices of the Royal Astronomical Society*, Vol. 452/3, pp. 2269-2291, Oxford Academic Press, Oxford, https://doi.org/10.1093/mnras/stv1428.

The Royal Society (2017), "After the Reboot: Computing Education in UK Schools", The Royal Society, London, https://royalsociety.org/~/media/policy/projects/computing-education/computing-education-report.pdf.

Russell, S., and P. Norvig (2016), "Artificial Intelligence: A Modern Approach. Global Edition", Pearson Education Limited, Harlow, England.

Schneider, G, (2017), "Automating drug discovery", *Nature Reviews Drug Discovery*, Vol. 17, pp. 97-113, Springer Nature, http://doi.org/10.1038/nrd.2017.232.

Segler, M.H.S., M. Preuss and M.P. Waller (2018), "Planning chemical syntheses with deep neural networks and symbolic AI", *Nature*, Springer Nature, http://doi.org/10.1038/nature25978.

Van Heerden, E., A. Karastergiou and S. Roberts (2016), "A Framework for Assessing the Performance of Pulsar Search Pipelines", *Monthly Notices of the Royal Astronomical Society*, Vol. 467/2, pp. 1661-1677, Oxford University Press, Oxford, https://doi.org/10.1093/mnras/stw3068.

Wakefield, J. (2017), "Google DeepMind's NHS deal under scrutiny", *BBC News*, webpage, 17 March, https://www.bbc.co.uk/news/technology-39301901.

Wilczek, F. (2006), *Fantastic Realities: 49 Mind Journeys and a Trip to Stockholm*, World Scientific Publishing, Singapore.

Williams, K. et al. (2015), "Cheaper Faster Drug Development Validated by the Repositioning of Drugs Against Neglected Tropical Diseases", *Journal of the Royal Society Interface*, Vol. 12/104, The Royal Society Publishing, London, https://doi.org/10.1098/rsif.2014.1289.

Chapter 6. Enhanced access to publicly funded data for STI

By

Alan Paic and Carthage Smith

Enhanced access to data can be a key enabler for science, technology and innovation (STI). It can support new scientific insights across disciplines, contribute to reproducibility of scientific results, and facilitate innovation. However, many countries have yet to develop comprehensive approaches to enhance access to data. This chapter focuses on policy concerns and potential policy action to enhance access to publicly funded research data for STI. It starts with an overview of public research data. It then outlines the specific policy dilemmas concerning enhanced data sharing. These include: (i) fostering data governance for trust and balancing the benefits and risks of data sharing; (ii) developing and implementing technical standards and practices; (iii) defining responsibility and ownership of data; (iv) changing recognition and reward systems to encourage scientists to share data; (v) implementing business models and long term funding for data provision; and (vi) developing human capital and skills to support data sharing and analysis. Finally, the chapter draws policy implications for the future by outlining two possible scenarios.

Introduction

Research is becoming increasingly data-intensive. "Big data" are no longer the prerogative of experimental physics and astronomy: they are spreading across all scientific domains. Access to data is a key enabler for science, technology and innovation (STI); not surprisingly, enhancing this access is a major priority for policy makers (OECD, 2006) (European Commission, 2014). As a critical element of open science, big data are expected to lead to new scientific breakthroughs, less duplication and better reproducibility of results, as well as bring about improved trust and innovation (OECD, 2015a, 2015b). The development of artificial intelligence (AI) further reinforces the importance of access to data, since AI algorithms need very large amounts of well-described data to "train", i.e. improve their performance.

Open data can be simply defined as "data that can be accessed and reused by anyone without technical or legal restrictions" (OECD, 2015a). This does not necessarily mean data is free of cost, although in the context of open science, it is normally assumed the user bears no charges. Openness is not a binary concept: data can be made more or less open, according to the specific nature of the data and the community of stakeholders involved. "As open as possible, as closed as necessary" is gradually replacing the "open-by-default" mantra associated with the early days of the open-access movement. Opening up data can help advance the STI agenda, but this needs to be balanced against issues of costs, privacy, security and preventing malevolent uses. Enhanced access to data is a term that is used increasingly in relation to public sector data and captures some of these important caveats around openness.

Enhanced access to data can be described as encompassing any practical and lawful means through which data can be effectively accessed by, and shared with an entity (individual or organisation) other than the data holder, for the purpose of fostering data re-use by the entity or a third-party chosen by the entity, while, at the same time, taking into account the private interests of individuals and organisations concerned (e.g. their intellectual property and privacy rights) as well as national security and public interests.

This chapter focuses on enhanced access to publicly funded research data for STI. It starts with an overview of public research data. It then develops the specific policy dilemmas concerning enhanced data sharing. Finally, it draws policy implications for the future. Much has already been written on this topic, and not all the important issues can be fully addressed in one short chapter. Hence, the chapter focuses on policy concerns and potential policy action. It builds on the recent OECD data-access survey of OECD countries regarding the OECD 2006 Recommendation on Access to Research Data from Public Funding (OECD, 2017a). It also draws on responses to the 2017 European Commission-OECD STI Policy Survey (STIP Compass) and the discussions held at an OECD expert workshop on principles for enhanced access to public data held in March 2018.

Public data for STI: An overview

Three broad categories of data are used for STI: 1) public-sector information (PSI), produced, curated and managed *by* or *for* governmental entities; 2) data resulting from publicly funded research; 3) privately owned or commercial data. This chapter covers only publicly funded data for STI, which includes both data produced by research and PSI used in research, such as meteorological or social survey data. These distinctions are somewhat artificial and partially overlapping, but they can be important in defining where policymaking responsibilities lie. For example, unlike data generated by research, access to PSI is not principally the remit of STI policymakers, and yet research data is sometimes treated as a subset of PSI, as is the case in the latest EC Directive on the re-use of PSI (European Commission, 2018a). Ensuring that PSI-related policies and practices that affect research are consistent with policies and practices affecting other research data requires co-ordination across policy communities.

Publicly funded research data are defined in the OECD 2006 Recommendation on Access to Research Data from Public Funding (OECD, 2006) as data "that are supported by public funds for the purposes of developing publicly accessible scientific research and knowledge". Research data can be further defined as: "factual records (numerical scores, textual records, images and sounds) used as primary sources for scientific research, and that are commonly accepted in the scientific community as necessary to validate research findings". We exclude from the scope of consideration here research data gathered for the purpose of commercialisation of research outcomes, or research data that are the property of a commercial sector entity. Access to such data is subject to a range of considerations that are beyond the scope of this document.

This chapter focuses on the *data* outputs of research, rather the *publication* outputs. It makes this distinction mainly for pragmatic reasons: because some policy issues – particularly the role of commercial publishers – are distinct, issues around open-access publications are normally considered separately from those concerning research data (e.g. OECD, 2015a). Nevertheless, research data and publications are widely recognised as part of a continuum and policies need to be connected accordingly. Access to data currently lags behind access to publications: more than 92% of universities in Europe have – or plan to have – open-access policies for publications, but under 28% have established guidelines concerning open access to data. The main institutional barriers to promoting research data management and/or open access to research data are: different "scientific cultures" within the university, absence of national guidelines or policies, limited awareness of benefits, legal concerns, and technical complexity (Morais and Borrell-Damian, 2018).

Rationales for sharing research data

At least six main rationales exist in favour of enhanced access to public research data (Borgman, 2012):

1. *New scientific insights*: Providing broader access to data allows more researchers (and citizens) to analyse and link those data to other data sources, to respond to different scientific questions. For example, biodiversity data are increasingly used by the health-research community working on emerging diseases.

2. *Reproducibility of scientific results*: Sharing access to the data underpinning scientific publications allows peers to test and reproduce scientific results. In practice, data alone are often insufficient for testing reproducibility, and enhanced access to analysis software is also necessary.

3. *Public research is a public good*: Data from publicly funded research should, in principle, be available to researchers, citizens and commercial actors who wish to use and derive value from them. This is sometimes also an issue of transparency and accountability.

4. *Promote innovation*: Allowing commercial companies to access public research data enables them to use the data to accelerate innovation on products (e.g. new drugs) or new data services (e.g. in weather forecasting). Data are an essential enabler for AI and related innovations.

5. *Support meta-analyses*: Enhancing access to and sharing of data encourages meta-analysis, which combines the results of different related studies (e.g. clinical trials of a drug) to provide greater statistical power.

6. *Avoid duplication*: Sharing datasets showing positive or negative results can avoid duplication of research efforts (Rothsteinet al., 2006).

When taken together, these rationales contribute to a more efficient and effective scientific enterprise. Access to data alone is insufficient to achieve all these expectations, but lack of access is a major barrier to achieving them.

There are also legitimate concerns about enhanced access to data, including privacy and intellectual property protection and national security and other public interests. These risks are discussed in the section on Future outlook. When, how and under what conditions public research data should be made accessible are important policy questions, which cut across the issues discussed in the rest of this chapter.

Policy action in favour of sharing research data

The OECD 2006 Recommendation (OECD, 2006) and the OECD Principles and Guidelines for Access to Research Data from Public Funding (OECD, 2007) represented an important step in multilateral efforts to foster open access to data in STI. A wide range of policies were implemented fairly quickly in response to these instruments: some countries introduced laws and comprehensive policies; others issued position statements and future plans (OECD, 2009).

A 2017 OECD survey on access-to-data policies among policy makers from 27 countries identified a total of 171 policy initiatives targeting enhanced access to data. Survey respondents were also asked to assess the relevance of the 13 principles cited in the 2006 OECD Recommendation (OECD, 2006). The principles considered the most pertinent today were openness, quality, security, interoperability, transparency, sustainability and legal conformity (Figure 6.1).

Building on earlier work by OECD (OECD, 2007), the findability, accessibility, interoperability and reusability (FAIR) data principles have been developed by a diverse set of stakeholders representing academia, scholarly publishers, industry and funding agencies, and are now becoming a mainstream reference for policy makers (Wilkinson et al., 2016) (Table 6.1).

Figure 6.1. An assessment of the relevance of the OECD principles concerning access to research data from public funding

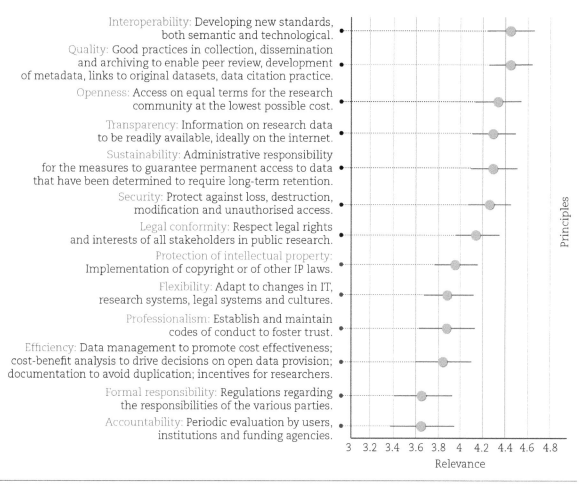

Additional principles quoted by the stakeholders but not included in the 2006 recommandations:

Discoverability is crucial if data is to be re-used. Therefore there is a need to establish ontologies and appropriate semantics to enable scientist from different scientific domains to find relevant data sets.

Machine-readability should be the norm, to facilitate treatment of large quantities of data

FAIR data principles. Findable, accessible, interoperable, re-usable.

Strict rules for the financial support of Open access to data: (i) responsibility for financing at international, national, regional and institutional level; (ii) definition of roles between financiers, operators and users; (iii) supervision mechanism for compliance with Open access to data rules, foresee legal sanctions

Definition of responsibility and ownership, including legal and ethical issues

Explicit recognition and reward system for data authorship

Publicly funded research data treated as Commons. Licencing under Creative Commons could be used to provide a framework which ensures openness while restricting re-use as needed.

Setting an embargo period for the exclusive use of data. The embargo could vary according to the nature of the output, and provide reassurance to authors.

Implications of blockchain technologies on open access to data should be investigated, since opinions of relevance are very different in the open science field. Blockchain technology is mentioned as a potential tool which could help re-allocate private monopoly rents from innovation back into the systemic network of public collaborative science and innovation (Soete, 2016).

Note: The 2017 survey asked respondents to assess the relevance of the 13 principles cited in the original OECD Recommendation (OECD, 2006) on a Likert scale (5 = very high relevance; 0 = no relevance). Responses were received from 55 organisations in 27 countries.

Table 6.1. Overview of FAIR principles

FAIR principles	Action items	Technical requirements
Findable – data should be easily found by humans and machines alike	Establish portals and open-science clouds	• Globally unique and persistent identifiers • Data indexed in a searchable database
Accessible – as open as possible, as closed as necessary	Use open licensing, whenever possible Establish trusted-user access for more sensitive datasets	• Machine readability • Standardised communication protocol • Metadata are accessible – even after the data are no longer available.
Interoperable – datasets need to be combinable with other datasets	Three aspects of interoperability: semantic (taxonomy), legal (rights) and technical (machine readability) Standard-setting	• Semantic interoperability – common vocabulary • Data include relevant references to other datasets
Reusable – it must be possible to re-use data in future research projects and then process these data further	Data curation Open Archival Information System (OAIS)-compliant repositories	• Metadata are exhaustive • Data describe multiple precise and appropriate properties • Data are released with a clear and accessible data licence • Data are connected to their origin • Data meet standards relevant to the field

Source: Author's analysis, based on Oxford Research (2018) and expert opinions (OECD, 2018).

Several other multilateral initiatives have been developed to foster data sharing, particularly at the European level (Box 6.1). At the national level, the 2017 edition of the EC-OECD STI Policy survey asked OECD member and partner countries to provide information about policy initiatives supporting open science and open access. Most of the 181 policy initiatives cited concern infrastructures and strategies, and a smaller number concern governance issues (Box 6.2).

Box 6.1. International policy initiatives to promote sharing of research data

The Transparency and Openness Promotion (TOP) guidelines were created by journals, funders and societies to align scientific ideals with practices. They include standards covering citations, data transparency, software, research materials, design and analysis, as well as preregistration of study and analysis plans, and replication. Journals select which of the eight transparency standards they wish to adopt, as well as a level of implementation for each standard (Center for Open Science, 2014).

In 2012, the European Commission issued a Recommendation on access to and preservation of scientific information, calling for co-ordinated open access to scientific publications and data, preservation and re-use of scientific information, development of e-infrastructures among EU Member States (European Commission, 2012). The Recommendation was updated in 2018 (European Commission, 2018).

In 2016, the European Commission published "Open Innovation, Open Science, Open to the World", a vision that incorporated its ambitious plans for a European Open Science Cloud (EOSC)[1] (European Commission, 2016). The EOSC is conceived to provide

> EU researchers an environment with free and open services for data storage, management, analysis and re-use across disciplines by connecting existing and emerging infrastructures, adding value and leveraging past infrastructure investment. The EOSC is expected to develop common specifications and tools to ensure data is FAIR and legally compliant with the European Union's General Data Protection Regulation (GDPR) and cybersecurity legislation. It also foresees mechanisms for cost recovery on cross-border access (European Commission, 2018b).
>
> Similar "cloud" initiatives include the National Institutes of Health (NIH) Commons in the United States (NIH, 2017), the Australian Research Data Cloud (eRSA, 2014), and the African Research Cloud (ARC). All of these initiatives aim to be interconnected and interoperable.

In addition to government policy, research funders and scientific journals are increasingly demanding open-access to data. Funders require data-management plans and have specific data-release policies; some (such as the UK Economic and Social Research Council) even require researchers wishing to collect new data to demonstrate that no existing data can be used for their purpose (Economic and Social Research Council, 2015). Many scientific journals require data statements and links; some (such as *Science*) require authors to share the computer code they used to create or analyse data.

In summary, since the OECD Recommendation drew international attention to the area in 2006, several multilateral initiatives to promote access to research data have been launched. The FAIR principles have de facto become an international norm, helping to guide policy actions. The majority of OECD countries are taking actions to promote open data, sometimes in association with plans to develop science clouds linking research data with services provided to the entire research community. However, several outstanding challenges need to be overcome before open data becomes a reality (the section on Future outlook addresses these challenges).

> **Box 6.2. Instruments concerning data access reported in the 2017 EC/OECD STI Policy Survey**
>
> Of the 181 policy initiatives reported as supporting open science and open access, 74 (42%) are concerned with research infrastructures, including portals offering open access to publications; repositories and archives for scientific data; search engines; virtual networks; and clouds connecting individual physical repositories. Examples include the European Open Science Cloud, and Research Data Infrastructure for Open Science in Japan. In some cases (Australia, Estonia, Finland and France), open-data infrastructure is treated within a national strategy on research infrastructures.
>
> 55 initiatives (33%) are national strategies and policies for open access to data and publications. These include:
>
> - dedicated strategies and policies for open access to data and publications at the policy-making level (Czech Republic, Korea, New Zealand, Norway, Slovenia and United Kingdom), as well as at the funding-agency level (Australia, Austria, Belgium-Federal, Canada, Lithuania, Nordic Council of Ministers, Netherlands, Norway, Portugal, Switzerland and United Kingdom)
> - open-data access within open-science policies (e.g. Chile, Colombia, Denmark, Estonia and the Netherlands); the Open Innovation Strategy (Austria); the Innovation and Science agenda (Australia); the Law on Scientific Activity (Latvia); and a specific Law 310/2014 for Public Research which focuses on co-operation between business and academia (Greece)
> - open-data access, integrated within open-government and public sector-information initiatives (Australia, Argentina, Brazil, Canada, Sweden)
> - bottom-up approaches through institutions (Centre national de la recherche scientifique and Institut national de la recherche agronomique in France; University of Malta; universities in Slovenia; and Concordat on Open Research Data in the United Kingdom).
>
> 13 initiatives (7%) aim to create or reform a governance body to foster open access. These include:
>
> - Etalab, a high-level, pan-governmental open-data platform in France co-ordinating open-data and open-government initiatives, which is chaired by the national chief data officer and reports to the French Prime Minister
> - a national focal point (chief science officer Canada, national chief data officer in France, point of reference in Slovenia) for access to and preservation of scientific data
> - an agency for information systems used in higher education and research (CERES – National Center for Systems and Services for Research and Studies, Norway)
> - The Datacite consortium, which enables researchers to attach a digital object identifier (DOI) to research data (Estonia)
> - the Data Archiving and Network Services institute, which facilitates data archiving and re-use, and provides training and consultancy (Netherlands)
> - open-data institutes (Canada and the United Kingdom) supporting economic, environmental and social-value creation opportunities arising from open data

> 8 initiatives (4%) are concerned with networking and collaborative platforms to facilitate open access to data. These include:
>
> - OpenAire Advance, a network of repositories with 34 National open science desks promoting open science as the default solution in Europe
> - library networks (HEAL Link in Greece, HAL and Persée in France)
> - think tanks sharing good practice and engaging in advocacy (EPRIST in France)
> - a data-analytics initiative linking disparate government datasets (Data61 in Australia)
> - cooperatives of research, educational and medical institutions (e.g. the SURF cooperative in the Netherlands), aiming to promote innovation in information technology
> - a commercialisation marketplace (Open Data Exchange in Canada)
>
> 5 initiatives (3%) undertake formal consultations of stakeholder groups, including expert groups. These include:
>
> - working groups and committees for open science and open access to scientific data (e.g. the European Commission Directorate General for Research, Technology and Innovation, and initiatives in France, Greece, Ireland, Japan, Slovenia, Turkey and the United Kingdom)
> - an open-data forum advocating the development of open-data policies (United Kingdom)
>
> Several initiatives aim to collect data about researchers, research projects and policies. For an overview of these initiatives, see Chapter 12 on digital science and innovation policy.
>
> *Source*: EC/OECD (2017)

Policy challenges to promoting enhanced access to data

The 2017 OECD data-access survey and a follow-up workshop in 2018 identified six key areas of policy concern with regard to enhancing access to public data for STI, as follows:

- *Data governance for trust – balancing the benefits of data sharing with the risks*: Opening up data can help advance the STI agenda, but this needs to be balanced against issues of costs, privacy, intellectual property, national security and other public interests.

- *Technical standards and practices*: Achieving FAIR goals hinges on the development and adoption of a common technical framework. The challenge is that technology development is now far outpacing standard-setting, creating regulatory gaps.

- *Defining responsibility and ownership*: Intellectual property rights and licensing arrangements associated with data need to be clearly defined and respected. IPR protection can be an important incentive for private sector investment in research and innovation. At the same time, enhanced access to data is also a driver for innovation. Public-private partnerships present a particular challenge, with the risk of "privatising" and preventing access to data derived from publicly funded research.

- *Incentives*: Recognition and rewards encourage researchers to share data. Current academic-reward systems mostly motivate researchers to publish their scientific results and do not attach enough value to the sharing of data.

- *Business models and funding*: The costs of providing open data are mostly borne by data providers, while the benefits accrue to users including those who develop "value added" data services. There are a variety of business models for providing data access and services, but these are often restrained by policy mandates and incentives.

- *Building human capital and institutional capabilities to manage, create, curate and re-use data*: A lack of skills breeds a lack of trust. It is important to ensure there are appropriate skills along the full data value chain, including data management skills of researchers, curation skills with data stewards, and data literacy among users.

The following subsections develop these six challenges.

Data governance for trust – balancing benefits and risks

Balancing the potential public benefits and risks of sharing research data is a critical issue for data governance. Sound data governance is needed to ensure trust from both data providers and users, and promote a culture of sharing, with the aim of making data "as open as possible and as closed as necessary".

Sharing data presents multiple risks related to: 1) individual privacy (e.g. in the case of clinical research data); 2) misuse (e.g. data about rare and endangered species, or rare minerals); 3) misinterpretation (particularly as concerns datasets of uncertain quality, and/or lacking the appropriate metadata); and 4) national security (e.g. data from research with potential military applications). More granular data often have higher potential research value, but the risk increases as well.

Providing access to personal data or human subject data is a particular challenge (OECD, 2013). Although anonymisation techniques can remove personally identifiable information from individual datasets, true anonymisation becomes very difficult as more and more data from different sources are integrated (President's Advisors on Science and Technology, 2014). Moreover, the research value of personal data often stems from the ability to link it back to individual characteristics. In the United Kingdom, for example, linking information from hospitals with the cancer-data repository, and data from various screening programmes, has made it possible to recommend changes in medical protocols that are likely to improve cancer survival rates. Rules and laws can be a disincentive to breaching anonymity, but the financial incentives to do so can be high in certain industries, and legal regimes are very difficult to implement across national jurisdictions.

Alongside anonymization, informed consent is the second pillar underlying the use of personal data in research. Consent is a right recognised in many countries and enshrined in legislation, such as the recent GDPR (European Commission, 2016). However, situations exist in research where consent for using data for specific research purposes is impossible or impractical to obtain, particularly if these purposes were not envisaged when the data were originally collected. For example, when analysing new forms of data from social networks in ways the collector had not anticipated, it might be unfeasible to go back to all the individuals to ask for consent. It is notable that the GDPR[2] makes exceptions for the use of data in research, where consent is one consideration, but is not prescribed as the legal basis for data use. Recent OECD work on the subject stressed the need for properly constituted independent ethics review bodies (ERBs), outlining their role in evaluating applications to access publicly funded personal data for research purposes. Well-functioning ERBs contribute to building trust (OECD, 2016). This recent work also emphasised the importance of public engagement in defining norms on the use of personal data in research. The approach adopted by the Australian Government, which aims to achieve value creation with open data while transparently managing risk, is one example (Box 6.3).

Box 6.3. In my view – Trust is the key to unlocking data

The Hon. Michael Keenan MP, Minister for Human Services and Digital Transformation, Australian Government

Data is the fuel powering our new digital economy. However, news of data breaches and misuse of personal information erodes trust and leads the public to believe that data is bad or something to be feared.

If these negative perceptions become entrenched, we risk missing out on the enormous opportunities and benefits data offers to improve people's lives, help grow the economy and become more successful as a nation.

As a Government, we have a responsibility to use data to make the best possible decisions to improve people's lives. In May 2018, the Australian Government announced reforms to simplify the way public data can be shared and used, and clarify accountabilities around the management of data. These reforms are made up of four components:

- A Consumer Data Right to give Australians greater access and control over their data, to enable them to get a better deal from their bank, energy and telecommunications companies;
- A National Data Commissioner to manage the integrity and improve how the Australian Government manages and uses data;
- A new National Data Advisory Council to provide advice on ethical data use, technical best practice, and industry and international developments; and
- Enabling legislation – the *Data Sharing and Release Act* – to improve the use and re-use of data while strengthening security and privacy protections for personal and sensitive data.

These reforms represent a tremendous opportunity to unlock national productivity. However, we will only seize this opportunity if public data is used in a safe and transparent manner and citizens trust their privacy and security is being valued and protected at all times.

To achieve that, we are working hard to secure the trust of the public at the core of our reforms.

This is the only way we can ensure the benefits of data and insights are driving effective outcomes for all people and organisations and indeed, for the entire economy and society.

Data is the fuel of growth and trust is the key that will enable us to get ahead.

If the full benefits of open data are to be realised, trust is required at multiple levels, not just as it relates to personal data. Power relations between individuals, institutions and countries are a critical component of trust, and need to be considered when developing data access policies. The reality is that open research data can be more readily exploited by more advanced companies, institutions and countries, which master the technology and the algorithms needed to analyse extract value from the data. Less empowered stakeholders can easily be reduced to simple data providers, while the (research and monetary) value is captured elsewhere.

In order to secure public trust and accountability, the socio-economic impacts of open research data need to be monitored. Over time, such impact assessments should help society evaluate the value of open-data initiatives. The 2006 OECD Recommendation suggested considering a few core aspects for external evaluation, including overall public investments, the management performance of data collection, and the extent to which existing datasets are used and reused (OECD, 2006). This provides useful starting guidance. Nevertheless, it must be noted that such assessments are quite challenging to implement, since the methodologies are not yet well developed and standardised.

Data integration is another major opportunity. For example, New Zealand's Integrated Data Infrastructure[3] allows registered researchers to access microdata about people and households, including data on education; income and work; benefits and social services; population; health; justice and housing. Such an integrated dataset enables social-science research on issues such as the life outcomes of socially disadvantaged groups, linking their educational attainment to income, health and crime outcomes.

Building on current experience and looking forward, some policy implications can be drawn for governments:

- Public data for STI should be "as open as possible, as closed as necessary". When it comes to accessing sensitive data, governance arrangements are critical. Ethics review boards that include data experts can play an important role in this respect.

- Governments should strive to enhance trust among different stakeholders, and create consensus around data sharing and re-use. Risks of privacy breaches cannot be completely avoided, but should be managed, and the procedures to this end should be clear and transparent.

- Specific initiatives can be launched to support data integration, exploring ways in which data from different sources can be combined transparently across different institutions. These initiatives should explore important issues relating to sensitive data, such as anonymization and informed consent.

- Socio-economic assessments should be undertaken to monitor the impact of open research data, with specific attention to where – and to whom – benefits accrue.

Technical standards and practices

As the volume and variety of research data increases, the resources required by data providers to make their data available, and the time invested by users to discover available data, also increase proportionally (OECD, 2015a). Insufficient information exists on what data are available, both for and from research. When data can be found, they are not always useable, because they do not conform to standards, lack metadata or are not machine-readable.

At the national scale, a large variety of institutional and domain-specific data catalogues, search engines and repositories are being established to enhance the findability of data (Box 6.1 and Box 6.2). At the international scale, increased efforts to co-ordinate and support global data networks are necessary (OECD, 2017c), to provide the foundation for developing open-science cloud initiatives that will facilitate data usage (Box 6.1).

Scientific publications are another major channel of discoverability. Many researchers first read about potentially interesting data in a journal article; the question then is how to obtain access to that data. Persistent links should appear in published articles, which should also include a permanent identifier for the data, code and digital artefacts underpinning the published results. Data citation should be standard practice. Broken links or inadequate metadata are common challenges, especially as journals tend to be lenient on data requirements for fear of losing good papers to competing journals. Several publishers have recently developed data journals, which can play an important role in promoting the use of published datasets.

Formal standard-setting through bodies, such as the International Standards Organisation, is a slow iterative process of negotiation that can take several years. As a result, pro-active commercial or public players in a position of power can set de facto standards. One example is Google's General Transit Feed Specification, a common format for public transportation schedules and associated geographic information (OECD, 2018).

The research community can turn this into an advantage if it takes the lead in developing appropriate standards and in so doing, consults fully with all concerned stakeholders. This is the approach taken by organisations that are helping to build the social and technical infrastructure to enable open sharing of data across national and disciplinary borders. For example, the Research Data Alliance produces recommendations – which can be adopted as standards – on a broad range of issues related to interoperability, data citation, data catalogues or workflows for publishing research data (Research Data Alliance, 2017).

Good metadata are essential for data interoperability and re-use (Table 6.1). Provenance information tracks the history of a dataset and is an essential part of metadata, necessary to understand both the source of the information and the history of the dataset (it is also important for incentivising data access, as discussed in the section 'A recognition-and-reward system for data producers'). In this regard, the Open Archival Information System (OAIS) reference model is of particular interest. OAIS was initially developed in the context of archival of data from space missions. It is designed to preserve information over the long term and disseminate it to a designated community that should be able to understand the data independently in the form in which it is preserved. OAIS covers the steps of ingesting, preserving and disseminating the data. It is universally accepted as the common language of digital preservation (Lavoie, 2014). An increasing number of repositories strive to be OAIS-compliant, since this ensures the possibility of re-using data in the long term.

Going forward, some policy implications can be drawn for governments:

- The development and adoption of community agreed standards is critical for FAIR data. Individuals and bodies (such as the Research Data Alliance) that work in this area should be supported accordingly.

- Good metadata are critical for data interoperability and re-use. The compliance of data controllers with standardised reference models (such as OAIS) should be encouraged.

Definition of responsibility and ownership

Issues of ownership and responsibility, including copyright and intellectual property need to be considered when enhancing access to public research data, as they can have important implications for how – and by whom – data can be used. Data creators may not necessarily hold the intellectual property rights (IPR) to the data they collect: in the case of human-subject data, for example, the participants themselves may hold those rights.

Most saliently, any IPR associated with research data, and the licensing arrangements for the use of that data, must be clearly specified. In the absence of such specification, data acquire the statutory IPR of the jurisdiction in which they are used. This may include copyright and *sui generis* database rights (e.g. as in Europe), which can greatly inhibit the further use of data. Such protections arise automatically unless expressly excluded, waived or modified (Doldirina et al., 2018).

Legislation and other rules for managing research data are not harmonised across organisations and countries. Data custodians often operate under various legal frameworks governing the collection and use of research data (e.g. Box 6.4 on South Africa). In the United States, for example, different research-funding agencies have different IPR policies (EARTO, 2016). In the European Union, copyright can be claimed on data that may not be copyrightable in other jurisdictions (such as the United States), with implications for the use of text and data mining in research. According to Hargreaves (2011), "Copyright, once the exclusive concern of authors and their publishers, is today preventing medical researchers studying data and text in pursuit of new treatments."

Tensions between public- and private-sector actors over access to research data are a concern, bearing in mind that one of the main drivers for open data is to improve knowledge transfer and innovation. Enormous potential exists for combining public research data with private-sector data (including social-media data); However, IPR and/or licensing arrangements ensuring both adequate protection of legitimate commercial interests, and the openness and transparency necessary to promote reproducibility and public confidence, are required (OECD, 2016). In this regard, the OECD Principles and Guidelines for Access to Research Data from Public Funding state that: "Consideration should be given to measures that promote non-commercial access and use while protecting commercial interests, such as delayed or partial release of such data" (OECD, 2007).

Going forward, there are a number of policy implications:

- Information about ownership and licensing should be contained within the metadata and specified for all prospective data products in research data management plans. Open-use licences, such as those developed by Creative Commons, should be used, wherever appropriate (OECD, 2015c).

- The implications of any amendments to copyright legislation and IPR regimes, as they relate to access to publically funded data for research, should be carefully considered. They should not inhibit research and innovation in new areas, such as text and data mining, and deep learning.

Box 6.4. In my view: Greater clarity in intellectual property (IP) and data-management policies can contribute to promoting open-data practice

Michelle Willmers, Curation and Dissemination Manager of the Global South Research on Open Educational Resources for Development (ROER4D) project, University of Cape Town, South Africa

The ability of researchers to legally share outputs arising from their work is dictated by institutional IP policies, which are in turn largely influenced by national copyright acts. In the African context, many universities have nascent policy environments, meaning that they may not have an IP policy, or it is out of date and inadequate to cover the intricacies of online content sharing – particularly as relates to open data transfer and publication. There are also instances in which policy environments provide conflicting or contradictory stipulations. This situation makes for confusion on the part of academics in terms of what their actual rights are in the context of data sharing … or, in some cases, may lead to flagrant disregard for policies and mandates.

Both the IP Policy and the Research Data Management Policy of the University of Cape Town (UCT) state that research data are owned by UCT, unless otherwise agreed in research contracts. This may lead many academics to assume they do not have the legal rights to share their data, which is not the case. UCT promotes the use of Creative Commons licensing in its IP Policy, and has a concerted campaign underway to promote responsible data sharing at all levels of the academic enterprise.

Possible confusion in this regard is compounded by the fact that the institutional terms of deposit for sharing data in repositories state that: "UCT grants the Principal Investigator (PI) of a research project the right to upload UCT research data supporting a publication required by a journal publisher or a funder and all UCT project data where this is a specific funder requirement, as long as the data complies with any ethics requirements (e.g. patient confidentiality, consent, etc.)."

This caveat raises questions around the rights of academics who are not operating in research contexts led by PIs, or are functioning in a context where there is no publisher or funder requirement in this regard. The fact that the caveat only exists on a website designed to promote data sharing and is not captured in any of the formal institutional policies regulating data sharing makes the institutional open data policy landscape confusing for academics to navigate, and may serve to build reluctance and confusion, rather than promote a culture of sharing where academics are certain of their legal rights.

Grant agreements and repository deposit terms do increasingly provide exceptions and caveats to restrictive or confusing IP policies, but these agreements are often not adequately scrutinised by academics, and the lack of cohesion between institutional policies, the dictates of funding entities and the intricacies of repository terms and conditions can ultimately amplify the distrust of – and therefore the reluctance to engage with – open-data practice.

National and regional initiatives to assess and revise institutional IP policies so that they are conducive to open data sharing and form part of a set of clear, cohesive institutional stipulations would be extremely valuable in terms of promoting open data practice, and ensuring a functional understanding of the legal and ethical aspects of the process – the uncertainty of which often inhibits academics' practice in this regard.

A recognition-and-reward system for data producers

Data sharing entails a cultural change among researchers in many fields of science. Appropriate acknowledgement and reward systems need to counterbalance perceived barriers and risks of providing open access to data. The emphasis on competition in research, including the way in which it is evaluated and funded, can be a strong disincentive to openness and sharing.

Researchers have incentives to publish (preferably positive) scientific results. Incentives to publish data are less developed, and usually seen as a constraint imposed by funding agencies and/or publishers. Data citation has not been widely implemented; although the prerequisites for achieving this (e.g. standard formats and citation metrics) already exist, they are not being broadly adopted. Data activities (including those relating to negative results) need to be embedded in evaluation systems, to ensure that researchers who provide high-quality research data are rewarded.

Despite the progress achieved, sharing of research data remains suboptimal. In a 2016 OECD Survey of scientific authors,[4] only 20-25% of corresponding authors had been asked to share data after publication. If asked, a significant share (30-50%) said they would grant access to the data, or at least undertake steps to grant them; about 30% of authors said they would seek to clarify the request. Depending on the discipline, 10-20% of authors would refuse to share data on legal grounds (Boselli and Galindo-Rueda, 2016). Authors of scientific papers are more reluctant to share their data openly than to access data from other research groups (Elsevier and Centre for Science and Technology Studies, 2017).

The TOP Guidelines (Box 6.1) recognise data citation as one of the levers to incentivise data sharing. They propose making data citation mandatory, and citing and referencing all datasets and the codes used in a publication with a DOI (Center for Open Science, 2014). The adoption of unique digital identifiers for researchers, such as the Open Researcher and Contributor ID (see Chapter 12), is also important in this context, as it would greatly simplify provenance mapping and related citation.

Adopting data citation as standard practice so that it can be used to incentivise and reward data sharing also requires developing appropriate data-citation metrics. These could then be used alongside other assessment measures, such as bibliometrics, in recruitment and evaluation processes (OECD, 2018). The approach adopted by the National Science Foundation (NSF) in the United States is an interesting example in this regard. The NSF has implemented an incremental strategy for accessing research data over the past decade. Since 2013, datasets and publications are treated equally as products in the context of an individual researcher's "biosketch". In 2016, the NSF added to the proposal section a requirement to discuss evidence of research products and their availability, including data, in prior NSF-funded research. In France, the newly published national Open Science Plan (Ministère de l'Enseignement supérieur, de la Recherche et de l'Innovation, 2018) adopts similar principles, pleading for a more qualitative rather than purely quantitative, approach to evaluating researchers. The Open Science Plan is based on the San Francisco Declaration on Research Assessment, which calls for a more holistic evaluation of scientists considering all their research outputs, including data and software (DORA, 2012).

Although recognising data citation and data products in academic evaluation processes may incentivise researchers, it will not necessarily value the critical contribution of data stewards. These are the people who curate and manage data, and ensure their long-term availability and usability (see the section on Future outlook). Career paths for this cohort of data professionals are unclear. Mechanisms to assess their performance should be distinct from the evaluation mechanisms applied to researchers, but should be linked to the data that they manage. New measures, incentives and reward systems will be required for data stewards.

Going forward, possible policy measures to incentivise and promote data sharing by researchers include:

- developing new indicators/measures for data sharing, and incorporating these into institutional assessment and individual researcher-evaluation processes

- promoting the use of unique digital identifiers for individual researchers and datasets, to enable citation and accreditation

- developing attractive career paths for data professionals, who are necessary to the long-term stewardship of research data and the provision of services.

Business models and funding for open data provision

"Open access" does not necessarily imply "free of charge". However, many experts agree that public research data should ideally be free at the point of usage (OECD, 2018), implying that the costs of the stewardship and provision will be assimilated by the data provider or repository. These costs can be substantial and require long-term financial commitment, often over several decades. Ultimately, most of the funding for open research data is likely to come from the public purse, although alternative revenue streams exist for some types of data (OECD, 2017b). A key question from the science-policy or funder perspective is how best to allocate this funding. The answer depends on a full understanding of the business models and value propositions of specific data repositories and of the networks in which they are integrated (Figure 6.2).

This must consider multiple factors, including the role of the repository, national and domain contexts; the stage of the repository's development or lifecycle phase; the characteristics of the user community; and the data product required by this community (influencing the level of investment necessary to curate and enhance the data). Business models are constrained by – and need to be aligned with – policy regulation (mandates) and incentives (including funding) (OECD, 2017b).

Many different kinds of data repositories provide a large variety of services, ranging from raw data to complex online analyses. Institutional repositories, national repositories, domain-specific repositories and international repositories are all parts of a complex landscape. This landscape is constantly changing as valuable new data resources arise from projects and transition into longer-term sustainable infrastructures, with longer-term funding requirements. At the level of the individual research system, potential economies of scale can be obtained by centralising or federating the management of data resources; this is common practice in some fields. However, not all data can be transferred across institutional or national boundaries for legal, proprietary or ethical reasons; a certain amount of redundancy in the system can also present some advantages, by making it more resilient. Federated networks can provide some of the benefits of scale, while respecting diversity (OECD, 2017c).

Even when business models are well-developed, and long-term funding is identified, there are limits on how data repositories can operate to provide FAIR access to increasing volumes of data. Priorities need to be established and choices made, e.g. between providing immediate online access or putting data into deep storage. With very big data from experimental facilities (such as the Square Kilometre Array telescope), it is impossible to provide open online access to all users; thus, tiered access systems have been developed. Prioritisation and data selection will be an increasingly significant challenge in the future. Addressing this challenge will require dialogue with data provider and users, as well as more systematic cost-benefit analyses (bearing in mind that data that may be of little value today can be very valuable tomorrow, and today's users may be different tomorrow).

Figure 6.2. Creating a value proposition for data repositories

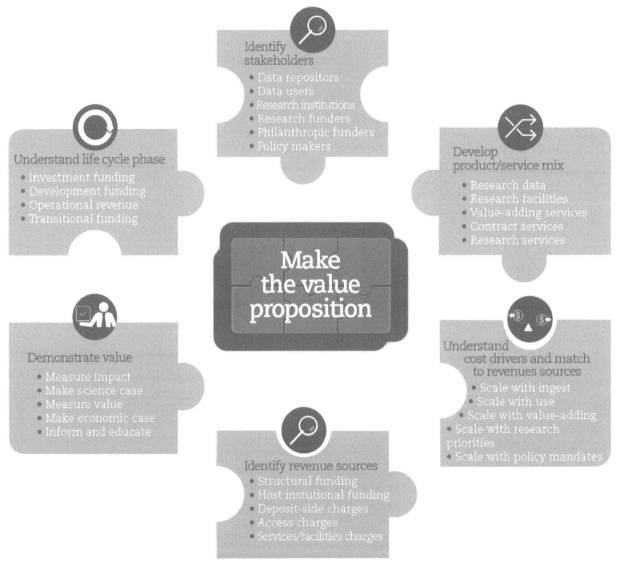

Source: OECD (2017b).

Research-data repositories and services can also be developed as public-private partnerships. Some private companies are opening their data for non-monetary gain (e.g. for recruiting, improving their image or exchanging data). For instance, medical researchers may want to combine data about people's medical history, genomics, food intake and mobility. Here, medical and genomic data may come from the public sector, while mobility and food data could depend on access to private-sector data. Provided that IPR and ethical issues can be agreed on, public-private partnerships built around such themes should be encouraged, as they can support the development of data infrastructure and the creation of value-added services. The governance arrangements of such public-private partnerships need to be carefully designed to promote trust among all stakeholders, and ensure transparency and accountability (OECD, 2016).

Going forward, some policy implications can be drawn for governments:

- Develop strategies and roadmaps, including long-term funding plans and business models, to build sustainable research-data infrastructure (i.e. data repositories and services).

- Explore how public investment in research data and infrastructure can be used to leverage private investment (as well as skills and data resources), while ensuring openness and accountability.

Building human capital

Depending on the scientific domain, researchers normally have some training in data analysis, but often lack data-management skills. Users (who may be from different academic sectors or from the private sector) do not always have the appropriate skills to interpret and analyse the data correctly. The effective operation of data repositories requires specialised skills in data curation and stewardship. Various other skills – related to ethical, legal and security issues, as well as risk management, communication and design – should be included in any well-functioning open-data ecosystem. A lack of these skills breeds lack of trust.

"Data science" and "data scientists" are overarching terms encompassing a wide range of skill needs. The National Institute of Standards and Technology Big Data Interoperability Framework (Volume 1)[5] defines a data scientist as "a practitioner who has sufficient knowledge in the overlapping regimes of business needs, domain knowledge, analytical skills and software and systems engineering to manage the end-to-end processes in the data life cycle." In reality, very few individuals exist in most scientific fields who fit this definition and are leaders in each of these skill areas. Research increasingly depends on collaboration and co-operation between individuals with different data skillsets. Defining the needs and gaps for these skillsets in different scientific fields is a challenge.

Several detailed analyses exist of the data-skill requirements for science, e.g. the Data Science Framework developed by the EDISON project funded by the European Commission[6] (Box 6.5).

> **Box 6.5. Data skills**
>
>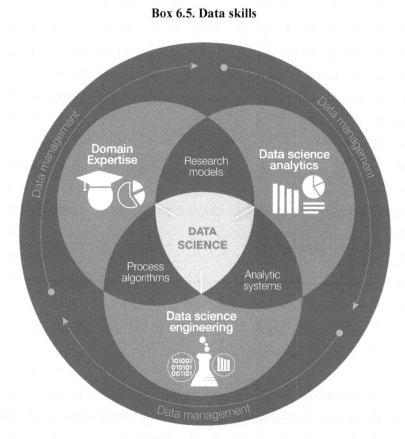
>
> This diagram illustrates the main competence groups within data science, as defined in the EDISON project: data-science analytics, data-science engineering, and domain knowledge and expertise. Data management, including curation and long-term stewardship, is sometimes classified as part of data science or as a separate competence group. These various competences need to be integrated into the different aspects of the research process, from design to experimentation, analysis and reporting.

Different scientific domains are equipped to varying degrees when it comes to data skills. Traditionally data-intensive fields, such as experimental physics or astronomy, are generally well-positioned (although competition for data scientists from commercial actors is affecting recruitment and retention in academia). Other areas, such as medical research, have significant skill gaps. Moreover, the additional burden of curating and stewarding data to make it available for secondary use creates a human-resource challenge that cuts across all areas of science.

Identifying skill needs and gaps across different research domains is a necessary first step. Meeting these needs is an even greater challenge, which requires a combination of retraining existing personnel (e.g. retraining librarians and archivists to perform data-stewardship functions), and providing new education and training opportunities for researchers and professional research-data support roles. Many initiatives are already taking place in this regard, presenting considerable opportunities for mutual learning across countries and different scientific domains.

Data scientists are in high demand in industry, and academic research competes for the best talent. An urgent need exists to develop recognition-and-reward structures and attractive career paths for all the specialists needed to realise the value of public research data. As in other research areas, workforce diversity will be an important determinant of success (Chapter 7) that should be considered at the outset when developing human-resource strategies for the digital research age.

Going forward, some policy implications can be drawn for governments:

- Develop a national data-skill strategy for STI, identifying specific skill gaps, and the education and training requirements needed to fill them.

- Facilitate co-operation across different education and research actors, to ensure coherence and complementarity in data-skill capacity-building activities.

Future outlook

The significance of data for STI will undoubtedly continue to increase over the next decade. The volume of data produced globally amounted to 16 zettabytes (ZB)[7] in 2016 and is expected to grow to 163 ZB by 2025 (Reinsel, Gantz and Rydning, 2017). The importance of artificial intelligence in assisting scientific discovery is also expected to grow significantly; access to well-managed data is a key enabler of this development (Kitano, 2016).

Enhanced access to research data holds considerable promise for increasing research productivity and innovation, and developing solutions to complex societal challenges. However, realising this potential, and minimising the potential risks, will require strategic planning and policy interventions. The OECD Recommendation (OECD, 2006) and the more recent FAIR principles for data access provide a broad guiding framework for policy development and co-operation across communities. Many countries have already taken up the challenge and have adopted open-science policies and/or open access to research-data strategies; at the European level, the European Commission has taken the lead in ensuring policy coherence across countries.

Box 6.6 references two possible scenarios. Successful implementation of open-data policies and strategies crucially requires establishing governance systems and processes that ensure transparency and foster trust across the research community and society at large. Mandates and incentives will need to be used judiciously to support and facilitate changes in research behaviour, without stifling creativity and innovation. Long-term investment in technical infrastructure and human capital will be required. Technical standards need to be developed, and legal and ethical concerns addressed.

A lot needs to be done, and a lot is already being done. Understandably, policy intervention focuses on realising the exciting opportunities presented by enhanced access to research data. Open data can help address issues related to the reproducibility and accountability of scientific research; it can help provide solutions to pressing socio-economic challenges; and it can unite the global scientific community around these issues. Looking to the future, however, it is also important to properly consider and mitigate some of the potential risks. The advent of data-driven science coincides with a crisis of confidence in science and the advent of the "post-truth" era.

Opening up public research data means that new actors will be able to analyse and interpret the data from their own perspectives, and not necessarily with the critical objectivity expected from scientists. The old adage that "if you have enough data, you can prove anything" is not unfounded.

In the new world of open science, the scientific community will need to work rigorously, communicate clearly the scientific method and limitations of its analyses, and engage in honest discourse and dialogue with public and policy makers. In a hyper-competitive research enterprise characterised by enormous pressure to succeed and growing hype around scientific breakthroughs, there is a need to ensure that open science and data can be trusted. Technological developments can assist in this regard. Ultimately, however, trust is a social construct that needs to be carefully nurtured over time.

Box 6.6. Possible future scenarios for access to data for STI

In a desirable future scenario, trust would be earned across society through strong governance initiatives. These would ensure robust risk management and mitigation, elaborated in transparent consultation with stakeholders. Ethics review boards would credibly represent individual interests and arbitrage consent issues. On the technical side, strong global standards would emerge, akin to Transmission Control Protocol/Internet Protocol for Internet communication, complemented by more specialised standards for specific applications. IPR and licensing provisions would promote responsible data access and re-use and be a standard part of machine-readable metadata. Data citation would become ubiquitous and an integral part of researcher evaluation. Financing of repositories would be based on long-term infrastructure strategies and sustainable models. Finally, digital skills would be addressed through a strategic approach encompassing initial education and lifelong learning for data producers, stewards and users.

A "worst-case" scenario is also possible, in which repeated security and privacy breaches would be inadequately managed, fostering a general level of mistrust. Standards would continuously lag behind technology development while IPR would be insufficiently defined to support widespread data re-use. Incentives for researchers to publish their data would remain weak, and initiatives to support data skills development would be poorly designed.

Notes

[1] "EOSC Declaration: European Open Science Cloud – New Research & Innovation Opportunities": https://ec.europa.eu/research/openscience/pdf/eosc_declaration.pdf#view=fit&pagemode=none.

[2] Regulation 2016/679 defines "consent" of the data subject means any freely given, specific, informed and unambiguous indication of the data subject's wishes by which he or she, by a statement or by a clear affirmative action, signifies agreement to the processing of personal data relating to him or her.

[3] http://archive.stats.govt.nz/browse_for_stats/snapshots-of-nz/integrated-data-infrastructure.aspx.

[4] http://oe.cd/issa.

[5] https://bigdatawg.nist.gov/V1_output_docs.php.

[6] http://edison-project.eu/.

[7] 1 ZB = 1 trillion gigabytes, or 10^{21} bytes.

References

Borgman, C. L. (2012). The conundrum of sharing research data. *Journal of the American Society for Information Science and Technology*, 63(6), 1059-1078.

Boselli, B. and F. Galindo-Rueda (2016), "Drivers and Implications of Scientific Open Access Publishing: Findings from a Pilot OECD International Survey of Scientific Authors", *OECD Science, Technology and Industry Policy Papers*, No. 33, OECD Publishing, Paris, http://dx.doi.org/10.1787/5jlr2z70k0bx-en.

Center for Open Science (2014), "Transparency and Openness Promotion (TOP) Guidelines", Center for open Science, Charlottesville, VA, http://cos.io/top.

Doldirina, C. et al. (2018), "Legal Approaches for Open Access to Research Data", *LawArXiv Papers*, Cornell University, Ithaca, NY, http://dx.doi.org/10.31228/OSF.IO/N7GFA.

DORA (2012), *San Francisco Declaration on Research Assessment (DORA)*, https://sfdora.org/ (accessed on 07 July 2018).

EARTO (2016), "EARTO Background Note: Overview of US Federal Agencies Data Sharing Policies", Defense Technical Information Center, Fort Belvoir, VA, http://www.dtic.mil/dtic/pdf/dod_public_access_plan_feb2015.pdf (accessed on 19 July 2018).

Elsevier and Centre for Science and Technology Studies (2017), *Open data: The researcher perspective*, Elsevier and Centre for Science and Technology Studies, Amsterdam and Leuven, https://www.elsevier.com/__data/assets/pdf_file/0004/281920/Open-data-report.pdf (accessed on 26 July 2018).

EC/OECD (2017), *STIP Compass: International Science, Technology and Innovation Policy (STIP)* (database), edition 2017, https://stip.oecd.org.

Economic and Social Research Council (2015), "Research Data Policy", webpage, https://esrc."ukri.org/funding/guidance-for-grant-holders/research-data-policy/ (accessed on 19 June 2018).

eRSA (2014), "The Australian Research Cloud: Your most powerful eResearch tool yet – eRSA", webpage, https://www.ersa.edu.au/the-australian-research-cloud-your-most-powerful-eresearch-tool-yet/ (accessed on 8 November 2018)

European Commission (2018a), "Proposal for a DIRECTIVE OF THE EUROPEAN PARLIAMENT AND OF THE COUNCIL on the re-use of public sector information (recast)", http://ec.europa.eu/transparency/regdoc/rep/1/2018/EN/COM-2018-234-F1-EN-MAIN-PART-1.PDF

European Commission (2018b), "COMMISSION RECOMMENDATION (EU) 2018/790 of 25 April 2018 on access to and preservation of scientific information", Official Journal of the European Union, Vol. L/134, pp. 12-18, https://eur-lex.europa.eu/legal-content/EN/TXT/PDF/?uri=OJ:L:2018:134:FULL&from=EN.

European Commission (2016), "REGULATION (EU) 2016/679 OF THE EUROPEAN PARLIAMENT AND OF THE COUNCIL – of 27 April 2016 – on the protection of natural persons with regard to the processing of personal data and on the free movement of such data, and repealing Directive 95/46/EC (General Data Protection Regulation)", Official Journal of the European Union, http://eur-lex.europa.eu/legal-content/EN/TXT/PDF/?uri=CELEX:32016R0679&from=EN.

European Commission (2014), "Validation of the results of the public consultation on Science 2.0: Science in Transition", European Commission, Brussels, http://ec.europa.eu/research/consultations/science-2.0/science_2_0_final_report.pdf.

European Commission (2012), "COMMISSION RECOMMENDATION of 17.7.2012 on access to and preservation of scientific information", European Commission, Brussels, https://ec.europa.eu/research/science-society/document_library/pdf_06/recommendation-access-and-preservation-scientific-information_en.pdf.

Hargreaves, I. (2011), "Digital Opportunity: A Review of Intellectual Property and Growth An Independent Report", ORCA Online Research @Cardiff, Cardiff University, https://orca.cf.ac.uk/30988/1/1_Hargreaves_Digital%20Opportunity.pdf.

Kitano, H. (2016), "Artificial Intelligence to Win the Nobel Prize and Beyond: Creating the Engine for Scientific Discovery", *AI Magazine*, Vol. 37/1, p. 39, Association for the Advancement of Artificial Intelligence, Palto Alto, CA, http://dx.doi.org/10.1609/aimag.v37i1.2642.

Lavoie, B. (2014). The Open Archival Information System (OAIS) Reference Model: Introductory Guide (2nd Edition) Principal Investigator for the Series Neil Beagrie. Retrieved 07 03, 2018, from https://www.dpconline.org/docs/technology-watch-reports/1359-dpctw14-02/file

Ministère de l'Enseignement supérieur, de la Recherche et de l'Innovation (2018), "Plan national pour la Science ouverte, mercredi 14 juillet 2018", Ministère de l'Enseignement supérieur, de la Recherche et de l'Innovation, Paris, http://cache.media.enseignementsup-recherche.gouv.fr/file/Actus/67/2/PLAN_NATIONAL_SCIENCE_OUVERTE_978672.pdf (accessed on 17 September 2018).

Morais, R. and L. Borrell-Damian (2018), "2016-2017 EUA SURVEY RESULTS", European University Association, Brussels, http://www.eua.be/Libraries/publications-homepage-list/open-access-2016-2017-eua-survey-results (accessed on 19 June 2018).

NIH (2017). ERA Commons - NIH. Retrieved 08 November 2018, from https://public.era.nih.gov/commons/public/login.do.

OECD (2018), "Towards New Principles For Enhanced Access To Public Data For Science, Technology And Innovation", Joint CSTP-GSF Workshop, March 13, 2018, Paris.

OECD (2017a), "Open access to data in science, technology and innovation – initial survey findings", OECD, Paris.

OECD (2017b), "Business models for sustainable research data repositories", *OECD Science, Technology and Industry Policy Papers*, No. 47, OECD Publishing, Paris, https://doi.org/10.1787/302b12bb-en.

OECD (2017c), "Co-ordination and Support of International Research Data Networks", webpage, http://www.oecd.org/going-digital (accessed on 16 July 2018).

OECD (2016), "Research Ethics and New Forms of Data for Social and Economic Research", *OECD Science, Technology and Industry Policy Papers*, No. 34, OECD Publishing, Paris, http://dx.doi.org/10.1787/5jln7vnpxs32-en.

OECD (2015a), "Making Open Science a Reality", *OECD Science, Technology and Industry Policy Papers*, No. 25, OECD Publishing, Paris, http://dx.doi.org/10.1787/5jrs2f963zs1-en.

OECD (2015b), *Data-Driven Innovation: Big Data for Growth and Well-Being*, OECD Publishing, Paris, http://dx.doi.org/10.1787/9789264229358-en.

OECD (2015c), "Assessing government initiatives on public sector information: A review of the OECD Council Recommendation", *OECD Digital Economy Papers*, No. 248, OECD Publishing, Paris, http://dx.doi.org/10.1787/5js04dr9l47j-en.

OECD (2013), "New Data for Understanding the Human Condition: International Perspectives – OECD Global Science Forum Report on Data and Research Infrastructure for the Social Sciences", OECD, Paris, http://www.oecd.org/sti/sci-tech/new-data-for-understanding-the-human-condition.pdf.

OECD (2009), "Access to Research Data: Progress on Implementation of the Council Recommendation", Science, Technology and Innovation for the 21st Century. Meeting of the OECD Committee for Scientific and Technological Policy at Ministerial Level, 29-30 January 2004 - Final Communique, OECD, Paris, http://www.oecd.org/document/15/0,3343,en_2649_34269_25998799_1_1_1_1,00.html.

OECD (2007), *OECD Principles and Guidelines for Access to Research Data from Public Funding*, OECD Publishing, Paris, https://doi.org/10.1787/9789264034020-en-fr.

OECD (2006), *Recommendation of the Council concerning Access to Research Data from Public Funding*, OECD, Paris, https://legalinstruments.oecd.org/en/instruments/OECD-LEGAL-0347.

President's Advisors on Science and Technology (2014), "Report to the President: Big data and privacy: a technological perspective", The White House, Washington DC, https://bigdatawg.nist.gov/pdf/pcast_big_data_and_privacy_-_may_2014.pdf.

Research Data Alliance (2017), "All Recommendations & Outputs – RDA Endorsed Recommendations", webpage, https://www.rd-alliance.org/recommendations-and-outputs/all-recommendations-and-outputs (accessed on 25 September 2017).

Wilkinson, M. et al. (2016), "The FAIR Guiding Principles for scientific data management and stewardship", Scientific Data 3, Article No. 160018 (2016), *Nature Communications*, http://dx.doi.org/10.1038/sdata.2016.18.

Chapter 7. Gender in a changing context for STI

By

Elizabeth Pollitzer, Carthage Smith and Claartje Vinkenburg

The under-representation of women in certain areas of science, technology and innovation (STI) has long been a concern. As the benefits of diversity in STI – both in terms of research excellence and relevance – become clearer, most countries are implementing policies to try to address gender equity. However, issues such as gender stereotypes and evaluation bias are embedded in research systems, and are resistant to simple interventions. This chapter begins by looking at the key issues affecting gender equity in science at different life stages. It starts with gender stereotypes that influence educational choices and career expectations in early childhood. It follows with a discussion of undergraduate and graduate education, as well as gender issues in research careers and the research system. It then considers the changing context for STI, and how this increases the emphasis on diversity. Finally, it lays out a future vision for a more diverse and productive scientific enterprise. While acknowledging that most countries have included gender diversity as one of the key objectives in their national STI plans, the chapter argues that policy initiatives remain fragmented. A more strategic and systemic long-term policy approach is necessary.

The statistical data for Israel are supplied by and under the responsibility of the relevant Israeli authorities. The use of such data by the OECD is without prejudice to the status of the Golan Heights, East Jerusalem and Israeli settlements in the West Bank under the terms of international law.

Gender equity: A persistent science, technology and innovation (STI) policy imperative

An estimated USD 12 trillion (US dollars) could be added to global gross domestic product by 2025 by advancing gender parity (McKinsey, 2015); this alone provides a strong rationale for including gender issues in STI policy. However, the benefits of tackling the under-representation of women in STI go well beyond economic gains and access to talent. In addition to the important issues of social justice and fairness, growing evidence suggests that diversity improves the quality of research and the relevance of its outcomes for society (Smith-Doerr, Alegria and Sacco, 2017). It is not surprising that gender has figured on STI policy makers' agendas for several decades and is now receiving even greater attention in most countries, with the expectation that STI will make a major contribution to the Sustainable Development Agenda 2030 and the 17 Sustainable Development Goals (SDGs). Not only does a specific goal (SDG 5) target gender, but diversity and inclusiveness in STI are considered a prerequisite for producing the types of knowledge and innovation required to respond effectively to all the SDGs.

In 2006, OECD published a report *Women in Scientific Careers: Unleashing the Potential* that took stock of the gender imbalances in different scientific fields and at different career stages in both the public and private sectors. It reviewed the policy actions taken by governments to address these imbalances, and concluded that "few countries appear to have a comprehensive approach to promoting the participation of women in scientific education and research careers". Since 2006, there has been some progress in some fields in some countries but the picture today remains largely the same as it was then (OECD, 2017a) and the same challenges prevail (Table 7.1). There are also some new issues related to gender bias in the selection of research topics and related innovations that were not much discussed a decade ago, but are increasingly recognised as important to STI policy.

Today, most OECD countries are implementing a variety of policy measures to address obvious gender inequalities (Box 7.1). Nevertheless, gender imbalances persist, and are particularly evident in some Science, Technology, Engineering and Mathematics (STEM) areas. While some policy measures – such as targeted support for individuals – are relatively easy to implement and assess, other areas such as changing gender stereotypes and eliminating implicit gender bias are much more resistant to intervention and require longer term cross-sectoral action.

Table 7.1. Gender issues and STI

Life/career stage	Issue	Cause	Policy options
Early childhood	Societal expectations of girls and boys are different	Gender stereotypes; cultural norms	Work with school teachers and media to address stereotypes; raise parental awareness of negative effects of stereotypes
Secondary education	Girls less likely to choose science, technology, engineering and mathematics (STEM) than boys	Gender stereotypes; parental expectations; peer pressure	Work with teachers to address career stereotypes; promote role models
Undergraduate	Women under-represented in certain STEM fields	Gender stereotypes; "exclusive" disciplinary cultures; bias in standardised selection tests	Design specific curriculum and reform pedagogic methods; highlight opportunities in STEM
Post-grad/PhD	Women continue to be under-represented in certain fields	Cumulative stereotypes; hyper-competition and bias in assessment of individual performance	Targeted individual support; innovative PhD training for careers beyond academia
Post-doc/early career	Women disproportionally drop out of STI	Precarity and hyper-competitivity; more attractive options outside of academia and research	Targeted individual support; more tenure-track positions; improve social and employment provisions for child care and parental leave
Career path	Women's career progress is slower than men's	Care responsibilities; gender bias in academic norms and evaluation; unfavourable cultures in technology-intensive sectors; unequal salaries	Targeted individual support; conditions promoting retention, e.g. flexibility in working hours and part-time leave; raise awareness of gender bias and shift norms accordingly
Senior appointments	Very few women in senior posts	Cumulative stereotypes and biases regarding career paths; lack of role models; bias in selection criteria and processes	Targeted individual support; legislate on pay discrimination; support female role models; raise awareness and provide training on evaluation and selection biases

---

Box 7.1. What are countries doing to address gender equity in STI?

The 2017 OECD/EC STI Policy survey reveals that almost all countries have implemented specific policy initiatives related to gender in STI (OECD, 2017b). Over 100 initiatives were reported, with several countries reporting multiple initiatives led by different institutions In almost all countries, gender equity is a strategic priority as identified in a variety of national plans. It is mainly addressed through targeted competitive funding – i.e. fellowships, project grants and prizes – for different stages of science training and career paths. Some countries have also attached conditions predicated on gender equity to the awarding of institutional funding. Several countries have also implemented public-awareness campaigns, e.g. schemes to support outreach by women scientists in schools. Fewer initiatives focus more precisely on the systemic issues – such as peer-review, reward and promotion mechanisms – that influence gender balance in academic institutions. The section immediately below provides some examples of how countries are combining multiple policy actions to address gender equity.

Australia: Gender equality is emphasised in the National Innovation and Science Agenda, 2015 and the Australian government is implementing a range of initiatives to support women's participation in STEM studies and careers. The funding schemes of the Australian Research Council (ARC) are underpinned by policies to support gender diversity. All ARC schemes take into account any career interruptions as part of the assessment processes and part time work and parental leave are enabled in some schemes. ARC also provides each year two named Laureate Fellowships for female researchers to undertake an ambassadorial role to promote women in research. Several other research-funding agencies have also implemented mechanisms to promote gender equality and diversity, including adjusting the criteria for awarding institutional block grants that support PhD training; establishing a dedicated Women in Health Science Working Committee to monitor gender-balance issues; and targeted training initiatives in mathematics. In 2016, the inaugural programme of the Homeward Bound leadership initiative culminated in the largest ever female expedition to Antarctica[1].

Germany: Only 24% of university lecturers and 15% of the country's 38 000 tenured professors in Germany are women. To improve those statistics, the German Education Ministry has introduced a Women Professors scheme, whereby the Ministry pays the salary of one to three female professors or lecturers at universities that are committed to redressing gender imbalances at the leadership level. The Government has committed substantial funds to its equal-opportunity programme for universities, with the aim of creating 200 additional posts for highly qualified female academics. Each post will be funded for five years, with the Federal Government and the individual states (*Länder*) splitting the costs. To secure the funding, the universities had to submit plans demonstrating their commitment to promoting more women to top academic positions and sustainably restructuring the university.

Ireland: Starting in 2013, the Ireland Research Council, the Higher Education Authority and Science Foundation Ireland have emphasised the need for a consistent and concerted approach to gender issues. Specific actions to this end include changing the eligibility criteria for the Starting Investigator national grant programme to increase the number of women applicants, and directing institutions to adopt the Athena Scientific Women's Academic Network (SWAN) award accreditation as a mandatory requirement to receive research funding. [The Athena SWAN Charter is an internationally recognised "quality mark" for gender equality administered by the Equality Challenge Unit in the United Kingdom. A similar accreditation, the Juno Excellence award, focuses on physics and is managed by the UK Institute of Physics.]

Japan: Women account for fewer than 15% of all researchers in Japan. The Japanese Government has taken several policy measures to address the cultural norms and practices driving this imbalance. Since 2006, a scheme to "support girl students to choose a science course" sponsors events where schoolgirls meet and talk with women scientists and engineers. In 2015, the Government implemented a new initiative to "realise diversity in the research environment", by supporting women researchers with family and care responsibilities.

Mexico: Mexico is implementing several initiatives to promote diversity and inclusion in STI, including the Mexican Mothers Heads of Households Scholarships, aimed at single, divorced, widowed or separated mothers who are pursuing professional studies (technical specialisation or third-level degree) in public higher education institutions; and the Scholarship Programme for Indigenous Women, which provides support for postgraduate studies in Mexico or abroad. Dedicated funding is also available for research projects that

> can generate knowledge, technological developments or innovations addressing women's issues and needs.
>
> Source: OECD (2017b), EC/OECD STI Policy survey, https://stip.oecd.org/stip.html.

Several trends – including globalisation, the internationalisation of higher education, increasing researcher mobility, and new paradigms of open science and inclusive innovation – are changing the landscape in which gender, STI and socio-economic conditions interact. This is generating both new challenges and new opportunities for women in STI. The shifting context heightens the need for a greater understanding and acknowledgement of how gender inequalities are created and perpetuated in science institutions, within scientific research, and when translating scientific knowledge into innovation. While the availability of sex-disaggregated data for socio-economic analyses has been improving, better data and new indicators are needed to monitor the evolving situation and inform appropriate policy interventions that address the causes as well as the symptoms of gender inequality in STI.

This chapter begins by looking at the key issues affecting gender equity in science at different life stages. It then considers the changing context for STI, and how this increases the emphasis on diversity. It starts with gender stereotypes that influence educational choices and career expectations in early childhood. It follows with a discussion of undergraduate and graduate education, as well as gender issues in research careers and the research system. Finally, it considers the main drivers for change in STI, and lays out a future vision for a more diverse and productive scientific enterprise. It is difficult to do justice to all the important aspects of gender equality in STI in a single short chapter; hence, some important issues are referenced, but not discussed in detail.[2] The chapter does not make a strong distinction between scientific careers in the public/academic sector and the private sector: although differences exist, the key issues affecting gender equity are very similar. The chapter also does not deal in depth with some specific issues relating to gender and innovation, although they are recognised as an increasingly important area for policy development.

Childhood and gender stereotypes

The relative over-representation of men in STEM starts at an early stage and is reflected in the numbers of men versus women in school subjects, types of education and degree programmes. While some debate is taking place as to what exactly causes these gender differences, the evidence points to stereotypes more than capabilities (Miller, Eagly and Linn, 2015). Interestingly, the numbers of men and women vary depending on disciplines, countries and cohorts. This indicates there are structural, cultural and socio-economic factors at play, rather than inherent, unchangeable factors.

Gender stereotypes are common expectations about the roles of men, women, boys and girls in society, at home and at work. These "received ideas" do not only reflect what men and women typically do, but also what they *should* do, and are therefore normative and prescriptive (Heilman, 2012). The main expectation is that men work and women care, and that men have a higher innate ability for most STEM fields than women (Leslie et al., 2015). The visible division of labour at work and at home is "justified" by inherent biological differences. Making counter-stereotypical "choices" is therefore harder and generates more disapproval than fitting the stereotype.

Stereotypes are acquired at an early age – even before schools starts. From age six or so, both boys and girls say that boys are more likely to be "really, really smart" than girls (Bian, Leslie and Cimpian, 2017), and both boys and girls are more likely to draw a man than a woman when asked to "draw a scientist" (Miller et al., 2018). These stereotypes generally intensify during adolescence (OECD, 2018a), and are reinforced at key stages over the life course, including marriage, childbirth and ageing/caring. Understanding the cumulative effects of stereotypes on individual choices and careers is one reason why the ability to measure the persistence rates of women in STEM is important (Section 4).

The finding that countries scoring highly on gender equality, as measured by the Gender Gap Index,[3] number fewer women in certain STEM areas is most likely related to internalised gender stereotypes (Charles, 2017). Gender stereotypes are socially and culturally embedded, and resistant to simple policy actions. Although a growing number of evidence-based policy measures are being developed, persistent gender stereotypes in the media - including social media - and advertising may counter or even cancel out the positive effects of these interventions (European Parliament, 2014). Children think they cannot be what they cannot see; reproduced stereotypes inhibit their motivation, ability and self-efficacy, and ultimately restrict their choices. Thus, gatekeepers to STEM careers – parents, teachers, career counsellors, future employers – need to work together with policymakers to prevent gender stereotyping of jobs and skills (Box 7.2).

Box 7.2. Overcoming gender stereotypes

Overcoming stereotypes of what men and women do at home and at work (particularly in STEM professions) is possible. However, the process needs to start early in life, and to be regular and consistent. Occasional exposure to the counter-stereotype may simply reproduce the norm, on the premise that "the exception proves the rule". Several examples of evidence-based interventions addressing gender stereotypes in STEM exist in both educational institutions and the media.

Primary education:

The #RedrawTheBalance campaign by Education and Employers (United Kingdom) started from a project asking children to draw their future profession, which revealed both a limited view of their possible future and the existence of gender stereotypes. Subsequently developed interventions include Inspiring the Future[4] and Primary Futures.[5,6]

The Young Scientists Australia programme is one of many initiatives that engage and challenge young children to explore and invent, with inspirational assignments from different STEM fields.[7] A similar programme, Let Toys be Toys, was developed in the United Kingdom to support teachers in challenging stereotypes in the classroom.[8]

Secondary education:

Efforts to engage girls in coding and other aspects of computer science are often driven jointly by information and communication technology (ICT) companies and non-governmental organisations (NGOs) or government agencies. One far-reaching example with a strong social media presence is the US-based Girls Who Code.[9]

Various initiatives are being developed in local settings to attract and sustain girls' interest in physics. These include: "girls' days" organised at German physics research institutes, such as Deutsches Elektronen-Synchrotron;[10] classroom interventions developed by the

> UK Institute of Physics;[11] and an Italian school competition for developing a video or interactive website that changes stereotypes of women in the natural sciences.[12]
>
> Finally, the Hypatia project has put together an evidence-based toolkit in various languages, with resources and national knowledge hubs to promote gender-inclusive STEM education in schools, science centres and museums across the European Union.[13]
>
> **Media representation and advertising**
>
> The Dutch NGO WOMEN Inc. has launched a campaign, *#BeperktZicht* ("limited vision") and established a coalition with media partners. The goal is to improve the representation and inclusion of women and ethnic minorities in the media, not only by increasing numbers, but also by featuring them in more counter-stereotypical roles.[14]
>
> In 2016, the US Government developed a factsheet for media partners depicting STEM opportunities (including for women) as part of STEM 2026. This "aspirational vision for STEM education" is inspired by the so-called "Scully effect" (in reference to the popular TV series "The X-Files") of growing interest from girls in science, medicine and especially forensics.[15]
>
> The United Nations recently launched a global initiative to eradicate harmful gender stereotypes from advertising. The Unstereotypes Alliance brings together leaders from business, technology and the creative industries around a joint commitment to foster inclusive communication.[16]

Higher education

Undergraduate level

Higher education systems have expanded considerably in recent years, leading to a growing number of students and graduates, including more women. In many countries, the share of women completing tertiary education has grown faster than the share of men. In Europe, the share of 30- to 34-year-olds having completed tertiary education grew steadily, from 24% to 39 % over 2002-16. Growth was considerably faster for women, who in 2016, were at 44 %. In contrast, for men the share was 34%.[17]

In science education, women and men are unequally distributed across academic courses. Women have traditionally dominated in the social sciences and humanities, and are increasingly dominating in the life sciences and medicine, whereas men prevail in other STEM areas (Figure 7.1). These differences seem to largely reflect cultural stereotypes.

Figure 7.1. Percentage of new female students entering tertiary education, by selected fields of education, 2015

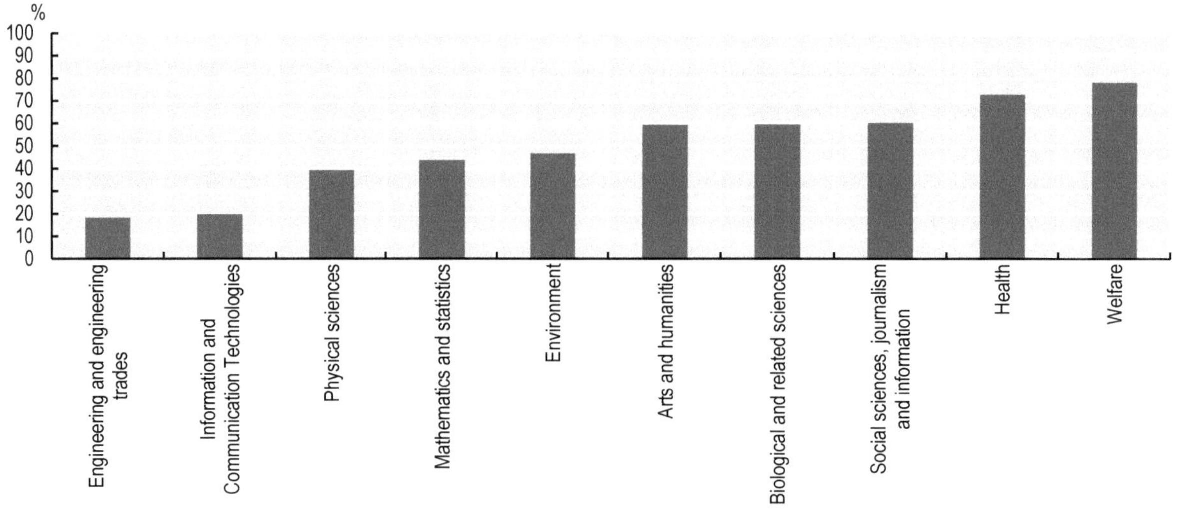

Source: OECD (2017d), Education at a Glance 2017: OECD Indicators, https://doi.org/10.1787/eag-2017-en.

StatLink https://doi.org/10.1787/888933858240

Historically, the under-representation of women in STEM has received much greater attention than the under-representation of women in philosophy or economics, or the under-representation of men in psychology, veterinary sciences or nursing. When assessing the effect of efforts to attract more girls into STEM bachelor programmes, it is important to consider the disciplines separately (Figure 7.1). "Stuffing the pipeline" only helps when there is a future to be found in these fields upon graduation (Miller and Wai, 2015).

Efforts to address women's under-representation in STEM are shaped and constrained by the manner in which the problem is framed, which needs to be sensitive to evolving contexts. For example, if it appears that more girls are not attracted to engineering because they are unaware of the opportunities or do not understand the nature of engineering work, then corrective actions should focus on outreach and informing them of the opportunities offered by engineering careers (Beddoes, 2011).

Furthermore, it is necessary to (re)consider how entry into certain higher education fields may be biased by reliance on standardised tests. For example, the Scholastic Aptitude Test scores commonly used for college admissions in the United States have been show to under-predict women's academic performance and over-predict men's, and test score differences do not necessarily translate into meaningful professional distinctions (Nature Editorial, 2005). Finally, if gender gaps in participation and performance are mutually reinforcing, educators seeking to promote women should address both factors simultaneously to maximise student achievement (Ballen, Salehi and Cotner, 2017). The small minority of female students who choose to enter some STEM fields may need mentoring and peer support – e.g. through networks – to perform optimally.

PhD level and early-stage careers

The doctoral level is the only educational level with near gender parity: all fields considered, 3.0% of men and 2.9% of women on average enter a doctoral programme across

EU countries (Figure 7.2). In practice, however, this means that the share of women in higher education declines at the postgraduate level, particularly in STEM fields. Nevertheless, the share of women in certain STEM fields has significantly increased over time, and the leaky pipeline between graduate and postgraduate education and training is no longer a major challenge (Miller and Wai, 2015). For example, only 14% of US doctoral degrees in biological sciences were awarded to women in 1970, compared to 49% in 2006. In 2015, more women in Europe received a PhD degree in life sciences than men. Entry into other STEM areas has been slower, but substantial. For example, 5.5% of US doctoral degrees in physical sciences were awarded to women in 1970, compared with 30% in 2006; 8% of US doctoral degrees in mathematics and statistics were awarded to women in 1970, compared to 32% in 2006 (Hill, Corbett and Rose, 2010).

There is growing concern that there are not enough jobs in academia for the rapidly growing population of PhD holders, although the potential scale of this issue varies across countries (Figure 7.2). Early-stage researchers often hold precarious positions in a very competitive environment: their academic careers begin with fixed-term contracts, often based on project funding. Hyper-competition, and its reinforcement of assertive and self-assured stereotypes, serves as an exclusionary mechanism for those who cannot or will not compete continually. The choice to enter this competition coincides with "the rush hour of life", i.e. establishing partnerships and families, which tends to reinforce gender imbalances.

Figure 7.2. Doctorate holders in the working-age population, 2016

Per 1 000 population aged 25-64.

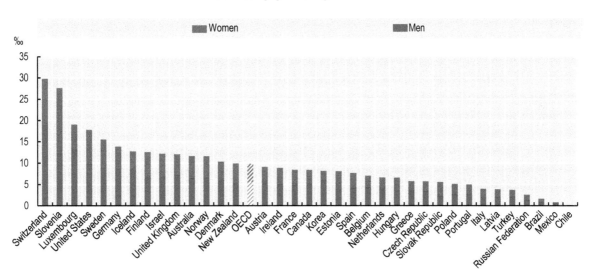

Source: OECD (2017c), OECD Science, Technology and Industry Scoreboard 2017: The digital transformation; https://doi.org/10.1787/9789264268821; based on OECD/UNESCO data collection in OECD/UNESCO (2017), Careers of Doctorate Holders 2017, http://oe.cd/cdh; OECD (2017d), Education at a Glance 2017: OECD Indicators, https://doi.org/10.1787/eag-2017-en; OECD (2009), Education at a Glance 2009: OECD Indicators, https://doi.org/10.1787/eag-2009-en.

StatLink https://doi.org/10.1787/888933858259

The precariousness of academic careers reduces the attractiveness of research for new and talented entrants, who can find more secure and better paid employment elsewhere (European Science Foundation [ESF] EUROAC, 2015; Janger et al., 2017). Women scientists, especially at the early career stage, are generally far less satisfied with their social and job security than men. This can be at least partially related to individual and societal expectations about motherhood and family structures. The share of female researchers with children is lower than for male researchers; this is especially true for researchers working in full-time positions. At the same time, the share of part-time working mothers in research is higher than the share of part-time working fathers (Janger et al., 2017). These divisions reflect the overall unequal distribution in society of care (including elder care) responsibilities between women and men. Targeted policies addressing employment conditions are required to address this, including through: flexible working practices; availability of parental leave, paternity leave and childcare facilities; dual-career opportunities for couples; flexible pension plans; and opportunities for career breaks (EC-PPMI, 2016). To be effective, however, flexible employment conditions need to be accompanied by "compensatory" measures with regard to performance evaluation, e.g. extended eligibility windows for tenure.

Careers in the research system

Efforts to increase the numbers of women studying STEM at the undergraduate and doctoral levels have not translated into equal or equitable representation of women in senior STEM positions. Various analyses within particular disciplines and/or national settings indicate this phenomenon will not simply resolve itself over time as more women enter STEM education. Other explanations need to be considered to understand why women's careers in STEM progress more slowly, stall more often and are more likely to be discontinued than men's (ESF, 2009; European Commission, 2015; National Science Foundation [NSF], 2017).

To identify levers to achieve gender equality, it is necessary to 1) track research careers across disciplinary, national and sectoral borders; and 2) gain a better understanding of the causes and consequences of different types of mobility between positions, both within and across institutions (Box 7.3). Researcher mobility is generally considered a good thing, which should be encouraged; the evidence shows that researchers who are mobile produce more highly cited research (OECD, 2017b). Yet it is easy to see how an over-emphasis on mobility could inadvertently disadvantage women at various life stages. In Europe, it takes an average of 17 years after obtaining a PhD to reach the most senior level in research (Janger et al., 2017). Despite some efforts to map research careers (ESF EUROAC, 2015; Janger et al., 2017), understanding exactly how they develop in terms of patterns or moves through positions, institutions, sectors and national borders remains largely uncharted territory. No quantitative gender-segregated data exist on the career paths and working conditions of researchers (MORRI, 2015).

Box 7.3. Tracking research careers

Recent overviews of the position of women in STEM such as European Commission's SheFigures (European Commission, 2015) and NSF (2017) use various statistical indicators to show the relative representation of women within a given professional category or career stage. The OECD STI Scoreboard also collects a number of STI indicators, which can be disaggregated by gender (OECD, 2017b). These combined sources of information give a good overall picture of the gender balance in STEM. As discussed in this chapter (Section 3.2), one of the major challenges for women who enter into scientific careers is career progression. However, current indicators are typically static and do not inherently indicate career progress. Even if representation changes at various

> career stages and/or over time, this difference or change cannot be said to unequivocally reflect actual *career development*, hence the need for an indicator that reflects career progress or development within STEM professions over time. Such indicators have already been developed for other professional fields and could be adapted to reflect the specificities of STEM careers (Dries et al., 2009).
>
> Developing a career-progression indicator for STEM is feasible, because research careers around Europe and North America reflect a highly comparable logic of four to five consecutive levels – from PhD to full professor or senior researcher, with minimal disciplinary and institutional variations*. Often, the dominant career system reflects an up-or-out dynamic, with a permanent contract only achieved after reaching a certain level within a limited amount of time.
>
> Such an indicator would make career progress quantifiable and comparable. It would help track gender, national or disciplinary differences and similarities between careers, and could shed light on the actual movement of researchers across positions and institutions. Together with existing statistical measures, it could inform policy making on research careers at the international, national and institutional levels.
>
> *The five career levels are described in the report of the ERCAREER project (2012-14), funded by the European Research Council to study "the paths and patterns, differences and similarities in the career paths of women and men ERC grantees"[18].

It is important to recognise that research careers do not necessarily progress within academia: research and development (R&D) positions in both the private and public sectors provide growing employment and career opportunities for PhD holders (Figure 7.3). In some countries, more women researchers are employed outside of academia than within (OECD, 2017a). Although industry generally lags behind the public sector as regards gender equity in the research workforce, there is growing recognition this has negative economic consequences (Peterson Institute, 2016). Several leading corporations have recently adopted strategies to boost the representation of women in engineering manufacturing, information technology and product management. In 2016, the Anglo-Australian mining company BHP announced its intention to reach 50/50 gender representation among its 65 000 employees; in February 2017, General Electric announced its goal of achieving 50/50 representation for all entry-level technical programmes and hiring 20 000 women to fill STI roles by 2020. For researchers – especially women – toiling in precarious post-doc positions in academic settings, these "outside opportunities" may offer better career prospects than academia, combined with more security and flexibility. Thus, creating opportunities for inter-sectoral mobility at all career stages (and monitoring this mobility) could be an important contributor to increasing overall research productivity, while at the same time promoting gender equity.

Figure 7.3. Women researchers as a percentage of total researchers in each sector (headcount)

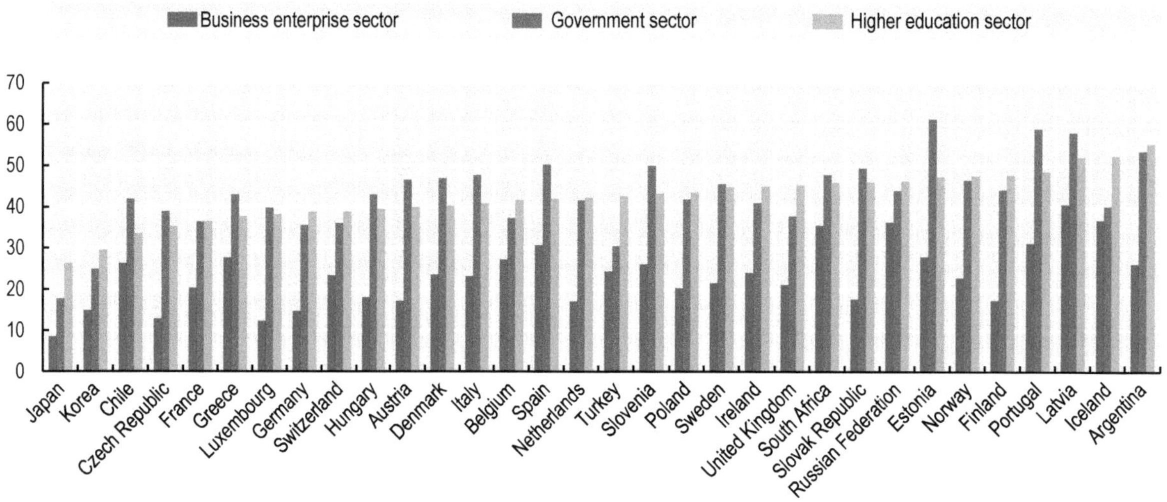

Source: OECD (2018b), OECD Main Science and Technology Indicators, www.oecd.org/sti/msti.htm (accessed on 18 June 2018).

StatLink https://doi.org/10.1787/888933858278

Gender differences in scientific careers are not only apparent in representation and advancement, but also in pay and decision-making power. As reported in European Commission (2014), the hourly wage difference in the scientific R&D area is around 18% and widens with age (see also OECD, 2017a). Similarly, only around 20-25% of board members and heads of research institutions are women. Sullerot's Law[19] seems to apply here: as the representation of women in particular STEM fields, professions and hierarchical levels rises, overall pay levels and status drop (Levanon et al, 2009).

The evaluation of performance plays a central role in the functioning of research systems. In academic settings, this principally includes peer review of publications and grant proposals, and national research and teaching assessment exercises. Often, such evaluation boils down to an assessment of individual rather than team performance, with principal investigators and corresponding authors endowed with the highest status, and obtention of individual grants and prizes considered more prestigious than participation in large-scale collaborations. These evaluations, in turn, help determine individual promotion and tenure awards.

Individual performance evaluation is very susceptible to gender bias – which, strictly defined, refers to a cognitive distortion that affects decision-making. Gender bias is linked to gender stereotypes, which perceive a better fit between men's innate abilities and STEM compared to women (Leslie et al., 2015). As a result, women working in STEM are "presumed inherently less competent" (Saini, 2017), leading to shifting standards in performance and merit evaluation. Gender bias affects progression in research careers, by limiting women's chances of being promoted. It is deepened by (impending) motherhood, effectively making it even harder for women to fit the stereotype (although men with care responsibilities also often suffer from "flexibility stigma" in research institutions). The relative absence of women in senior positions helps reinforce the stereotype. As gender bias is often implicit and subtle, it is more difficult to recognise and acknowledge – and thus harder to counter than blatant and explicit discrimination (Biernat, Tocci and Williams, 2011).

Bias is prominent in the construction, operationalisation and application of evaluation criteria (Vinkenburg, 2017). It is especially pronounced in systems that rely on peer review (from recommendation letters to evaluations of grant proposals), but citations, student evaluations, journalists' quotes and questions asked at conferences tend to be equally biased in favour of men (Saini, 2017). In a research system that is inherently founded on merit, it is hard to prove that reward allocation and performance evaluation practices often result in an unequal distribution of success in favour of some compared to others, *regardless* of the actual distribution of merit. Nevertheless, the development and adoption of interventions to effectively mitigate bias is growing (Box 7.4).

Box 7.4. Overcoming gender bias in decision-making and performance evaluation in STEM

While many research organizations turn to implicit or unconscious bias training, there is only limited research evidence on the impact of this kind of training. However, there are some evidence based design specifications for systemic diversity interventions engaging "gatekeepers" or decision makers in mitigating the effects of gender bias in performance evaluation (Vinkenburg, 2017). These specifications take into account the target group, length, focus, behavioural nature, and structured nature of the interventions. Examples are given below.

Breaking the bias habit®, University of Wisconsin (United States) and funded as part of the NSF ADVANCE programme: this short training programme has had a proven impact on attitudes and behavioural intentions among faculty members, with the result that significantly higher numbers of women have been hired and promoted[20] (Devine and Carnes, 2017).

Monitoring gender equality, Swedish Research Council (Vetenskåpsradet): the Council performs active monitoring, using participant observation of research panels and feedback on meeting practices, to improve the application and success rates of women in research funding; two reports were published in 2013 and 2015.[21]

Training video for selection committees and panel members in research organisations: developed by CERCA Institute (Spain),[22] this training video is now used by the European Research Council.[23]

Bias interrupters, CWLL (United States): this website features practical tools to help organisations and individuals interrupt bias in performance evaluation, recruiting, assignments and compensation[24] (based on Williams, Philipps and Hall, 2016).

Gender-blindness in research and innovation is both a symptom and a cause of the under-representation of women in STEM, particularly at senior levels. Research priorities and agendas are largely established by men, and research design and resultant innovations may fail to consider gender specificities (Box 7.5). In extreme cases, the lack of attention to gender considerations when translating scientific knowledge into products or actions can actually be harmful to women.

> **Box 7.5. Gender-blind STI**
>
> The issue of gender in the context of STI goes beyond improving the number of women scientists, inventors and innovators: it influences the relevance and quality of research and innovation outcomes for women and men.
>
> The traditional "gender-blind" approach to STI assumes that research results are applicable independently of the researcher or intended end-user's gender. However, there exists a strong gender dimension to the choice of STI topics, and the way research is conducted and translated into innovations. Gender bias permeates important fields of scientific knowledge, with more evidence relating to men than to women. For example, much medical research has been done exclusively on male experimental models and men, and then extrapolated at the level of medical practice to the whole population, with little regard for the biological differences between the sexes. Innovations, including new technologies, tend to be determined by masculine norms.[25] "Gender-blind" research and innovation can be harmful to women. It can also miss important opportunities to create new markets for products – e.g. in health care – that utilise scientific knowledge of sex/gender differences (Pollitzer and Schraudner, 2015).
>
> A particular concern in relation to gender-blindness is the growing use of artificial intelligence (AI) and algorithms in STI. AI systems depend on training based on very large data sets (e.g. of images), which themselves often reflect societal gender stereotypes and inherent biases. There is a widespread assumption that algorithms – as opposed to human judgements – are objective and free from discrimination. In practice, the use of algorithms and AI can inadvertently perpetuate existing biases, and discriminate more consistently and systematically on a larger scale (O'Neil, 2016; Bolukbasi et al., 2016). Collecting evidence and raising awareness of the potential for digital discrimination in STI are important first steps in ensuring it does not propagate. On a more positive note, the potential also exists to harness "gender-neutral" AI to help eliminate gender bias in research evaluation (Erel et al., 2018).
>
> Most OECD countries recognise that the topics and conduct of research, and its translation into innovation, are often gender-blind. New governance arrangements, rules, guidelines, regulations and targeted funding schemes are being introduced to redress these inherent biases. In Korea, amending the Key Framework Act on Science and Technology (2009) to formally acknowledge the importance of gendered innovation is one of several measures being taken to implement a new science and technology plan "with and for the people". In Spain, a new national award for gender equality is being launched in 2018, with the dual aims of promoting gender equity, and integrating the gender dimension into research and innovation content in public research institutions. France has a similar dedicated financial-support mechanism for institutions that incorporate gender in their research content. Denmark has developed a broad agenda to systematically include relevant gender perspectives in research. Finally, the United Kingdom is funding multidisciplinary research on digital discrimination and gender.

The changing context for STI and the importance of diversity

The scale and pattern of international collaboration in STI has grown massively over the past two decades, driven by digitalisation and the emergence of new scientific powerhouses outside of the OECD. Similarly, higher education is increasingly a global enterprise, with international universities in many parts of the world educating large numbers of overseas students. As discussed, gender balance in science varies considerably across countries, despite their increasing interconnectivity and interdependence. Policy actions to promote international exchanges of female STEM students and the mobility of female researchers are one mechanism to redress some of the current imbalances. Within Europe, this is facilitated by dedicated European Commission funding schemes. Such mechanisms also exist at the bilateral scale: for example, a France-Morocco partnership has recently been established to strengthen the role of women in scientific research.

Globalisation, interconnectivity and technological development are not just affecting science and education, but are also fundamentally changing socio-economic systems. This leads to new and complex challenges, which in turn require new scientific approaches, as illustrated by the Paris accord on Climate Change and the United Nations SDGs (Section 1). Responding effectively to environmental change and meeting the SDGs will require integrating knowledge from many distinct scientific domains, and applying transdisciplinary research approaches that engage end users in the co-design and co-production of research. Natural and social scientists will increasingly need to work together, often in large transnational teams.

Another factor that is dramatically affecting research practice is digitalisation, which has enabled open science and data-driven science, with major implications for the future scientific workforce. Not surprisingly, policy often focuses on the core ICT professions or disciplines; the gender imbalances in these areas are certainly very substantial and need to be addressed (OECD, forthcoming). At the same time, digitalisation is transforming professions such as librarianship and archiving, where women are better represented. Opportunities exist to raise the status and reward for these professions, which are essential to developing the digital data services on which science will increasingly depend. Australia, for example, has identified re-training librarians as a major pillar of its digital skills for science strategy. The policy emphasis on open science and collaboration also implies that science communication, team building, ethics and legal knowledge will become more important to the scientific endeavour. This will present opportunities to design and reward academic careers differently, and provide more options for women (and men) wishing to contribute to science.

Globalisation, complex societal challenges, open science and digitalisation all have one commonality – they emphasise the need for greater diversity in STI. While diversity considerations in science are not limited to gender, it is a cross-cutting issue that applies to all population groups. Women and men may differ in biology and behaviour, but they are also similar in many respects. Beyond the "binary" classification, the concept of "gender diversity" encompasses the differences deriving from the interactions between the biological, ethnic, cultural or psychological characteristics that individual women and men develop over their life course. These interactions are the subject of active scientific research, including on developing methods to measure and compare differences between individuals and groups. Nevertheless, it is generally agreed that gender equality, combined with cultural and cognitive diversity, improves the quality of research and innovation outcomes (Abbasi and Jaafari, 2013; Campbell et al., 2013; Hinnant et al., 2012; Jeppesen and Lakhani, 2010). Hence, solutions-focused research and innovation needs to reflect the diversity of the societies in which the solutions will be situated.

Future vision and how to achieve it

The many reports on women in science tend to share a future vision for a world in which there are equal opportunities for women to enter, contribute and progress in all scientific disciplines without prejudice or bias. This implies a more diverse, productive and attractive research enterprise, which fully recognises and rewards the equivalent and distinct contributions of both men and women. Clearly, achieving this vision is still a long way off, which is further complicated by the major transitions that science and innovation are currently undergoing.

As discussed throughout this chapter, almost all countries are taking policy actions to promote gender equity in STI. These focus on feeding the pipeline for STEM subjects and providing support for individual women scientists at various career stages; some seek to address the underlying causes of gender imbalance, including gender stereotypes and inherent gender bias in science and innovation systems. However, the overall picture shows a fragmented approach, characterised by multiple institutions acting independently, and limited co-ordination between education, science and innovation actors. There is little systematic evaluation of the effectiveness and sustainable impact of the many interventions under way. In some cases, this will require developing new indicators and measures, presenting important opportunities for mutual learning across different countries and developing communities of practice.

Addressing gender inequalities in STI will require a strategic and systemic long-term approach. Policy actions are necessary on several fronts to: 1) continue to monitor and address long-term challenges in scientific education, training and careers; 2) ensure that digital education and training strategies provide full and equal opportunities to girls and women, and do not enforce traditional gender stereotypes or introduce digital discrimination; and 3) ensure that the contribution of all disciplines and supporting professions is fully recognised, valued and rewarded in the transition to open science and greater transdisciplinary research. There is a need for strategic thinking and targeted interventions that will create positive feedback loops to strengthen the position of women within STI systems as a whole (Figure 7.4). Co-ordinated actions engaging multiple actors – governments, research funders, academia, public research organisations, educational institutions and corporations – must be implemented at multiple levels, from local to global.

Figure 7.4. Gender inequality in research careers: A system-dynamic model

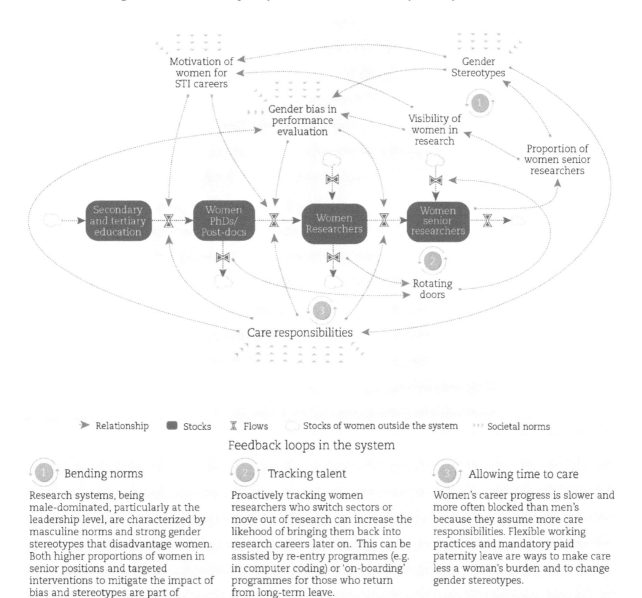

Source: Based on Bleijenbergh and Van Engen (2015)'s participative system-dynamic model of gender inequality at a technical university.

Looking to the future, diversity and inclusiveness will be critical to improving research productivity, and the relationship between science and society. Those countries, institutions and firms that achieve gender equity will be well placed to emerge as leaders in their fields. Policy makers have an important role to play in establishing and implementing the necessary regulatory and normative frameworks to achieve this, and will themselves need to fully embrace gender equality and diversity.

Notes

[1] https://homewardboundprojects.com.au/

[2] For instance, a comprehensive overview of the digital gender divide, including innovation-related aspects, can be found in "Bridging the Digital Gender Divide: Include, Upskill, Innovate" (OECD, forthcoming).

[3] The Gender Gap Index was introduced by the World Economic Forum as an overall measure of gender inequality at the national level. See: https://www.wikigender.org/wiki/global-gender-gap).

[4] https://www.inspiringthefuture.org/.

[5] https://primaryfutures.org/.

[6] See the 2018 report: https://www.educationandemployers.org/drawing-the-future/.

[7] http://www.youngscientists.com.au/.

[8] http://lettoysbetoys.org.uk/ten-ways-to-challenge-gender-stereotypes-in-the-classroom/.

[9] https://girlswhocode.com/.

[10] http://www.desy.de/schule/mint_fuer_maedchen/index_ger.html.

[11] http://www.iop.org/publications/iop/2015/file_66429.pdf.

[12] For the winning video, see: https://youtu.be/bDZF62gd1L4.

[13] http://www.expecteverything.eu/hypatia/.

[14] https://www.womeninc.nl/beperktzicht/.

[15] https://obamawhitehouse.archives.gov/sites/default/files/microsites/ostp/imageofstemdepictiondoc_02102016_clean.pdf.

[16] http://www.unstereotypealliance.org/en.

[17] http://ec.europa.eu/eurostat/statistics-explained/index.php/Europe_2020_indicators_-_education.

[18] https://cordis.europa.eu/project/rcn/105764_en.html.

[19] Evelyne Sullerot (1924-2017) was a French feminist, philosopher and writer. Sullerot's Law states that if women become the majority in a certain vocation, then prestige and salary will be lower than if men are the majority.

[20] http://wiseli.engr.wisc.edu/breakingbias_gender.php.

[21] https://www.vr.se/english/calls-and-decisions/assessment-of-applications/gender-equality.html.

[22] http://cerca.cat/en/women-in-science/bias-in-recruitment/.

[23] https://erc.europa.eu/thematic-working-groups/working-group-gender-balance.

[24] http://biasinterrupters.org/.

[25] http://genderedinnovations.stanford.edu/index.html.

References

Abbasi, A. and A. Jaafari (2013), "Research impact and scholars' geographical diversity", *Journal of Informetrics,* Vol. 7/3, pp. 683-692, Elsevier, Amsterdam, http://dx.doi.org/10.1016/j.joi.2013.04.004.

Ballen, C.J., S. Salehi and S. Cotner (2017), "Exams disadvantage women in introductory biology", *PLoS ONE,* Vol. 12/10, Public Library of Science, San Francisco, http://dx.doi.org/10.1371/journal.pone.0186419.

Beddoes, K.D. (2011), "Engineering Education Discourses on Underrepresentation. Why Problematization Matters", *International Journal of Engineering Education*, Vol. 27/5, pp. 1117-1129, Dublin Institute of Technology Tempus Publications, Dublin, http://www.sociologyofengineering.org/wp-content/uploads/2016/01/Beddoes-IJEE-2011.pdf.

Bian, L., S.-J. Leslie and A. Cimpian (2017), "Gender stereotypes about intellectual ability emerge early and influence children's interests", *Science*, Vol. 355/6323, pp. 389-391, American Association for the Advancement of Science, Washington, DC, http://dx.doi.org/10.1126/science.aah6524.

Biernat, M., Tocci, M.J. and Williams, J.C. (2011), "The Language of Performance Evaluations: Gender-Based Shifts in Content and Consistency of Judgment", *Social Psychological and Personality Science*, Vol. 3/2, pp. 186-192, https://doi.org/10.1177/1948550611415693.

Bleijenbergh, I. and M.L. Van Ergen (2015), "Participatory modeling to support gender equality", *Equality, Diversity and Inclusion: An International Journal*, Vol. 34/5, pp. 422-438, Emerald Group Publishing, Bingley, United Kingdom, http://dx.doi.org/10.1108/EDI-06-2013-0045.

Bolukbasi, T. et al. (2016), "Man is to Computer Programmer as Woman is to Homemaker? Debiasing Word Embeddings", 21 July 2016, arxiv.org, Cornell University, Ithaca, NY, arXiv:1607.06520v1 [cs.CL].

Campbell, L.G. et al. (2013), "Gender-Heterogeneous Working Groups Produce Higher Quality Science", *PLoS ONE*, Vol. 8/10, Public Library of Science, San Francisco, http://dx.doi.org/10.1371/journal.pone.0079147.

Charles, M. (2017), "Venus, Mars, and Math: Gender, Societal Affluence, and Eighth Graders' Aspirations for STEM", *Socius*, Vol. 3, Sage Publications, Thousand Oaks, CA, http://dx.doi.org/10.1177/2378023117697179.

Devine, P.G. and M. Carnes (2017), "A gender bias habit-breaking intervention led to increased hiring of female faculty in STEMM departments", *Journal of Experimental Social Psychology*, Vol. 73, pp. 211-215, Elsevier, Amsterdam, http://dx.doi.org/10.1016/j.jesp.2017.07.002.

Dries, N. et al. (2009), "Development and validation of an objective intra-organizational career success measure for managers", *Journal of Organizational Behavior,* Vol. 30/4, May 2009, pp. 543-560, Wiley-Blackwell, Hoboken, NJ, https://doi.org/10.1002/job.564.

EC-PPMI (2016), *Research Careers in Europe: Final report. Prepared by the PPMI Group, CARSA and INOVA for the European Commission*, European Commission and Public Policy Management Institute, Brussels and Vilnius, https://blacksea-horizon.eu/object/document/618/attach/NC0614200ENN_002.pdf.

Erel, I. et al. (2018), "Selecting Directors Using Machine Learning", *NBER Working Paper*, No. 24435, National Bureau of Economic Research, Cambridge, MA, http://papers.nber.org/papers/w24435?utm_campaign=ntw&utm_medium=email&utm_source=ntw.

ESF (2009), *Research Careers in Europe Landscape and Horizons*, European Science Foundation, Strasbourg, http://archives.esf.org/fileadmin/Public_documents/Publications/moforum_research_careers.pdf.

ESF EUROAC (2015), "Career tracking of doctorate holders. Detailed reports", in Fumasoli, T., G. Goastellec and M.B. Kehm (eds.), *Academic Work and Careers in Europe: Trends, Challenges, Perspectives*, Springer International Publishing, Basel.

European Commission (2015), *She Figures 2015*, Directorate-General for Research and Innovation, European Commission, Brussels, https://ec.europa.eu/research/swafs/pdf/pub_gender_equality/she_figures_2015-final.pdf.

European Commission (2014), *The EURAXESS initiative: Mobilisation of research careers*, European Commission, Brussels, http://ec.europa.eu/programmes/horizon2020/en/h2020-section/euraxess-initiative-mobilisation-research-careers.

European Parliament (2014), *A new strategy for gender equality post 2015: Report to the Committee on Women's Rights and Gender Equality of the European Parliament*, European Parliament Publications Office, Strasbourg, http://www.europarl.europa.eu/RegData/etudes/STUD/2014/509984/IPOL_STU(2014)509984_EN.pdf.

Heilman, M.E. (2012), "Gender stereotypes and workplace bias", *Research in Organizational Behavior*, Vol. 32/0, pp. 113-135, http://dx.doi.org/10.1016/j.riob.2012.11.003.

Hill, C., C. Corbett and A.S. Rose (2010), "Why so few? Women in Science, Engineering, Technology and Mathematics", AAUW, Washington, DC, https://www.aauw.org/files/2013/02/Why-So-Few-Women-in-Science-Technology-Engineering-and-Mathematics.pdf.

Hinnant, C.C. et al. (2012), "Author-team diversity and the impact of scientific publications: Evidence from physics research at a national science lab", *Library & Information Science Research*, Vol. 34/4, pp. 249-257, Elsevier, Amsterdam, https://doi.org/10.1016/j.lisr.2012.03.001.

Janger, J. et al. (2017), "MORE3 – Support Data Collection and Analysis Concerning Mobility Patterns and Career Paths of Researchers. Final Report – Task 4: Comparative and Policy-relevant Analysis", WIFO Studies, No. 60891, Austrian Institute of Economic Research, Vienna, https://ideas.repec.org/b/wfo/wstudy/60981.html.

Jeppesen, L.B. and K. Lakhani (2010), "Marginality and problem solving effectiveness in broadcast search", *Organization Science*, Vol. 21/5, pp. 1016-1033, HBS Scholarly Articles, Harvard Business School, Boston, MA, http://nrs.harvard.edu/urn-3:HUL.InstRepos:3351241.

Leslie, S.-J. et al. (2015), "Expectations of brilliance underlie gender distributions across academic disciplines", *Science*, Vol. 347/6219, pp. 262-265, American Association for the Advancement of Science, Washington, DC, http://dx.doi.org/10.1126/science.1261375.

Levanon, A., P. England and P. Allison (2009), "Occupational feminization and pay: Assessing causal dynamics using 1950-2000 U.S. Census data", *Social Forces*, Vol. 88, pp. 865-892, Oxford University Press and University of North Carolina Press, Oxford and Chapel Hill, NC, http://dx.doi.org/10.1353/sof.0.0264.

McKinsey (2015), The Power of Parity: How Advancing Women's Equality can add $12 trillion to Global Growth, McKinsey & Company, New York, www.mckinsey.com/featured-insights/employment-and-growth/how-advancing-womens-equality-can-add-12-trillion-to-global-growth.

Miller, D.I., A.H. Eagly and M.C. Linn (2015), "Women's Representation in Science Predicts National Gender-Science Stereotypes: Evidence From 66 Nations", *Journal of Educational Psychology*, Vol. 107/3, pp. 631-644, American Psychological Association, Washington, DC, http://dx.doi.org/10.1037/edu0000005.

Miller, D.I. et al. (2018), "The Development of Children's Gender-Science Stereotypes: A Meta-analysis of 5 Decades of U.S. Draw-A-Scientist Studies", *Child Development,* Vol. 0/0, Wiley and Society for Research in Child Development, Hoboken and Ann Arbor, http://dx.doi.org/10.1111/cdev.13039.

Miller, D.I. and J. Wai (2015), "The bachelor's to Ph.D. STEM pipeline no longer leaks more women than men: a 30-year analysis", *Frontiers in Psychology*, Vol. 6/37, Frontiers Media, Lausanne, http://dx.doi.org/10.3389/fpsyg.2015.00037.

MORRI (2015), *Analytical report on the gender equality dimension*, Monitoring the Evolution and Benefits of Responsible Research and Innovation (MoRRI), http://morri-project.eu/reports/2015-04-01-d2.3.

NSF (2017), *Women minorities and persons with disabilities in science and engineering. Interactive dataset and report*, National Science Foundation, Arlington, VA, https://www.nsf.gov/statistics/2017/nsf17310/digest/about-this-report/.

Nature Editorial (2005), "Separating Science from Stereotype", *Nature Neuroscience*, Vol. 253/8, Nature Publishing Group, London, https://doi.org/10.1038/nn0305-253.

OECD (forthcoming), "Bridging the Digital Gender Divide: Include, Upskill, Innovate", OECD Publishing, Paris.

OECD (2018a), Empowering women in the digital age: where do we stand?, High-level event

at the margin of the 62nd Session of the UN Commission on the Status of Women, 14 March 2018, New York City, OECD, Paris, https://www.oecd.org/social/empowering-women-in-the-digital-age-brochure.pdf.

OECD (2018b), OECD Main Science and Technology Indicators, www.oecd.org/sti/msti.htm (accessed on 18 June 2018).

OECD (2017a), The Pursuit of Gender Equality: An Uphill Battle, OECD Publishing, Paris, https://doi.org/10.1787/9789264281318-en.

OECD (2017b), EC/OECD STI Policy survey, OECD, Paris, https://stip.oecd.org/stip.html.

OECD (2017c), OECD Science, Technology and Industry Scoreboard 2017: The digital transformation, OECD Publishing, Paris, https://doi.org/10.1787/9789264268821-en.

OECD (2017d), Education at a Glance 2017: OECD Indicators, OECD Publishing, Paris, https://doi.org/10.1787/eag-2017-en.

OECD (2009), Education at a Glance 2009: OECD Indicators, OECD Publishing, Paris, https://doi.org/10.1787/eag-2009-en.

OECD (2006), Women in Scientific Careers: Unleashing the Potential, OECD Publishing, Paris, https://doi.org/10.1787/9789264025387-en.

OECD/UNESCO (2017), *OECD/UNESCO Institute for Statistics/Eurostat Careers of Doctorate Holders (CDH) project*, OECD/UNESCO, Paris, http://oe.cd/cdh.

O'Neil, C. (2016), *Weapons of Math Destruction: How Big Data Increases Inequality and Threatens Democracy*, Crown Publishing, New York.

Peterson Institute for International Economics (2016), "Is Gender Diversity Profitable? Evidence from a Global Survey", *Peterson Institute for International Economics Working Paper*, Vol. 16/3, February 2016, Peterson Institute for International Economics, Washington, DC, https://piie.com/publications/working-papers/gender-diversity-profitable-evidence-global-survey.

Pollitzer, E. and M. Schraudner (2015), "Integrating Gender Dynamics into Innovation Ecosystems", *Sociology and Anthropology*, Vol. 3/11, pp. 617-626, Horizon Research Publishing Corporation, San Jose, CA, http://doi.org/10.13189/sa.2015.031106.

Saini, A. (2017), INFERIOR – How Science Got Women Wrong and the New Research That's Rewriting the Story, Beacon Press, Boston.

Smith-Doerr, L., S.N. Alegria and T. Sacco (2017), "How Diversity Matters in the US Science and Engineering Workforce: A Critical Review Considering Integration in Teams, Fields, and Organizational Contexts", *Engaging Science, Society and Technology*, Vol. 3/17, pp. 139-153, https://doi.org/10.17351/ests2017.142.

Vinkenburg, C.J. (2017), "Engaging Gatekeepers, Optimizing Decision Making, and Mitigating Bias: Design Specifications for Systemic Diversity Interventions", *The Journal of Applied Behavioral Science*, Vol. 53/2, pp. 212-234, Sage Journals, Thousand Oaks, CA, https://doi.org/10.1177/0021886317703292.

Williams, J. C., K.W. Phillips and E.V. Hall (2016), "Tools for change: Boosting the Retention of Women in the STEM Pipeline", *Journal of Research in Gender Studies*, No. 6/1, p. 11, Addleton Academic Publishers, New York, https://repository.uchastings.edu/faculty_scholarship/1434.

Chapter 8. New trends in public research funding

By

Philippe Larrue, Dominique Guellec and Frédéric Sgard

Public research is expected to fulfil a widening set of objectives, from scientific excellence and economic relevance to contributing to a variety of societal challenges (inclusiveness, gender diversity, sustainability, etc.). Policy makers in ministries and funding agencies have broadened their portfolio of funding instruments and design variants to respond to this demand. However, little is known about the potential effects of the various funding instruments on research outcomes. This chapter aims to provide policy makers with analytical tools to help them decide upon what types of funding mechanisms and instruments should finance what types of research and for what effects. It examines recent changes in the modes of allocation of research funding that have blurred the formerly well-established boundaries between competitive and non-competitive funding instruments. It then proposes a simple conceptual framework to present the portfolio of research-funding instruments available to policy makers along multiple and continuous – rather than unique and binary – dimensions. The chapter then analyses the "purpose fit" of this growing set of funding instruments – i.e. their ability to fulfil different policy objectives – to help policymakers design and utilise them in a way that best corresponds to the expected impacts of public research. The chapter concludes with a forward-looking discussion that draws implications in terms of future analytical work and how emerging long-term trends (e.g. digitalisation and societal challenges) might influence the volume and types of research funding.

The statistical data for Israel are supplied by and under the responsibility of the relevant Israeli authorities. The use of such data by the OECD is without prejudice to the status of the Golan Heights, East Jerusalem and Israeli settlements in the West Bank under the terms of international law.

Introduction

What types of funding mechanisms and instruments should finance what type of research and for what effects? Despite progress in understanding the underlying dynamics, research funding is still the subject of lively discussions in academic and policy arenas.[1]

The various positions in these debates, often revolving around the two models of competitive and non-competitive funding, are entrenched in different conceptual views on how new knowledge is generated and used in the innovation process. They also reflect various communities' vested interests since the responses given to this question influence the allocation of funds to different actors. Finally, they are strongly related to the national institutional set-ups in which the funding systems are embedded, adding a further layer of complexity to the debate.

These policy debates have become more intricate as the boundaries between the formerly two well-established modes of research funding – competitive and non-competitive – have become increasingly blurred and porous. On the one hand, competitive funding can be allocated to certain institutions – particularly centres of excellence – for a period of several years; on the other hand, institutional funding increasingly integrates performance-based components, introducing a degree of competition into these funding mechanisms.

Reflecting changes in the policy arena, an extensive academic and grey literature has progressively moved away from the usual dichotomy between competitive and non-competitive funding instruments, introducing more nuanced measurement and comparison of national funding patterns. Scholars and experts also scrutinise the operational/technical aspects of the different funding instruments (e.g. the components of the funding formula for institutional funding, and the criteria and selection modes for competitive funding). This body of work now offers a richer understanding of the funding landscape, more closely related to the reality experienced by policymakers.

However, little is yet known about the effects of funding instruments. What are the merits of the various instruments (and their multiple design variants) in achieving certain policy objectives, including supporting research excellence, steering research in certain directions or triggering breakthroughs? Although they do not provide systematic responses to this question, various country reviews, evaluations of schemes and programmes supporting research, and research works provide some useful insights on this matter. Together, they help shed light on the "purpose fit" of instruments, i.e. how certain instruments are more or less adapted to specific policy objectives. They also provide a significant – though scattered – evidence base on the various factors influencing the desired effects at the different stages of the funding process, from high-level strategic orientation to research implementation in Higher education institutions (HEIs) and public research institutes (PRIs).

Connecting the technical ("how to fund?") and political ("for what desired effects?") aspects of research funding is essential, to help policy makers design and use funding instruments in a way that best corresponds to their objectives. This chapter builds on recent progress in the academic and empirical literature, analysing the policy objectives and desired effects underlying the different types of government research funding. The OECD has recently resumed work in this field (OECD, forthcoming a) and future OECD work on research funding is planned for the 2019-20 biennium.

The chapter takes stock of recent changes in the allocation modes of research funding. It examines the increasingly complex set of funding instruments designed to convey a widening set of policy objectives, and proposes a simple analytical framework of the mix of these funding instruments as a continuum. Regarding the purpose fit of funding instruments, the chapter pays particular attention to performance-based institutional funding instruments, which have undergone recent reforms in many countries and offer new policy levers to accommodate a wide set of policy objectives. It concludes with a forward-looking view, drawing implications for future analytical work and discussing how emerging long-term trends (e.g. digitalisation and societal challenges) might influence the volume and types of research funding.

Recent changes in research-funding

Innovation, particularly at the knowledge frontier and in emerging sectors, depends heavily on scientific progress (OECD, 2015a). HEIs and PRIs – which in 2016 represented just under 18% (HEIs) and 11% (PRIs) of gross domestic expenditure on research and development (GERD) in OECD member countries, far below business (69%) – perform more than three-quarters of total basic research.

HEIs play a growing role in research and development (R&D), surpassing PRIs, whose importance has decreased in many countries. In addition to providing higher education, universities are strongly engaged in the production of longer-term and higher-risk scientific knowledge, and increasingly in applied research, knowledge transfer and innovation activities.

Despite considerable country differences, government sources finance the bulk of academic research activities: in 2015, public funds supported 67% of academic research by HEIs and 92% of research by PRIs (OECD, 2017a). Budgetary restrictions in the aftermath of the 2008 global financial crisis negatively affected R&D funding (Box 8.1). However, research will remain an important component of public budgets, as the level of knowledge embedded in products and services keeps increasing, and the number of global challenges calling for radically new technological and social innovation also keeps rising.

Box 8.1. How has public funding of R&D evolved in recent years?

The share of government funding in the budget of PRIs has remained relatively stable since the 1980s. However, it has decreased steadily for HEIs, which have successfully sought third-party funding. A closer look at the more recent period (Figure 8.1, Panel A) reveals a significant increase in public research funding (as well as sharp increases in business R&D funding) immediately after the onset of the financial crisis, as many countries used research and innovation programmes in their stimulus packages (e.g. the 2010 Investments for the Future Programme [*Programme d'investissements d'avenir*] in France and the 2009 American Recovery and Reinvestment Act in the United States). However, this increase was short-lived: as early as 2010-11, increases in public R&D budgets slowed or reversed. Government spending on research in HEIs and PRIs slightly decreased, both in real terms and as a percentage of gross domestic product (GDP), as economic growth resumed without an increase in government funding of public research (Figure 8.1, Panel B).

This decrease cannot be attributed only to budgetary pressures: public funding for research also decreased as a share of total government expenditures. This is consistent with anecdotal evidence suggesting a certain "frustration", owing to the absence of sufficient tangible innovation results from past funds allocated to research. In such a context, advocates of science, technology and innovation (STI) activities are less well positioned to defend their budgets when negotiating with finance ministries and representatives of other policy areas.

Figure 8.1. Components of GERD financed by government, OECD, 2005-2015

In million USD 2010 PPP (Panel A), % of GDP (Panel B), % of total government expenditures (Panel C)

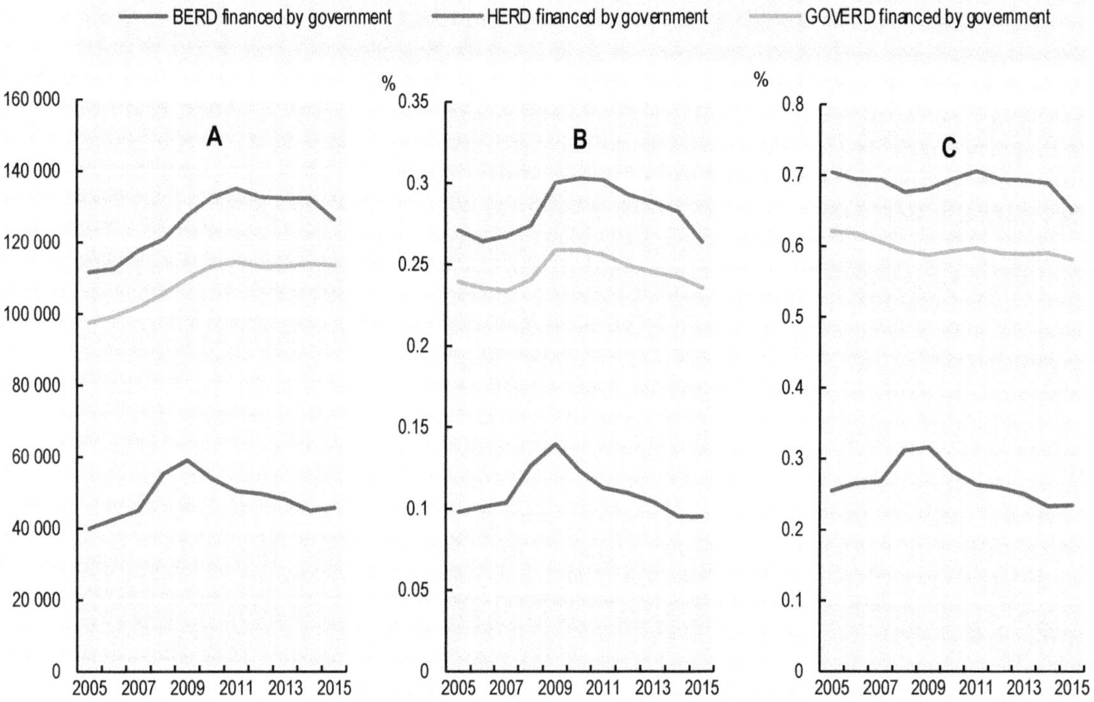

— BERD financed by government   — HERD financed by government   — GOVERD financed by government

Note: PPP: purchasing power parity; BERD: business enterprise expenditure on R&D; HERD: higher education expenditure on R&D; GOVERD: government intramural expenditure on R&D.
Source: Calculations based on OECD (2018a), Main Science and Technology Indicators, http://www.oecd.org/sti/msti.htm; and OECD (2018b), Research and Development Statistics, http://www.oecd.org/innovation/inno/researchanddevelopmentstatisticsrds.htm (accessed on 25 June 2018).

StatLink ⟶ https://doi.org/10.1787/888933858297

Research funding is allocated in very diverse ways, reflecting the institutional settings of national research systems. The earliest and simplest typology distinguishes between competitive and non-competitive funding mechanisms:

- Competitive project funding encompasses the programmes or instruments of funding agencies, research councils or ministries that allocate resources for a research activity limited in scope, budget and time, based on formal contests or competitions, in which applicants apply for funding. Financial awards can be of variable size and length, and may be allocated to individuals, projects or centres (OECD, forthcoming a).

- Non-competitive institutional funding includes institutional core or block funding, i.e. the general funding of research-performing institutions, without direct selection of R&D projects or programmes. It is generally allocated as a yearly government contribution to HEIs or PRIs (not to a specific sub-component or research group) to fund their day-to-day operations, such as staff salaries, infrastructure and maintenance related to education or research activities. While institutional funding was earmarked in the past for specific activities, it is now mostly allocated as a lump sum (block grant) that research institutions can spend as they see fit (OECD, 2015b; Jongbloed and Lepori, 2015).

The changing ways in which most governments allocate research funding have increasingly blurred the formerly well-established boundaries between the two major funding mechanisms in the two last decades. First, the gradual spread of new public management (NPM) thinking in many public administrations (including HEIs and PRIs in the 1980s and 1990s), and the growing pressure on budgets, have led public authorities to increase the share of research funds distributed through competitive project funding (Hicks, 2010). Furthermore, not only did NPM reforms further increase project-based funding, they also introduced performance-based variables and different conditions for institutional funding allocated to HEIs and PRIs (Lepori, Geuna and Mira, 2007). In some countries (e.g. Sweden) and institutions (e.g. PRIs in Norway), attempts have been made to include strategic components in institutional funding, in order to better align research activities and national priorities while preserving institutional autonomy. As a result of these changes, institutional funding (which still often retains a strong historical component) can no longer be considered non-competitive and non-oriented.

An even more recent trend has also challenged the previously binary typology of funding mechanisms. Governments increasingly use competitions to allocate multi-year funding to institutions (or part of them) through different types of research excellence initiatives (REIs). These initiatives aim to encourage outstanding research by allocating large-scale, long-term funding directly to designated research units; hence, they feature elements of both institutional and project funding. In 2014, over two-thirds of OECD countries were operating such schemes, mostly established within the past decade (OECD, 2014a). The 2017 edition of the European Commission EC-OECD science, technology and innovation policy (STIP) survey showed similar results: 31 countries (i.e. 61% of a total of 51 countries)[2] reported 84 initiatives using these funding instruments (EC/OECD, 2017).

An analytical framework of funding instruments

The evolution of the funding landscape has challenged the boundaries between competitive and non-competitive funding instruments, requiring "nuanced" conceptual frameworks. Several initiatives – mainly commissioned by the European Commission and the OECD since the early 2000s – have attempted to clarify the definitions of instruments in this moving landscape and reflect these observations in precise statistics (Box 8.2).

> **Box 8.2. Measuring national patterns of research funding allocation**
>
> One of the first significant attempts at measuring national patterns of research funding was conducted by the PRIME European network (2004-08), with support from the European Commission (Lepori, Geuna and Mira, 2007). PRIME developed a conceptual framework and definitions, which were then applied to existing data on a subset of six European countries. Using these results as a stepping-stone, the OECD Working Party on Science and Technology Indicators (NESTI) made a first attempt in 2012 to collect data differentiating different modes of funding (van Steen, 2012). Building on this seminal work, EUROSTAT started to collect voluntary information from European countries on the share of project and institutional funding. The European Commission sponsored another research consortium, Public Funding of Research (PREF), which collected new data and obtained results that are "broadly consistent" with EUROSTAT (Jonkers and Zacharewicz, 2016).
>
> These projects yielded the following main results:
>
> - The studies show a wide diversity of national configurations and evolution patterns of research-funding flows. One key indicator, the share of project funding of R&D in total domestic R&D, ranged from above two-thirds (in New-Zealand, South Korea and Ireland) to less than one-third (in Austria, Switzerland and the Netherlands) in the NESTI study. EUROSTAT data suggest that the relative importance of project funding typically ranges between 25% and 50% in the European countries analysed, with some noticeable exceptions (e.g. above 65% in Ireland).
> - The studies show a general trend of increasing project funding from 1970 to 2000, in real terms and relative to GDP. EUROSTAT data show a relative stability since the mid-2000s (again with exceptions, e.g. a strong increase in Greece since 2010 and a decline in Iceland until 2014). Although most of these studies focus on HEIs, funding sources for PRIs have also moved towards higher competitive and contractual funding in most countries. Despite the growth of project funding, institutional funding is still the main instrument for financing public research.
>
> Considerable progress has therefore been made, but the measurement agenda is still open. Ongoing work provides a more granular and less binary analysis than analysis based on the divide between institutional (or organisational) funding and project funding. To that aim, PREF authors defined "mixed models" of the two. They have also recently developed synthetic indicators of the degree of competition and "proximity to performance", rather than considering these notions as absolute features of funding instruments. For instance, the performance-based indicator for institutional funding ranges from 0 if it is allocated historically to 1 if it depends entirely on past research outputs. Out of the 14 European countries scrutinised in this research, 3 (Poland, Portugal and the United Kingdom) appear to have a strong orientation towards performance-based allocation of block funding, while 8 (Austria, Denmark, France, Italy, Sweden, Switzerland and Germany) have a lower dependence on performance (Reale, 2017). Despite increasingly introducing performance criteria in block-funding allocation formulae, most countries still award institutional funding mainly on a historical basis, with scaling parameters (often related to higher education activities, such as the number of students or teaching staff).
>
> Taking a broader perspective that considers both institutional and competitive funding, Lepori, Reale and Orazio Spinello (2018) developed a synthetic indicator of the

performance orientation of public research funding, tracking its evolution over reforms of the funding system. One key result is that although the wide variations in countries' performance orientation stem from the relative shares of project funding, the recent significant increases in performance orientation in some countries (Finland, Norway, Poland, Portugal and Sweden) followed the introduction or reform of institutional funding instruments.

No such systematic initiatives are known to have been undertaken outside of Europe recently. Despite some measurement issues (notably breaking down the block grant between research and education activities), general university funds (GUF),[1] considered as "a proportion of university block grants" (OECD, 2017a), provide some insights about the level of institutional funding on a broad international scale, while not reflecting the many variants of this type of funding. This indicator confirms wide dispersion in national use of institutional funding, from 0% in the United States (at the federal level; institutional funding is allocated at the US state level) to above 50% of government support for civil R&D in several countries, such as Iceland, the Netherlands and Austria (Figure 8.2). It also confirms that in many countries, the decrease in institutional funding was particularly significant in the 1990s and has plateaued since then (e.g. in Japan).[2]

Figure 8.2. GUF as a percentage of civil GBARD, 2016

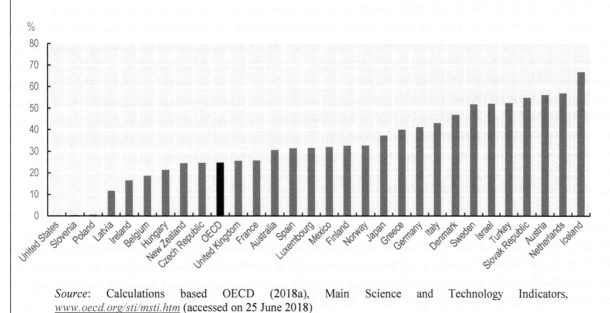

Source: Calculations based OECD (2018a), Main Science and Technology Indicators, www.oecd.org/sti/msti.htm (accessed on 25 June 2018)

StatLink https://doi.org/10.1787/888933858316

Considering these changes and looking more closely at the diverse variants of funding instruments requires reconsidering the dichotomy between non-competitive and competitive funding as a continuum (Dialogic and Empirica, 2014). Based on progress made over the last decade in understanding research funding, this chapter proposes a simple analytical framework to present the portfolio of research-funding instruments available to policy makers along multiple and continuous – rather than unique and binary – dimensions.

These dimensions, as well as the main parameters influencing the positioning of the different funding instruments along them, are discussed below:

- *Competition intensity*: competition is more intense when the number of applicants is large for a given total available budget. Since funders themselves often have little margin to augment the overall budget dedicated to a given funding stream, the size of the targeted population will be the main lever in their hands to manage competition intensity. Hence, the scope of the calls for proposals in project funding, the eligibility rules for institutional funding (e.g. targeting only research universities), together with factors affecting the selection rate and concentration of the distributed funds, are key determinants that intensify competition.

- *Granularity*: the selection/allocation unit can be an entire institution, part of an institution (e.g. a faculty), or a project or programme of different sizes and scope. This has important implications in terms of the scope and flexibility of the allocation, its stability, the level of fragmentation of the funding, etc.

- *Level*: competition can also involve different levels within a single organisation, depending on the elementary units of allocation and assessment. These two units may not coincide, e.g. in the case of institutional funding, where the assessment is performed at the level of departments or research groups, with funding allocated to the organisation as a whole. Depending on internal allocation rules, competition between institutions can translate internally into rivalry between and within parts of these institutions.

- *Type of assessment and selection criteria*: competition can be based on a wide array of criteria, using different timeframes for assessment. Selection/allocation criteria range from publications and citations, to third-party funding and expected social impact. Simplistically, a distinction can be made between input and output-related performance criteria. These criteria can be considered within timeframes with different durations (number of years) and directions (ex ante and/or ex post).

- *Orientation/directionality*: funding allocation can be open, or targeted towards priority areas or issues (e.g. scientific disciplines, economic or societal problems). The more granular and ex ante the allocation, the easier it is for policymakers to steer funding in selected directions.

Figure 8.3 schematises the mix of funding instruments as a continuum along three of the above dimensions. Although not all countries have implemented the full range of instruments, many of these overlap or accumulate. For instance, performance-based funding is almost always provided on top of historical block funding, to allow some stability in funding allocation over time. Similarly, performance contracts are most often coupled with an (ex-post) performance-based component. Therefore, the relative weights of the different funding components (e.g. the research performance-based component in Norway only affects 15% of the total block funding), and their possible synergistic effects, are an important variable when defining a national funding portfolio.

Figure 8.3. Classification of research-funding instruments by intensity, granularity and assessment type

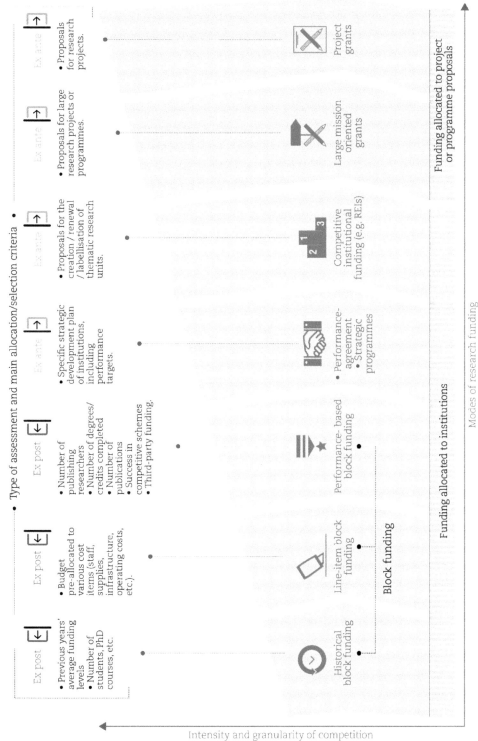

The purpose fit of research-funding instruments

What are the different funding instruments, with their multiple design variants, "good at"? As previously shown (Box 8.2), considerable conceptual, data-collection and case-study work has generated important progress in characterising and measuring research-funding trends over the last two decades. However, knowledge and evidence on the effects of research-funding mechanisms is much scarcer. A key preliminary step in assessing the effects of funding instruments consists in analysing their purpose fit, i.e. determining what policy instruments fit what objectives. This also reconnects the knowledge gained on instruments with the challenges facing policymakers as they attempt to respond to the mounting societal expectations of public research, far beyond a sole focus on scientific excellence.

Each instrument conveys an ever-widening range of policy objectives as new social needs arise, with more programmes stating multiple goals. A recent OECD project identified the desired effects most frequently stated in a dedicated questionnaire covering 75 competitive funding programmes from 21 countries (OECD, forthcoming a). The study distinguished between two sets of "internal" and "external" desired effects (Figure 8.4). Although they were not covered in the study, a similar array of objectives would probably apply to institutional funding instruments, albeit in different proportions.

Figure 8.4. Most frequently stated desired effects of research funding

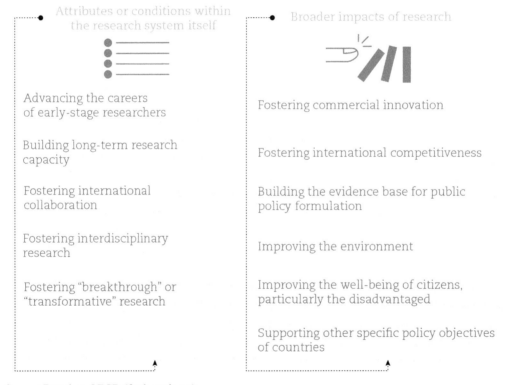

Source: Based on OECD (forthcoming a).

This trend toward more programmes stating multiple goals results in more complex policy-instrument designs to accommodate these various objectives (Jongbloed and Lepori, 2015). For instance, new "mixed" or "hybrid" funding model instruments have been introduced, either by adding competition and performance requirements to formerly "fixed" instruments, or adding more strategic and longer-term components in competitive schemes (e.g. REIs).

The increasingly complex design of instruments also offers many levers to make them more "amenable" to fulfilling different policy objectives. Table 8.1 describes how the design features of three main "families" of instruments can be fine-tuned to accommodate different policy objectives.

Table 8.1. Funding-instrument design parameters to fit specific purposes

	Enhancing research excellence	Steering research towards specific priorities	Creating the conditions for research breakthroughs
Project-based funding	• More weight to proposal selection criteria like the quantity/quality of publications, citations, etc. • Low selection rate and highly skewed allocation ("only the best")	• Specific performance target/missions in the specifications of calls for proposal • Narrower scope of the calls (e.g. thematic calls, mission-oriented calls) • Addition of criteria for selecting proposals related to certain priorities (e.g. extra points for project aligned with national priorities) • Larger (predetermined) budget per disciplines, in line with national priorities • Eligibility criteria to target certain population/projects (e.g. requirement for collaboration with business)	• Scope of the calls (open calls) • Selection criteria in calls for proposal (e.g. novelty of research, interdisciplinary research, targeting of young scientists) • Innovative design of panels ("sandpits") • Extended duration/time horizon of projects
Institutional funding	• More weight to variables in the formulae, e.g. quantity of publications and citations • Performance-based assessment restricted to a few "top" projects selected by the institution	• More weight to variables in the formulae, e.g. societal impacts and alignment with national priorities • Strategic component/programme as part of the institutional funding • Link of institutional funding with performance agreement (emphasis on developing profiles relevant to national priority)	• Absence of "innovation" performance-based conditions • Strategic "moonshot" component/programme as part of the institutional funding (preferably linking different institutions)
Mixed models: research excellence initiatives	• Longer duration of funding • More weight to ex-ante and ex-post evaluation criteria, e.g. quantity of publications and citations	• Targeted calls (e.g. thematic calls, mission-oriented calls) • Number of centres/units per theme determined ex ante • Ex-post evaluation criteria (impacts in line with objectives) • Industry involvement requirements	• Longer duration of funding for the centre • Selection criteria in the initial REI call for proposal (e.g. novelty of research, interdisciplinary research, targeting of young scientists) • Evaluation criteria (e.g. novelty of research)

It focuses on the three most frequent and comprehensive types of desired effects: enhancing research excellence; steering research towards specific priorities; and creating the conditions for breakthroughs.

The section below briefly reviews the main elements of institutional funding, project-based funding and funding of REIs against these three types of desired effects.

- **Institutional funding** focuses on maintaining a stable research infrastructure and underpinning longer-term "excellent" research. As "formal" selection is generally absent from this type of allocation, and academic institutions are entitled to use the funding as they see fit (serving the principle of academic freedom), it is generally not considered amenable to steering research towards specific national priorities. However, some of the initiatives presented in Box 8.3 show that institutional funding can create the appropriate conditions and incentives for researchers to engage in targeted research, providing the necessary strategic capabilities are present at the top level of the beneficiary institutions. They illustrate three main ways to steer research activities through institutional funding: "top-slicing" block grants to target specific priorities; providing additional earmarked institutional funding (either through direct negotiation or competitive awards) for large multi-year projects aligned with national priorities; and using performance contracts to help research institutions build up their profile in fields of national interest. If these initiatives are designed appropriately, and specific conditions are in place to promote co-operation between institutions (as with the Swedish Strategic research areas [SFO] programme), they could also serve the objective of creating breakthrough research.

- **Project funding** consists of allocating funds to groups or individuals to perform specific R&D activities, mostly based on a project proposal subjected to a competitive process. Project funding is considered a better policy tool to steer research, particularly with a view to producing higher-quality research and (to a lesser extent) research that is more relevant to socio-economic objectives (Hicks, 2010). By contrast, many studies have highlighted that an increasing reliance on competitive funding can result in shorter-term, lower-risk projects, rather than longer-term, higher-risk research, although the evidence for this is mixed[1]. Moreover, the resource and time burdens of applying for and reviewing competitive grants can deter some of the best researchers from participating. Finally, project funding hinders the ability of researchers and institutions to engage in long-term planning, because of uncertain future funding. This is especially true for project-based funding with low success rates. Policymakers have experimented with a few alternatives, such as "lotteries" and "sandpits" (OECD, forthcoming a).

- **REIs** provide the selected centres with relatively long-term resources, thereby allowing them in principle to carry out (as their name suggests) excellent research. REIs often include researchers and infrastructures from different institutions, hence promoting the interdisciplinary and co-operative context necessary for high-impact, high-risk "breakthrough" research (OECD, 2014a).

> **Box 8.3. Examples of institutional funding supporting strategic/targeted research**
>
> While the performance-based component of institutional funding has been widely documented, the strategic steering component remains understudied, primarily because it is used less frequently. However, mission-oriented research is attracting renewed interest in the academic and policy arenas. A few countries provide interesting examples of this trend (Mazzucato, 2018).
>
> **Norway** maintains a dual-tier institutional funding system comprising a fixed amount and a performance-related amount, which is complemented by separate funding for relatively large multi-year projects. These "strategic institutional initiatives" (SIS) are negotiated between the institutes, the ministries and the research council, and their budget is added to the envelope of the block grant. SIS aim to develop long-term expertise in the institutes' research fields that are deemed to be of high national interest, but are difficult to realise through competitive funding. SIS represented about 40% of the institutional funding of "environmental" research institutes and 30% of the "primary" research institutes' institutional funding in 2016 (overall institutional funding itself represented about 15% of these two types of institutes' total revenues) (OECD, 2017b).
>
> **Sweden** launched the SFO programme to increase the share of institutional research funding in universities' funding mix and strengthen co-operative university research in areas of national strategic relevance. SFO grants were allocated on a competitive basis for five years, based on proposals from university partnerships in priority areas. Once awarded, the selected universities could add the funds to their institutional funding and use them with total freedom in the priority areas determined by the Swedish Government according to the proposals' relevance to Swedish industry, as well as their capacity to reach the highest international quality levels, and solve important societal needs. The three selected areas were medicine and the life sciences, technology and climate change (OECD, 2016).
>
> In the **Netherlands**, TO2 Applied Research Institutes have seen a triple evolution in their block funding since the mid-2000s, with significant cuts in direct government funding, a greater share tied to performance, and stronger conditions for using the funding to better align research with the national priorities formalised as "top sectors". This change has been implemented in multi-year performance contracts, connected to specific public-private partnerships in the priority areas (OECD, 2014b).
>
> In **Austria**, performance agreements determine around 95% of the 22 universities' block funding for research (compared to 7% in the Netherlands, 10% in Ireland and 100% in Finland). First implemented in 2007, the agreements define a concrete set of measures and services to be fulfilled over three years, based on development plans individually negotiated between each university and the Federal Ministry of Education, Science and Research (BMBWF). These development plans are informed by the National Development Plan for Higher Education, which is formulated by the BMBWF and sets national objectives for a period of six years. The University Act (2002) also sets priorities to be addressed in institutional plans (OECD, forthcoming b).

Among the different policy objectives, the issue of how the different funding instruments support breakthrough research is attracting growing attention, particularly in light of rising concerns about the seemingly decreasing productivity of research (Bloom et al., 2017). As previously mentioned, the research community has expressed concerns that competitive funding mechanisms could disadvantage risky, potentially transformative, or transdisciplinary research proposals in favour of applied, incremental, or disciplinary proposals. Indeed, reconciling both a desire for more efficient and transparent research funding with the need to support more innovative (but also riskier) projects poses a real challenge.

Studies on this topic provide recommendations on how to design instruments to fund breakthrough research (e.g. Laudel and Gläser, 2014; Wang, Lee and Walsh, 2018). Some studies recommend tailoring funding mechanisms to the need for creativity in science, rather than simply adding criteria to existing project-funding schemes. Others claim that competitive funding can support breakthrough research, providing it is specifically adapted to this strategic objective (Heinze, 2008; Goldstein and Narayanamurti, 2018). The Japanese Government, for instance, announced that the number of selection panels in the main competitive instrument (the Grants-in-Aid for Scientific Research programme, "*kakenhi*") will drop from close to 500 to around 375, to foster research originality and creativity (Hornyak, 2017). The increase in competitive funding has been blamed for a markedly increased concentration of basic-research funding in the hands of a small number of Japanese institutions; this loss of diversity is detrimental to novelty and alternative scientific ideas (Matsuo, 2018).

Advancing the research-funding agenda

Considerable conceptual, data-collection and case-study work has generated important progress in characterising and measuring research-funding trends over the last two decades. The increasing diversity of design variants for funding instruments offers policymakers new levers to accommodate a widening set of policy needs. However, knowledge on the effects of research-funding mechanisms is far scarcer, notably owing to many methodological problems (Butler, 2010). This chapter has proposed a conceptual framework to represent the new research-funding landscape and analyse which policy instruments (and their design variants) can theoretically fulfil different policy objectives. However, this analysis of the 'purpose fit' of funding instruments is still in its infancy and will be the object of more work in the near future to assess how policy makers can best fund research to realise their priorities.

Pushing this research agenda further will require going beyond an "instrument-by-instrument" analysis, to examine the instruments' combined effects and interactions with the institutional environment:

- Competitive and non-competitive funding interact in several ways, exhibiting both positive complementarities and tensions. For instance, a project grant generally only covers part of the costs of the research activities and requires matching funds that might be found in the block funding for university research (often under the form of research staff time). Implementing the project also requires services and equipment financed through past and present institutional funding. Typically, institutional funding provides money to build and maintain basic capacity (i.e. skills and the work environment) and finance day-to-day operations, whereas project funding supports more targeted research (Lepori, Geuna and Mira, 2017). However, this traditional model is becoming blurred, as rules (not least concerning overheads and eligible expenses) are changing and vary among countries. As a result, making a clear-cut distinction between longer-term institutional funding and competitive-project respective contributions to the steering of research is even more difficult.

- The institutional environment is essential to understanding the funding landscape. Some important parameters to consider are the existence, size and scope of funding agencies, and their type of relationships with ministries; the existence of "umbrella organisations", to which government can delegate some programming and funding roles (e.g. the National Centre for Scientific Research [CNRS] in France); and universities' internal organisation (e.g. the internal funding-allocation mechanisms) and strategic management capabilities.

Research funding is a complex, staged and multifaceted issue, which calls for a systemic view in order to understand its dynamics and assess its effects. The "In my view" box below provides some guidelines to pursue a holistic analysis of research funding.

Box 8.4. In my view : A systems world needs systemic thinking about research funding

Erik Arnold, Chairman of Technopolis Group and Adjunct Professor in Research Policy, Royal Institute of Technology (KTH), Stockholm.

Funding research involves a range of actors, influences and policies, each with limited reach, which tend to be managed separately. If we look at the whole picture, it becomes clear that a range of policy levers exist to improve system performance (not all of which are accessible to all policy actors), and that a co-ordinated approach provides an opportunity to steer the whole system in a way that helps it develop, and supports the implementation of national research and innovation policy.

Fundamentally, funding instruments serve specific policy intentions and should be considered in the context of the overall system of rewards (and punishments) policy offers to research performers, such as universities.

Figure 8.5 provides a bird's-eye view of that system. The central box focuses on research funding. Traditionally, education ministries provided universities with institutional funding in the form of block funding – lump sums they could use to produce teaching and research. Some countries provided a detailed budget to indicate the intended uses of the block fund, but the principle of university autonomy meant (and still means) that there was a distance between what the education ministry could decide, and what the universities would actually do. More recently, education ministries have started not only to distinguish between institutional funding for education and institutional funding for research, but also to base parts of this funding on performance, introducing an unprecedented element of inter-university competition for institutional funding. Performance-based research funding systems have received increasing policy attention in recent years as policymakers try to manage national research systems more effectively. These systems can be contentious (academics hate them, university research managers love them), and a growing literature studies the role of performance assessment in their operation.

Recognising the difficulty of assuring the quality of research by autonomous universities, education ministries also tend to fund research councils offering competitive "external" (i.e. non-block) funding on a project basis. This is normally "bottom-up" and investigator-initiated research lacking any predetermined societal relevance. This "excellence" funding is expected to assure quality, as well as increase the volume of research. However, with academics controlling the research councils and the committees prioritising projects, it is the academics – not the rest of society – who are firmly in charge of the nature and quality of research.

Backed by "sector" or "mission" ministries, innovation agencies and sector funders (e.g. covering health, transport and the environment) offer other funding incentives for the research system to address societal problems.

However, the direct operation of these incentives is far from the only policy influence on the development of the research system. The overall amount or growth of research funding is one positive factor (for example, Denmark's dramatic surge in scientific performance in recent years builds on substantially increased funding). Internationalisation raises quality in lagging countries (international co-publications are more highly cited than single-author or national ones). University governance and management also have a big impact. It is widely believed that the competition involved in having a high share of external money in universities' research income drives up quality). Finally, there is increasing faith in performance-based research funding systems, as well as significant disagreement about whether it should govern a high proportion of institutional funding for research (there is evidence that both high and low proportions affect researcher behaviour.)

Figure 8.5. Research funding in a policy context

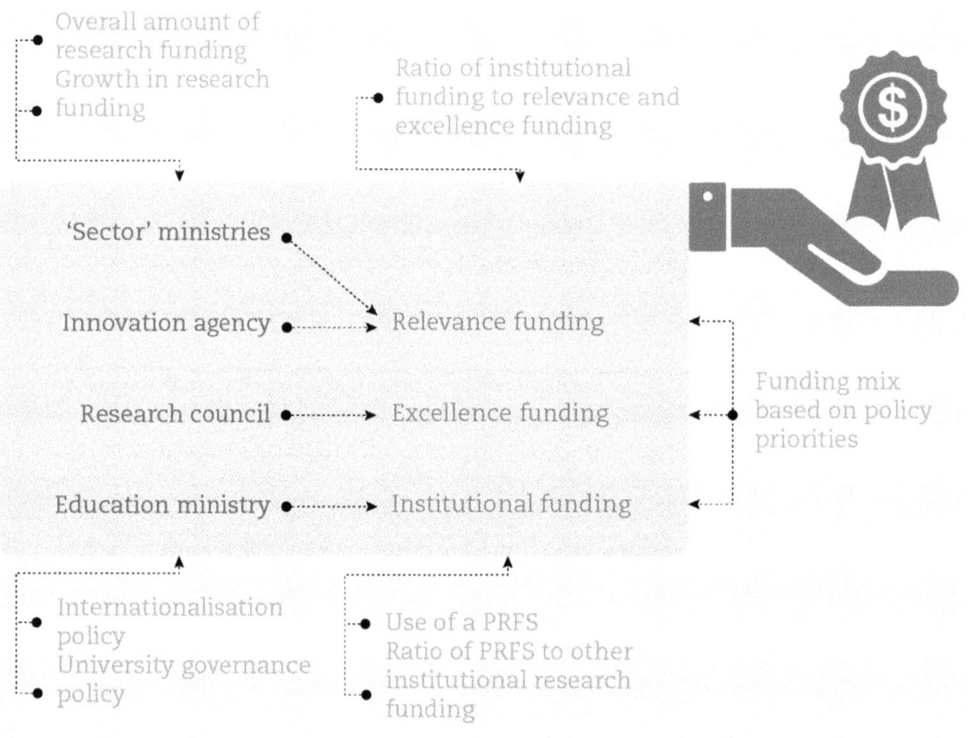

Statistically, it is very difficult to connect observed patterns in national performance to most of these policy levers. Multiple policies are at play. Their effects are hard to untangle; contextual factors, such as history and culture, are also important. Often, good performance seems to result from changes in one or more of the "levers" discussed above, rather than from the presence of particular ratios among funding streams. As with much else in innovation systems, policymakers need to adopt a systemic perspective of their specific national situation when analysing needs and using policy instruments. Ultimately, a single ministry cannot do this – a higher power, such as a research and innovation council or the

> government itself, needs to co-ordinate the different components of a research and innovation system.

A forward-looking view on research funding

Emerging or ongoing trends are already changing funding practices and landscapes; the future evolutions of research funding are therefore uncertain. With the growing importance of innovation in all human activities, the pressure will grow for research to deliver workable solutions to real-world problems. A likely scenario is that research will continue to evolve as a demand-driven activity, favouring mechanisms in which research users – rather than researchers alone – increasingly shape the research agenda. Such an evolution could not only promote competitive mechanisms, but also different forms of institutional funding that steer research. It could also result in a multiplication of the expected objectives underlying any research activity, as shown in the growing list of project-evaluation criteria and the expanding formula for performance-based institutional funding. This trend could jeopardise the ability of a given research project to excel in a specific dimension, e.g. scientific excellence, high-risk research or economic/social relevance. The modes of research support will most likely continue to evolve to deal with this issue, either by segmenting funding according to types of objectives or creating new modes of "customised" project evaluation.

The growing recognition of the Sustainable Development Goals (SDGs) as challenges to be addressed in research and innovation is a salient trend (Chapter 4 on the SDGs). The literature has widely documented that research relevant to SDGs will need to be transformational, hence ambitious, interdisciplinary and performed with a mid- to long-term horizon. While this does not in principle imply project-based funding, the pressure for greater accountability and cost efficiency will clearly favour competitive-funding approaches. Designing new instruments and programmes (such as different forms of mission-oriented programmes) will be key to juggling the competing requirements of strategic steering, competitive allocation and risk-taking.

The articulation between instrument design and policy objectives is also changing as digitalisation transforms the research and innovation enterprise (as evidenced in this Outlook). Digitalisation is improving the ability of policymakers and funders to monitor research: more up-to-date information is available, which can be analysed more in-depth, hence facilitating evaluation (see Chapter 12 on digital science and innovation policy). Information useful for resource allocation could be accessed directly through data processing, reducing the need for costly competitions. At the same time, digitalisation can lower the cost of competitive funding (project-preparation work can be subjected to versioning and re-used, and panels can be organised online), which could enhance its appeal.

Needless to say, research to address the SDGs and/or reap the opportunities of digitalisation will require ever-increasing financial inputs, in the context of the rising costs of research and budget pressure in indebted states. Tensions over budgetary negotiations will undoubtedly grow between policy fields. Research – which is both an increasingly costly policy field and a key enabler of the transformational agenda – will be at the heart of these debates.

Notes

[1] As shown, for instance, in the analysis of responses to the questions on the main public-research policy debates in the 2017 edition of the EC-OECD STIP survey, covering more than 50 countries (EC/OECD, 2017). See also Zdravkovic and Lepori (2018) for an analysis of the academic literature.

[2] Including 21 OECD of 36 member countries.

[1] The OECD Frascati Manual defines GUF as the share of R&D funding from the general grants universities receive from the central government (federal) ministry of education or the corresponding provincial (state) or local (municipal) authorities to support their overall research/teaching activities (OECD, 2015b).

[2] Part of the country differences relate to the relative weights of research activities performed in HEIs and PRIs.

[1] Similar criticisms can be also directed towards some forms of performance-based institutional funding.

References

Bloom, N. et al. (2017), "Are Ideas Getting Harder to Find?", *NBER Working Paper*, No. 23782, National Bureau of Economic Research, Cambridge, MA, http://www.nber.org/papers/w23782.

Butler, L. (2010), "Impacts of performance-based research funding systems: A review of the concerns and the evidence", in *Performance-Based Funding for Public Research in Tertiary Education Institutions: Workshop Proceedings*, OECD Publishing, Paris, pp. 127-165, https://doi.org/10.1787/9789264094611-en.

Dialogic and Empirica (2014), "The effectiveness of national research funding systems", Dialogic and Empirica, Utrecht/Bonn, https://www.dialogic.nl/wp-content/uploads/2016/12/2013.109-1422.pdf.

EC/OECD (2017), *STIP Compass: International database on STIP policies* (database), April 2018 version, https://stip.oecd.org.

Goldstein, A.P. and V. Narayanamurti (2018), Simultaneous pursuit of discovery and invention in the US Department of Energy, *Research Policy*, Vol. 47, pp. 1505-1512, Elsevier, Amsterdam, https://doi.org/10.1016/j.respol.2018.05.005.

Heinze, T. (2008), "How to sponsor ground-breaking research: A comparison of funding schemes", *Science and Public Policy*, Vol. 35/5, pp. 302-318, Oxford Academic Press, Oxford, https://doi.org/10.3152/030234208X317151.

Hicks, D. (2010), "Overview of models of performance-based research funding systems", in *Performance-Based Funding of Public Research in Tertiary Education Institutions*, OECD Publishing, Paris, http://dx.doi.org/10.1787/9789264094611-en.

Hornyak T. (2017), "Japan shakes up research funding system", *Nature Index*, 1 August 2017, Springer Nature, https://www.natureindex.com/news-blog/japan-shakes-up-research-funding-system.

Jongbloed, B. and B. Lepori (2015), "The Funding of Research in Higher Education: Mixed Models and Mixed Results", in Huisman, J. et al., *The Palgrave International Handbook of Higher Education Policy and Governance*, pp. 439-462, Palgrave Macmillan, London.

Jonkers, K. and T. Zacharewicz (2016*), Research Performance Based Funding Systems: a Comparative Assessment*, European Commission, Publications Office of the European Union, Luxembourg, http://publications.jrc.ec.europa.eu/repository/bitstream/JRC101043/kj1a27837enn.pdf.

Laudel, G. and J. Gläser (2014), "Beyond breakthrough research: Epistemic properties of research and their consequences for research funding", *Research Policy*, Vol. 43, pp. 1204-1216, Elsevier, Amsterdam, https://doi.org/10.1016/j.respol.2014.02.006.

Lepori, B., E. Reale and A. Orazio Spinello (2018), "Conceptualizing and measuring performance orientation of research funding systems", *Research Evaluation*, Vol. 1/13, Oxford University Press, Oxford, https://doi.org/10.1093/reseval/rvy007.

Lepori B., A. Geuna and A. Mira (2017), "Money matters, but why? Distribution of resources and scaling properties in the US and European higher education", Presentation at Leiden University, June 2017.

Lepori, B. et al. (2007), "Comparing the evolution of national research policies: What patterns of change?", *Science and Public Policy*, Vol. 34/6, pp. 372-388, Oxford University Press, Oxford, https://doi.org/10.3152/030234207X234578.

Mazzucato, M. (2018), *Mission-Oriented Research & Innovation in the European Union – A problem-solving approach to fuel innovation-led growth*, Directorate-General for Research and Innovation, European Commission, Publications Office of the European Union, Luxembourg, https://ec.europa.eu/info/sites/info/files/mazzucato_report_2018.pdf.

Matsuo, K. (2018), "The structure and issues in Japan's STI funding", Presentation at the Euroscience Open Forum (ESOF) Conference, Toulouse, 11 July 2018.

OECD (forthcoming a), *Effective Operation of Competitive Research Funding Systems*, OECD Publishing, Paris.

OECD (forthcoming b), *OECD Reviews of Innovation Policy: Austria 2018*, OECD Publishing, Paris.

OECD (2018a), *Main Science and Technology Indicators* (database), https://www.oecd.org/sti/msti.htm (accessed on accessed on 25 June 2018).

OECD (2018b), Research and Development Statistics, database, http://www.oecd.org/innovation/inno/researchanddevelopmentstatisticsrds.htm (accessed on 25 June 2018).

OECD (2017a), *OECD Science, Technology and Industry Scoreboard 2017: The digital transformation*, OECD Publishing, Paris, https://doi.org/10.1787/9789264268821-en.

OECD (2017b), *OECD Reviews of Innovation Policy: Norway 2017*, OECD Publishing, Paris. http://dx.doi.org/10.1787/9789264277960-en.

OECD (2016), *OECD Reviews of Innovation Policy: Sweden 2016*, OECD Publishing, Paris, http://dx.doi.org/10.1787/9789264249998-en.

OECD (2015a), *The Innovation Imperative: Contributing to Productivity, Growth and Well-Being*, OECD Publishing, Paris, https://doi.org/10.1787/9789264239814-en.

OECD (2015b), *Frascati Manual 2015: Guidelines for Collecting and Reporting Data on Research and Experimental Development, The Measurement of Scientific, Technological and Innovation Activities*, OECD Publishing, Paris, https://doi.org/10.1787/9789264239012-en.

OECD (2014a), *Promoting Research Excellence: New Approaches to Funding*, OECD Publishing, Paris, https://doi.org/10.1787/9789264207462-en.

OECD (2014b), *OECD Reviews of Innovation Policy: Netherlands 2014*, OECD Reviews of Innovation Policy, OECD Publishing, Paris, https://doi.org/10.1787/9789264213159-en.

OECD (2011), *Public Research Institutions: Mapping Sector Trends*, OECD Publishing, Paris, http://dx.doi.org/10.1787/9789264119505-en.

Reale, E. (2017), "Analysis of National Public Research Funding (PREF) – Final Report", JRC Technical report, European Commission, Publications Office of the European Union, http://publications.jrc.ec.europa.eu/repository/bitstream/JRC107599/kj0117978enn.pdf.

van Steen, J., (2012). "Modes of Public Funding of Research and Development: Towards Internationally Comparable Indicators", *OECD Science, Technology and Industry Working Papers*, Vol. 2012/04, OECD Publishing, Paris, http://dx.doi.org/10.1787/5k98ssns1gzs-en.

Wang J., Y.-N. Lee and J.P. Walsh (2018), "Funding model and creativity in science: Competitive versus block funding and status contingency effects", *Research Policy*, Vol. 47, pp. 1070-1083, Elsevier, Amsterdam, https://doi.org/10.1016/j.respol.2018.03.014.

Zdravkovic, M and B. Lepori (2018), "Mapping European Public Research Funding Studies: Selected results and some open questions", Presentation at the EU-SPRI conference, 7 June, ESIEE, Marne-la-Vallée.

Chapter 9. The governance of public research policy across OECD countries

By

Caroline Paunov and Martin Borowiecki

Good governance of public research policy can boost the effectiveness of public investment in research. This chapter describes the governance of public research policy across 35 OECD member countries and its evolution over 2005-17. It sheds light on different research-policy contexts that explain why a "one-size-fits-all" approach is inappropriate. The chapter successively addresses four core governance dimensions with important implications for research sector performance. It first discusses the objectives of national STI strategies for higher education institutes (HEIs) and public research institutes (PRIs), which are increasingly expected to contribute to raising national R&D intensity and to address societal challenges. It then describes the variety of organisations allocating funding and evaluating performance. The section that follows discusses the growing autonomy of HEIs and PRIs and the use of associated policy tools, such as performance contracts. The last of the four core governance dimensions relates to the modes of stakeholder involvement in policy decision-making. The chapter concludes with a review of potential future developments.

The statistical data for Israel are supplied by and under the responsibility of the relevant Israeli authorities. The use of such data by the OECD is without prejudice to the status of the Golan Heights, East Jerusalem and Israeli settlements in the West Bank under the terms of international law.

Introduction

The contributions to innovation of research conducted by higher education institutions (HEIs) and public research institutions (PRIs) are well recognised, as is the need for public support for such research. In the emerging globalised knowledge economy, where the best innovations are key success factors, research is more important than ever. Yet many countries struggle to increase public budgets for research. Consequently, countries deploy a battery of policy instruments to orient investments in public research. Each national policy mix is shaped by the mechanisms and institutional arrangements governing policy action on publicly funded research in HEIs and PRIs. More effective policy governance arrangements can enhance the effectiveness of research funding. For instance, involving all stakeholders in policy design can help identify better policies to overcome obstacles hindering public research activities.

This chapter describes the governance of public research policy across 35 OECD member countries and its evolution over 2005-17. It sheds light on different research-policy contexts that explain why a "one-size-fits-all" policy approach is inappropriate. It outlines institutional choices countries are in a position to change.

More specifically, the chapter addresses four core dimensions that shape the policy mix regarding HEIs and PRIs and provides findings (Figure 9.1), with important implications for the research sector's performance (e.g. Aghion et al., 2010; Breznitz, 2007).

Figure 9.1. Four core dimensions that shape the policy mix

National science, technology and innovation strategies	Institutions allocating funding and evaluating performance	Autonomy of HEIs and PRIs	Stakeholder involvement
The objectives of national science, technology and innovation (STI) strategies for HEIs and PRIs: these objectives establish how governments expect HEIs and PRIs to contribute to national STI agendas.	The institutional arrangements in place to allocate public funding for HEIs and PRIs, and evaluate their performance: these arrangements include which institutions are in charge of funding and evaluation. They articulate the incentive structures for HEIs and PRIs to reach established objectives.	The autonomy of HEIs and PRIs in deciding on their budgetary expenses, human resources and industry relations: opportunities to shape the design of research policy to reach objectives differ according to which decisions individual institutions can take independently.	The stakeholders involved in policy making on HEIs and PRIs: the extent to which HEIs and PRIs, civil society and industry participate in public research policy making affects how different interests and demands are considered in objective-setting.

This chapter identifies a number of common characteristics and trends across these dimensions of the governance of public research policy (Table 9.1) using the results of an OECD survey on the governance of research policy conducted after a three-year process of in-depth data collection and validation[1] (Borowiecki and Paunov, 2018).

The chapter is structured as follows: The first section discusses the objectives of national STI strategies for HEIs and PRIs. This is followed by a description of the institutions allocating funding and evaluating performance. The third section discusses the autonomy of HEIs and PRIs, followed by a section devoted to an overview of stakeholder involvement in policy decision-making. The final section concludes with a review of potential future developments.

HEIs and PRIs in national STI strategies

Public research features prominently in national STI plans or strategies, which are in place in 33 (i.e. 94%) of the 35 OECD countries surveyed. They outline national priorities for research and innovation, and define the expected contributions of HEIs, PRIs, industry and civil-society actors (e.g. non-governmental organisations [NGOs] and foundations). Policy demands across OECD countries include finding solutions to societal challenges (e.g. demographic change and sustainable growth); developing key technologies (e.g. digital technologies) for competitiveness; and increasing national research and development (R&D) intensity. Countries' STI strategies differ in terms of the national priorities they set (i.e. societal challenges, research fields and/or industries), the targets they define (i.e. R&D intensity) and how they monitor progress in reaching these targets.

Table 9.1. Common characteristics across OECD countries

Dimension	Common characteristics of how public research is organised across OECD countries
HEIs and PRIs in national science, technology, and innovation (STI) strategies	• National STI strategies set out prominently the expected contributions of higher education and public research to technology development (incl. of digital technologies), raising national R&D intensity and addressing societal challenges, such as the Sustainable Development Goals. • STI strategies often set measurable targets for HEIs and PRIs, such as increasing the number of tenure positions for young researchers, the share of female researchers and the number of collaborative research projects with industry.
Institutions allocating funding and evaluating performance of HEIs and PRIs	• Specialised agencies are in charge of competitive, project-based funding to HEIs and PRIs. Where several agencies provide such funding they are specialised by research field, provide either funding for research or innovation, or there are separate agencies for the national and regional level. • Performance contracts between ministries/agencies and individual HEIs have been adopted in several OECD countries over the past decade. They set goals and link them to the block funding of HEIs. • Countries have invested substantially in evaluation and monitoring the performance of HEIs and PRIs. Several new institutions have been created for this purpose over the past decade.
Autonomy of HEIs and PRIs	• Reforms over the past decade have increased HEIs' autonomy with regard to budget allocations, recruitment and promotions of researchers, as well as industry relations, including the creation of technology transfer offices, spin-offs, and industry partnerships. • Most national restrictions to autonomy apply to the setting of researchers' salaries.
Stakeholders' involvement in policy-making	• Stakeholder involvement in university boards has increased across the OECD. Civil society and industry shape policy decisions of HEIs – particularly where these have substantial autonomy – by sitting on HEI governing boards or councils. • National research and innovation councils often offer opportunities to shape policy directions for stakeholders from civil society – including members of labour unions and non-profit organisations (NGOs) – and industry – often large firms but also in some cases SMEs. • New tools such as online consultations to solicit input from civil society have been used more widely in combinations with traditional consultation methods, such as working groups and roundtables.

Looking at the data collected by the OECD governance of research policy survey, three main observations can be made. First, most strategies (i.e. in 31 of 33 countries, plus the Brussels-Capital Region, the Flemish Region and the Walloon Region in Belgium) identify specific scientific research areas, technologies and economic fields, e.g. energy and energy technologies; health and life sciences; information and communication technologies; and nanotechnology and advanced materials. A growing number of strategies place digital-transformation objectives at the core of their strategic orientations, as discussed in Chapter 3.

Second, STI strategies also define the expected contributions of HEIs and PRIs to overcoming socio-economic challenges. In 30 (i.e. 91%) of the 33 countries surveyed, and in the Brussels-Capital Region, the Flemish Region and the Walloon Region, STI strategies address major societal challenges, including demographic change, health, environment, smart transport and cities. The STI strategies of 25 (i.e. 76%) of 33 countries, the Brussels-Capital Region and the Walloon Region stress the need for research and innovation to develop a sustainable economy. The strategies of 13 (40%) of 33 countries emphasise the importance of STI in addressing demographic change. Finally, the STI strategies of 15 (45%) of 33 countries, as well as the Flemish Region and the Walloon Region, also encourage investment in STI to improve transport systems.

Third, most national STI strategies include quantifiable benchmarks for policy outcomes (Figure 9.2).

Figure 9.2. Quantitative targets included in national STI strategies

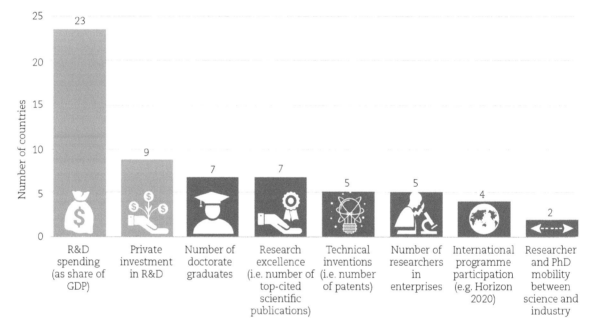

Note: The figure corresponds to question 2.6.e of the OECD survey on the governance of research policy ("Does the national STI strategy or plan address any of the following priorities? Quantitative targets for monitoring and evaluation [e.g. setting as targets a certain level of R&D spending for public research?]). It showcases only countries where the national STI strategies have quantitative targets. Israel and Luxembourg do not have national STI strategies. Australia, Chile, the Czech Republic, Denmark, Finland, France, the Netherlands, Sweden Switzerland, and the United Kingdom have not set quantitative targets.
Source: Borowiecki and Paunov (2018).

The national STI strategies of 23 (70%) of 33 countries, and the Brussels-Capital Region, the Flemish Region and the Walloon Region have a national R&D intensity target. The data also shows that 11 (or 33%) of national STI strategies and the STI strategies of Brussels-

Capital Region and the Walloon Region have targets HEIs and PRIs at the core of policy attention. These targets include raising funding for doctoral students (7 of 33 strategies, plus the Brussels-Capital Region), and increasing job placements for researchers and PhDs in industry (5 out of 33 strategies, plus the Brussels-Capital Region). Japan's Fifth Science and Technology Basic Plan for 2016-20 features targets for increasing the number of tenure positions for young academic researchers and raising the share of female researchers among newly hired university personnel. It also sets quantitative benchmarks for knowledge transfer between universities and industry. These include increasing private funding for university research, the amount of collaborative research funds received from industry and the number of licence agreements on university patents.

Institutions allocating funding and evaluating performance

Institutions allocating project-based funding

Project-based funding – i.e. funding mostly allocated by agencies to a research group or researcher to perform a specific item of research and/or innovation – is an important tool to incentivise HEIs and PRIs to contribute to national STI objectives. Together with institutional block funding, it accounts for the bulk of funding for public HEIs and PRIs, complemented (to a lesser degree) by funding from industry and other segments of the private sector. The governance setting, notably which institution provides such funds, also contributes to raising the effectiveness of project-based funding.

The evidence shows that in 31 (i.e. 89%) of 35 OECD countries, national agencies decide on project-based funding allocations for HEIs. In most countries (30 countries out of 35, plus Wallonia and Flanders, for HEIs; 25 out of 34 countries, plus Wallonia and Flanders, for PRIs), ministries provide institutional block funding. The main roles of these agencies is to fund research and innovation projects; among other responsibilities, they also provide expert advice on related policy.

The institutional landscape for project-based funding is a dynamic one. Between 2005 and 2016, 10 OECD countries created new project-funding institutions. They include the French National Research Agency (ANR), created in 2006; the Innovation Fund Denmark, created in 2014; and the State Research Agency (AEI) in Spain, created in 2015.

Several countries use multiple agencies to allocate project-based funding. In 12 of 31 OECD countries, a single agency provides project-based funding, compared to 2 or more specialised agencies in the remaining 19 countries (Figure 9.3).

Agencies specialising in research fields usually exist where such research has very special features (e.g. health and medical research) and are an important research base in the country. In Australia, for instance, the National Health and Medical Research Council manages funds for health and medical research, whereas the Australian Research Council handles competitive calls for all other research fields. Canada has several such specialised agencies, including the National Research Council, the Natural Sciences and Engineering Research Council; the Canadian Institutes of Health Research; and the Social Sciences and Humanities Research Council of Canada.

In several countries featuring multiple agencies, research and innovation tasks are separate, reflecting the divided responsibilities across different ministries. In Austria, the Austrian Science Fund (FWF) is responsible for basic research, whereas the Austrian Research Promotion Agency (FFG) and the CDG-Christian Doppler Research Association fund applied research. This reflects the ministerial division of responsibilities, whereby the

Federal Ministry of Science, Research and Economy is responsible for research, and the Federal Ministry for Transport, Innovation and Technology is in charge of innovation.

In countries with federal structures, education, research and innovation tasks are shared between the national level and the federal state or subnational level. In Germany, the federal states oversee education policy (including teaching at HEIs), whereas nationwide PRIs and the national German Research Fund (among other national and regional players) provide financing for research and innovation. In addition, a variety of competitive funding tools for project-based research funding of HEI and PRI have been implemented. In Belgium, five regional funding agencies provide project-based research funding.

Over 2007-17, some countries reduced the number of funding agencies to simplify funding applications (creating a "one-stop-shop"), reduce funding fragmentation and increase efficiency. Denmark, for instance, created the Innovation Fund Denmark in 2014 by merging the Danish Council for Strategic Research, the Danish National Advanced Technology Foundation and the Danish Council for Technology and Innovation. The merger's objective was to simplify grant applications for researchers and businesses. Estonia created the Estonian Research Council in 2012 by consolidating the functions of three agencies to reduce fragmentation in public research funding.

Figure 9.3. Number of public agencies in charge of project-based funding allocations in countries with agencies in place

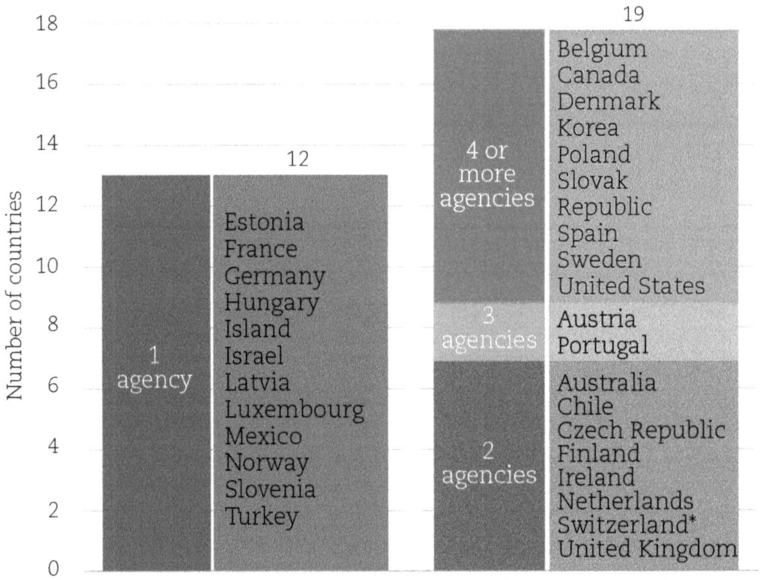

- **1 agency** (13 countries): Estonia, France, Germany, Hungary, Island, Israel, Latvia, Luxembourg, Mexico, Norway, Slovenia, Turkey
- **2 or more agencies** (19 countries total, 12 shown in detail):
  - **2 agencies**: Australia, Chile, Czech Republic, Finland, Ireland, Netherlands, Switzerland*, United Kingdom
  - **3 agencies**: Austria, Portugal
  - **4 or more agencies**: Belgium, Canada, Denmark, Korea, Poland, Slovak Republic, Spain, Sweden, United States

Note: The figure corresponds to question 1.2.c of the OECD survey on the governance of research policy ("Name of the institution in charge of project-based funding"). Information is displayed for 31 countries where at least 1 national agency allocates project-based funding. * The Swiss funding agency Innosuisse started operating in 2018.
Source: Borowiecki and Paunov (2018).

Agencies specialised in evaluation and monitoring

Specialised agencies in charge of evaluating and monitoring the performance of HEIs and PRIs are in place in 19 (56%) of 34 countries, in Wallonia in Belgium, and in Massachusetts in the United States. The agencies' objective is to conduct high-quality, independent evaluations, to inform policy on funding programmes for HEIs and PRIs. The High Council for the Evaluation of Research and Higher Education (HCERES) in France is one such agency. In Ireland, the Higher Education Authority (HEA) is responsible for system governance and institutional block funding for HEIs, whereas the Quality and Qualifications Ireland (QQI) oversees quality assurance. Both the HEA and QQI conduct quality and strategic evaluations of HEIs and PRIs, based on criteria set by the government. In the Netherlands, the Higher Education and Research Review Committee is an independent committee that evaluates the attainment of performance targets set in performance contracts. Evaluation and monitoring is performed by ministries in 11 (32%) of 34 countries; and by HEI\PRIs in the Netherlands and Spain. In Belgium and the United States, regions/federal states are in charge of evaluations of HEIs and PRIs.[2]

Examples of recently established agencies and independent committees for evaluation and monitoring include the Agency for Assessment and Accreditation of Higher Education (A3ES) in Portugal (2007); the Higher Education and Research Review Committee in the Netherlands (2012); and the National Agency for Evaluation of Universities and Research Institutes (ANVUR) in Italy (2010).

Performance contracts

The move towards stronger performance evaluation has also increased the importance of performance contracts and performance-based funding instruments. Performance contracts are set up between national ministries/agencies and individual HEIs; they define goals and link them to block funding of HEIs. Performance contracts are in place in 13 (37%) of 35 OECD countries and several regions/federal states (e.g. Scotland in the United Kingdom; and Baden Württemberg, Brandenburg and North Rhine-Westphalia, among other federal states in Germany). Nine countries introduced performance contracts during the past decade (Figure 9.4).

Performance contracts vary across countries in several respects, including the shares of HEI budgets they cover. Among the nine countries and four regions/federal states for which such information is currently available, the shares subject to performance contracts vary from 1% in Denmark and 7% in Latvia and the Netherlands, to 94-96% in Austria and 100% in Finland and Korea. At the regional/federal level, performance contracts affect 50% of institutional funding of HEIs in Scotland, for instance. In the German federal states for which information is available, performance contracts apply to 2% of HEIs institutional funding in Brandenburg and 23% of block funding of HEIs in North Rhine-Westphalia.

Figure 9.4. Year of introduction of performance contracts and shares of HEI institutional block funding involved

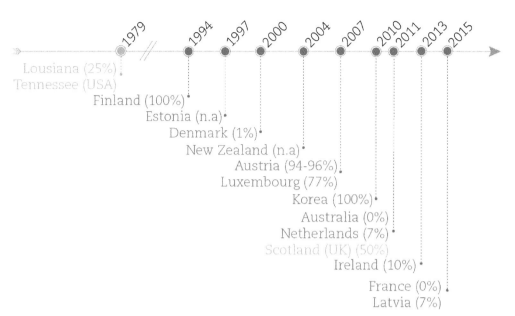

Note: This figure corresponds to questions 1.3.a ("Do performance contracts determine institutional block funding of HEIs?") and 1.3.b. ("Share of HEI budget subject to performance contracts") of the OECD survey on the governance of research policy. Values in parentheses show the share of institutional block funding of HEIs subject to performance contracts. Information on the year of introduction of performance contracts is missing for Japan. Information on the share of the budget of HEIs subject to performance contracts is missing for Estonia, Japan and New Zealand. At the regional/federal state level, Scotland; the US states of Louisiana and Tennessee; and several German federal states, including Baden Wurttemberg, Brandenburg, and North Rhine-Westphalia have performance contracts in place. Performance contracts are in place in most Federal States in Germany, e.g. Brandenburg and North Rhine Westphalia. Some Federal States introduced them in the 2000s while others had introduced them earlier. The share of HEI institutional block funding involved also varies, e.g. 2% in Brandenburg and 23% in North Rhine Westphalia.
Source: Borowiecki and Paunov (2018).

Other differences in performance contracts relate to their targets. Targets are used to monitor the performance of HEIs and assess whether they have met their objectives. As expected, education and research targets are the main criteria, used in the 12 countries and 2 regions (Scotland and North Rhine-Westphalia) with performance contracts in place and for which target-related information is available; 10 of these countries and 2 regions (Scotland and North Rhine-Westphalia) focus on the role of HEIs in supporting innovation performance; 5 countries and Scotland address socio-economic challenges and include targets to support the local economy. Differences also exist in how targets are defined. Some countries use qualitative indicators, while others rely more on quantitative indicators. Table 9.2 describes the cases of Austria, Finland and Scotland.

Table 9.2. Performance contracts in Austria, Finland and Scotland

Country	Targets	Process
Austria	Qualitative and quantitative criteria used in performance contracts set education, research and innovation targets for universities. Education indicators include the number of students who complete full credits per academic year, the number of graduates and the quality of teaching. Research indicators pay specific attention to the generation of basic research, as well as young academics' career paths. Innovation-outcome indicators vary across institutions. The University of Vienna, for instance, commits to increasing the number of patents and providing courses on technology transfer (University of Vienna, 2015).	Each of the 22 Austrian institutions signs a specific performance agreement for a period of 3 years, based on institutional development plans. The National Development Plan for Higher Education, formulated by the Federal Ministry of Science, Research and Economics for a period of six years, sets national objectives that inform the universities' development plans. These goals include increasing the number of students in different disciplines, increasing the number of graduates, and improving student-staff ratios. The University Act (2002) also fixes a set of issues to be addressed in institutional plans, such as strategic goals, co-operation with other universities and knowledge transfers.
Finland	Quantitative indicators for education include: - the number of bachelor's, master's and PhD degrees awarded - the percentage of students awarded more than 55 study credits per academic year - the number of employed graduates. Research indicators include: - scientific publications - the percentage of competitive funding in the institution's total funding. Several indicators focus on the degree of internationalisation, including: - the number of international teaching and research personnel - the number of master's degrees awarded to foreign nationals - student mobility to and from Finland. Other education-and-science policy indicators include strategic development efforts, field-specific funding and contributions to "national duties" (e.g. teacher-training schools). A different formula applies to universities of applied science, with criteria focusing on education (79%), R&D (15%) and strategic development (6%).	A funding formula serves as a basis for each university to negotiate its performance agreement with the Ministry of Education and Culture (MEC) at the beginning of every four-year term. Each performance agreement contains specific institutional targets. Universities participate in the monitoring and evaluation process. The evaluation process also involves on-site visits by MEC staff. Performance reviews are conducted jointly by representatives from the MEC and individual institutions. To enable the MEC to monitor performance, HEIs provide information to a central statistical database maintained by the Ministry. An assessment of the performance of HEIs is published every year.
Scotland	Qualitative and quantitative criteria used in performance contracts include: - equality: admission targets for students from diverse backgrounds. - innovation: the number of research grants and contract income received; the share of income from the UK Competitive Research Council; and the use of innovation vouchers for specific science-to-business collaborations. - graduate employability: the number of first-degree qualifiers; the number of undergraduate entrants in science, technology, engineering and mathematics curricula; the development of an on-campus "employability and enterprise hub"; and the development of an employability award as part of an alumni mentoring programme.	Outcome agreements are made between the Scottish Funding Council and individual HEIs, and run for three years. These agreements also set annual targets for institutional priority areas. In 2014-15, four main priority areas were selected: equality (opportunity); innovation; graduate employability and enterprise; and sustainable institutions. Universities have defined quantitative indicators to help monitor their performance.

Source: De Boer, H. et al. (2015), "Performance-based funding and performance agreements in fourteen higher education systems, http://doc.utwente.nl/93619/7/jongbloed%20ea%20performance-based-funding-and-performance-agreements-in-fourteen-higher-education-systems.pdf.

Performance contracts are only one measure introduced over the past decade. Among other reforms, 9 (27%) of 33 countries introduced performance indicators in the formula for allocating university block grants.

Several countries have also strengthened their programmes for research excellence. In 2005, Germany established the "Excellence Initiative", a competition among German research universities for top-up funds from the Federal Government to make German universities more competitive internationally. In 2007, three "excellence universities" were selected, based on criteria of research excellence. Each university received USD 26 million (US dollars) (EUR 21 million [euros]) annually. Another 18 universities received funding to establish international graduate schools and "excellence clusters", i.e. research hubs bringing together different research groups from within and across universities in the region. The competition's second round in 2012 expanded funding to 11 elite universities. In 2018, the Initiative for Excellence was renamed the "Excellence Strategy", providing support only for created excellence clusters and the selected excellence universities.

Autonomy of HEIs and PRIs

Institutional autonomy is an important, much-discussed issue in the governance of public research. Institutional autonomy allows HEIs and PRIs to decide for themselves how best to meet the objectives set in national STI strategies and to select the most relevant funding criteria for their specific contexts. This can be useful, for example, when considering the commercialisation of public research, since their opportunities to collaborate with industry differ according to the type of research conducted, the relations with industry, their local economic context, etc.

Reforms implemented over the past decades have increased the autonomy of HEIs. In many OECD countries, HEIs can take their own decisions regarding industry relations, budget allocation, recruitment and promoting researchers (Figure 9.5). In 29 (85%) of 34 OECD countries, HEIs are free to create legal entities, such as technology offices and spinoffs, and decide on the conditions for collaborating with industry. In many cases, autonomy is the outcome of the reforms implemented over 2005-17. In France, for instance, HEIs have been free to establish their own for-profit entities and joint R&D ventures with industry since 2011 (Freedom and Responsibilities for Universities Act 2011). In Portugal, Law 62/2007 of 10 September 2007 on Higher Education Institutes (RJIES) granted some HEIs more autonomy.

HEIs do not have full autonomy, however, to decide salaries, which also depend on the funding sources and institutional conditions. HEIs can decide on the salaries of their academic staff in 12 (34%) of 35 OECD countries. In some countries (e.g. Denmark and France), national laws regulate salary bands for academic personnel; in other countries (e.g. Austria and the Netherlands), collective bargaining agreements are in place.

When it comes to internal budget allocation decisions, public HEIs in 23 (68%) of 34 OECD countries can decide on the share of institutional block funding to allocate to teaching, research and innovation activities. PRIs in 23 (79%) of 29 countries providing this information can freely decide their budget allocations.

Figure 9.5. Autonomy of HEIs across the OECD-34

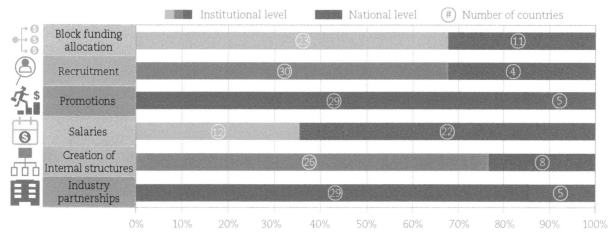

Note: The figure corresponds to questions 3.4.a of the OECD survey on the governance of research policy ("Who decides about allocations of institutional block funding for teaching, research and innovation activities in HEIs?"); 3.5.a ("Who decides about recruitment of academic staff in HEIs?"); 3.5.c ("Who decides about salaries of academic staff in HEIs?"); 3.5.e ("Who decides about reassignments and promotions of academic staff in HEIs?"); 3.6.a ("Who decides about the creation of academic departments, such as research centres in specific fields, and functional units, e.g. technology transfer offices in HEIs?"); and 3.6.c ("Who decides about the creation of legal entities and industry partnerships in HEIs?"). Information on HEI autonomy is missing for New Zealand.
Source: Data on university autonomy for Australia, Canada, Chile, Israel, Japan, Korea, Mexico, New Zealand and the United States, as well as data on the autonomy of PRIs, were collected by the authors (Borowiecki and Paunov, 2018). Data on university autonomy for Austria, Czech Republic, Denmark, Estonia, Finland, France, Germany (BB, Hesse and NRW), Greece, Hungary, Iceland, Ireland, Italy, Latvia, Lithuania, Luxembourg, the Netherlands, Norway, Poland, Portugal, the Slovak Republic, Spain, Sweden, Switzerland, Turkey and the United Kingdom are based on a survey conducted by the European University Association (EUA) between 2010 and 2011. The answers were provided by the secretaries general of national rectors' conferences and can be found in the EUA report (Estermann et al., 2015).

Stakeholder involvement in policy decision-making

The final structural dimension strongly shaping governance is how HEIs and PRIs themselves, as well as civil society (including citizens, labour unions, NGOs and foundations) and industry, participate in decision-making on research policy.

Stakeholder participation in research councils and university boards

The first important way for stakeholders to shape research and innovation policy is to participate in research and innovation councils (particularly those with strong policy mandates), which are in place in 31 of 35 OECD countries. Councils are permanent public bodies outside of ministries and agencies, which are mandated to engage in one or several of the following activities: provide policy advice (28 of the 31 countries, i.e. 90%); develop strategic priorities (23 countries, 74%); evaluate policy reforms (15 countries, 48%); co-ordinate within both government and non-public stakeholders (15 countries, 48%); and allocate research and innovation budgets (7 countries, 23%).

Stakeholders outside of government are often represented in research and innovation councils. Civil society (including members of labour unions and NGOs) is active in 15 (48%) of the existing 31 councils. Private-sector representatives – often large firms, but also some small and medium-sized enterprises (SMEs) – are present in 26 councils (84%).

Foreign experts participate in 6 (19%) of the 31 OECD countries with councils, i.e. Austria, France, Germany, Greece, Switzerland and the United Kingdom. Foreign experts come mostly from academia; a few (e.g. in Austria and the United Kingdom) come from industry or the public sector (Figure 9.6).

Figure 9.6. Who formally participates in the research and innovation council?

		SWE	LVA	FIN[1] Old council (86-16')	FIN[1] New council (since 16')	FRA	JPN	EST	KOR	AUS	ISL	MEX	SVK	TUR	DEU[2] Innovation Dialogue	DEU[2] Council of Science and Humanities	DEU[2] Expert Commission for Research and Innovation	AUT	SVN	LUX[3]	PRT[4] National Council for ST	PRT[4] National Council for Entrepreneurship and Innovation	CHL	ESP	ISR	POL	GRC	BEL[5] BEL-FED	BEL[5] BEL-BC	BEL[5] BEL-FL	BEL[5] BEL-WA	CAN[6]	GBR	HUN	CZE	CHE	NLD	DNK	USA	Share of countries with Councils (%)
Government representatives	Head of State/Prime Minister	x	x	x	x	x	x	x		x	x	x	x	x	x																										39% 12 of 31
	Ministers	x	x	x	x	x	x	x	x	x	x	x	x	x	x				x	x	x	x		x	x	x	x	x													68% 21 of 31
	Other national government officials		x	x	x	x		x	x			x	x				x						x	x	x				x			x	x	x							45% 14 of 31
	Funding agencies representatives			x	x				x			x	x	x									x	x								x	x								29% 9 of 31
	Local and regional government	x			x												x									x		x				x	x	x							26% 8 of 31
Stakeholders	HEI representatives	x	x	x	x	x	x	x	x	x	x	x	x	x	x	x	x	x	x	x	x	x	x	x		x		x	x	x		x	x	x	x	x	x	x	x	x	94% 29 of 31
	PRI representatives	x	x	x	x	x		x	x		x	x	x	x	x	x	x	x	x	x	x	x	x					x	x	x		x	x	x	x	x	x	x	x	x	77% 24 of 31
	Private sectors representatives	x	x	x	x	x	x	x	x	x	x	x	x	x	x		x		x	x	x	x	x					x	x	x		x	x	x	x			x	x	x	84% 26 of 31
	Civil society	x		x			x		x		x	x							x	x								x	x	x	x	x					x	x			48% 15 of 31
	Foreign experts			x													x	x	x										x			x	x								19% 6 of 31

← High-level policy representation (Head of State/Prime Minister, Ministers) → Stakeholders only

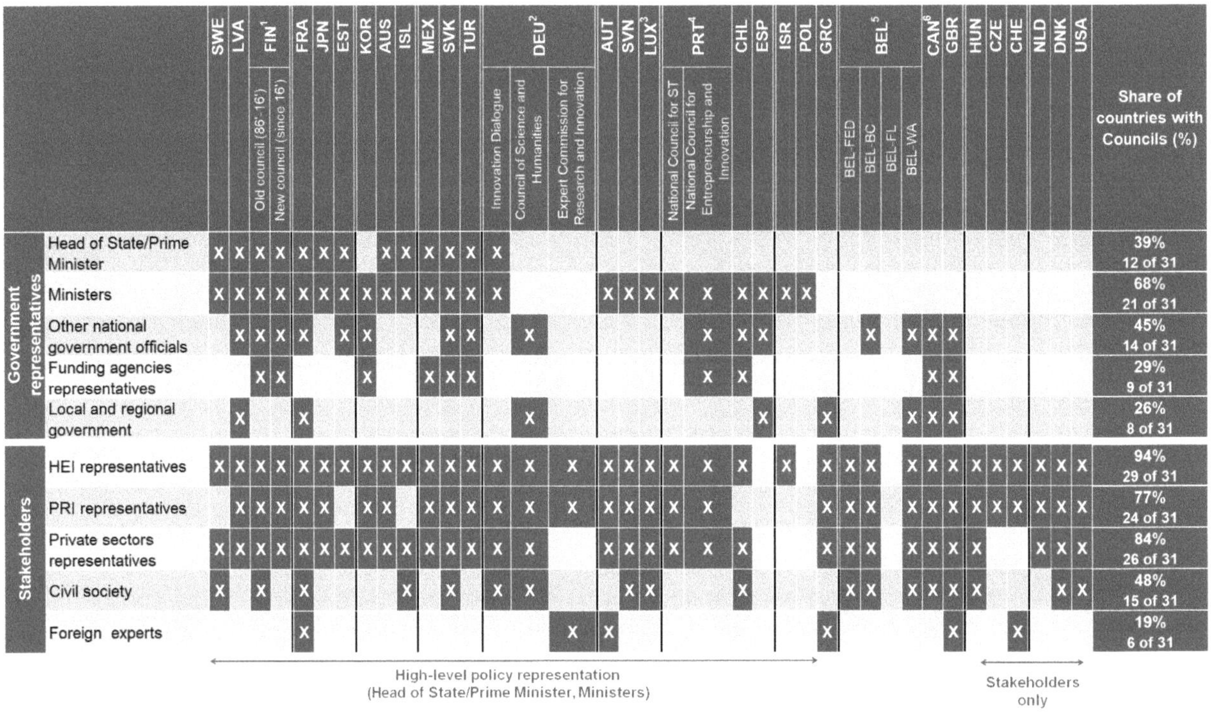

Note: This figure corresponds to question 2.3 of the OECD survey on the governance of research policy ("Who formally participates in the research and innovation council?"). Ireland, Italy, Norway and New Zealand do not have a research and innovation council. Percentages are expressed as a share of countries with a council in place (N=31).
1. The Finnish Research and Innovation Council was dissolved in 2016, and a new council was established under the same name in the same year. Owing to changes to the composition and mandate of the Council, this analysis treats them as two separate entities.
2. Germany has three main councils: the Council of Science and Humanities, the Expert Commission for Research and Innovation, and the Innovation Dialogue. Information provided by all three councils was used for the cross-country comparison. All three councils' mandates include policy co-ordination and policy advice. The mandates of the Council of Science and Humanities, and the Expert Commission for Research and Innovation, also include developing strategic priorities and policy evaluation.
3. In Luxembourg, the Superior Committee for Research and Innovation has not convened since 2014.
4. Portugal has two main councils: the National Council for Science and Technology, and the National Council for Entrepreneurship and Innovation. They have not convened since 2015.
5. Belgium has a federal council (Federal Science Policy Council), a council for the Brussels-Capital Region (Council of Science Policy), a council for the Flemish Community (Flemish Council for Science and Innovation), and a council for the Walloon Region (Science Policy Pole).The Federal Science Policy Council comprises experts from academia, the private sector and policy circles, who participate in their own capacity.
6. In Canada, the Minister of Science and Sport has announced that the Science, Technology and Innovation Council, which provided confidential advice to the government on issues related to science, technology and innovation policy, is being replaced by a new council that will be more open and transparent. Interpretation of the figure: the last row shows that in France, Austria, Greece, Switzerland, the United Kingdom and Germany, the Expert Council for Research and Innovation includes foreign experts.
Source: Borowiecki and Paunov (2018).

A second way for stakeholders from civil society and industry to shape the policy decisions of HEIs (particularly those with substantial autonomy) is to sit on their governing boards or councils. In most OECD countries, the university governance structure include a board (also known as a senate). The university board is the main decision-making body and is responsible for setting priorities. Stakeholder representation is important, in that it helps HEIs understand and answer public demands on their teaching and research activities.

University boards in 28 (82%) of 34 countries have outside stakeholder representation (Figure 9.7). In 25 (90%) of these 28 countries, the boards include private-sector representatives – mostly from large firms, but sometimes from SMEs. University boards in 23 (68%) of 34 countries include representatives from civil society – i.e. citizens, NGOs and foundations. In 21 (62%) of the 34 countries, the boards include representatives from both the private sector and civil society. In 10 (29%) of these countries, foreign experts sit on university boards. In 4 (12%) of these 34 countries, only the private sector is represented (Figure 9.7).

Figure 9.7. Who formally participates in public university boards?

	AUS	CHE	GBR	IRL	ISR	NZL	USA-MA	USA-CA	DNK	AUT	BEL-FL	CAN	ESP	FIN	ISL	NLD	NOR	PRT	SWE	POL	DEU-BW	DEU-NW	DEU-BB	FRA	HUN	JPN	SVN	SVK	ITA	KOR	GRC	CHL	CZE	LUX	LVA	MEX	TUR	Share of countries with boards
Private sector	x	x	x	x	x	x	x	x	x	x	x	x	x	x	x	x	x	x	x	x	x	x	x	x	x	x	x	x										74% 25 of 34
Civil society	x	x	x	x	x	x	x	x	x	x	x	x	x	x	x	x	x	x	x	x	x	x	x	x	x	x			x	x								68% 23 of 34
Foreign experts	x	x	x	x	x	x	x	x	x	x																			x		x							29% 10 of 34
No formal representation																																x	x	x	x	x	x	18% 6 of 34

← Private sector and civil society: 59% →   ← No external stakeholder representation: 18% →

Note: This figure corresponds to question 3.1 of the OECD survey on the governance of research policy ("Do stakeholders participate as formal members in governing boards of HEIs?"). There is no formal stakeholder participation in HEI boards in Chile, the Czech Republic, Luxembourg, Latvia, Mexico or Turkey. Information on participation in university boards is missing for Estonia. Percentages are expressed as a share of countries with information on the composition of HEI boards (N=34). Interpretation of the figure: the first row shows that in all countries except Italy, Korea, Greece, Chile, the Czech Republic, Luxembourg, Latvia, Mexico and Turkey, the private sector is represented in HEI boards.
Source: Borowiecki and Paunov (2018).

In some countries, external stakeholder representation on university boards is fairly new. In Portugal, for instance, university reforms introduced stakeholder representation on the governing boards of HEIs in 2007. In France, the Law on Higher Education and Research introduced the representation of business and local actors in the governing bodies of HEIs and PRIs in 2013.

New forms of stakeholder involvement

Online public consultations are a new policy instrument, devised to include civil society more fully in policy formulation (see Chapter 10 on technology governance). Online platforms were used to develop national STI plans in Australia (National Research Infrastructure Roadmap 2016); Canada (Innovation and Skills Plan 2017); France (National Research Strategy 2015); Japan (Fifth Science and Technology Basic Plan 2016-20);

Mexico (National Development Plan 2013-18); and the Netherlands (Dutch National Research Agenda, 2016). In 2016, the UK Department for Business, Energy and Industrial Strategy issued an online consultation to prepare for the National Innovation Strategy. In 2017, Finland introduced an online consultation to develop the national Vision for Higher Education and Research 2030, along with a roadmap.

Other more traditional, yet still important stakeholder-investment methods include working groups, roundtables and calls for inputs. Like online consultations, these temporary methods allow broader consultation and sectoral targeting. For example, the Scientific and Technological Research Council of Turkey (TÜBITAK) conducted an open-ended survey to identify priorities in the biomedical technology sector, gathering over 1 200 ideas from 300 researchers and experts. Technology roadmaps and policy programmes were developed based on these inputs (OECD, 2016a). New and established mechanisms were used jointly to engage stakeholders in the development of a 'Made-in-Canada' Athena SWAN programme. It will be aimed at supporting the careers of under-represented groups, including women, Indigenous peoples, members of visible minorities, and persons with disabilities, across all disciplines in higher education and research. Similarly, the Estonian Ministry of Education and Research formed a strategy preparation committee, convening over 200 specialists from research, business (including entrepreneurs) and government, to help prepare the Estonian Research and Development and Innovation Strategy 2014-20, "Knowledge-based Estonia". These exercises are flexible instruments that engage stakeholders in policy making, complementing the more permanent consultations already in place.

Future outlook

This chapter described some of the characteristics of public research policy across OECD member countries and recent trends in its organisation. The evidence shows OECD countries use formal instruments to evaluate the performance and contributions of HEIs and PRIs to achieving national STI priorities. Specialised agencies in charge of evaluating and monitoring the performance of HEIs and PRIs are an important component, together with strong stakeholder involvement in the policy process governing the publicly funded research conducted in these institutions. Reforms implemented over the past decades have increased the autonomy of HEIs and PRIs, allowing them to take their own decisions regarding industry relations, budget allocation and recruitment, and promotions.

Based on the trends evidenced by the data, the following four factors are expected to shape the future organisation of public research policy:

- National STI strategies will increasingly solicit the contributions of HEIs and PRIs to achieve a wider set of socio-economic objectives, including technology development (e.g. digital technologies) and the societal priorities described in the Sustainable Development Goals. National STI strategies are also likely to go beyond traditional R&D intensity targets, with new objectives placing HEIs and PRIs at the core of policy attention. These will include raising funding for doctoral students, and securing job placements in industry for researchers and PhDs.

- With increased pressure on public budgets and demands to account for spending, OECD countries will likely further invest in and consolidate the evaluation and monitoring structures for HEIs and PRIs. Specialised agencies are already in place in many OECD countries. New forms of evaluation, exploiting big-data analysis and digital platforms, will play an important role in these efforts (see discussion in Chapter 12).

- Efforts to expand and enhance multi-stakeholder consultations will greatly contribute to organising public research policy and identifying societal needs. National research and innovation councils, which provide platforms for engaging with civil society and industry, are already part of the standard national policy toolkit. University outreach already takes the form of stakeholder engagement in university boards, and linkages between HEIs and PRIs and wider society will grow stronger. Online consultations soliciting input from the population at large will likely expand further. The use of big-data and semantic-analysis tools will also increase, making it possible to process unstructured stakeholder inputs.[3]

- HEIs and PRIs will become more autonomous. This will afford them more opportunities to decide how they can best meet the objectives of national STI strategies, likely resulting in a diversification of approaches. More autonomy also means that the contributions of HEIs and PRIs to national STI strategies will increasingly depend on the amounts and modalities of the public funding contracts established between them and their government.

Notes

[1] The resulting database is publicly available at https://stip.oecd.org/resgov.

[2] There are regional agencies in place in Wallonia (Belgium), and in Massachusetts and California (United States), while there is a regional Ministry in charge of evaluations of HEIs and PRIs in Flanders (Belgium).

[3] For a discussion of the potential of semantic analysis for innovation policy analysis, see: www.innovationpolicyplatform.org/semantics.

References

Aghion, P. et al. (2010), "The governance and performance of universities: Evidence from Europe and the US", *Economic Policy*, Vol. 25, pp. 7-59, Oxford University Press, Oxford, http://www.jstor.org/stable/40603192.

Borowiecki, M. and C. Paunov (2018), "How is research policy across the OECD organised? Insights from a new policy database", *OECD Science, Technology and Industry Working Papers*, No. 2018/55, OECD Publishing, Paris, https://doi.org/10.1787/235c9806-en.

Breznitz, D. (2007), Innovation and the State, Yale University Press, New Haven, CT.

De Boer, H. et al. (2015), "Performance-based Funding and Performance Agreements in Fourteen Higher Education Systems", Centre for Higher Education Policy Studies, No. C15HdB014I, Enschede, Netherlands, http://doc.utwente.nl/93619/7/jongbloed%20ea%20performance-based-funding-and-performance-agreements-in-fourteen-higher-education-systems.pdf (accessed on 18 October 2016).

Estermann, T., T. Nokkala and M. Steinel (2015), "University Autonomy in Europe II – The Scorecard", European University Association, Brussels, https://eua.eu/resources/publications/401:university-autonomy-in-europe-ii-the-scorecard.html (accessed on 19 September 2016).

University of Vienna (2015), Performance agreement of the University of Vienna (Leistungsvereinbarung 2016–2018), website (German), https://www.univie.ac.at/mtbl02/02_pdf/20151222.pdf (accessed 22 November 2016).

Chapter 10. Technology governance and the innovation process

By

David E. Winickoff and Sebastian M. Pfotenhauer

Innovation reaps major benefits for economies, but some emerging technologies carry public concerns and risks. However, governing and steering emerging technologies to achieve good outcomes, while important, remains difficult. This chapter first examines how governance of emerging technologies should be recast from post-hoc regulation to approaches that engage the process of innovation itself. It then successively discusses three policy instruments that show promise as a means of addressing societal goals, concerns and values during the innovation process: participatory agenda-setting, co-creation (e.g. in the form of test beds), and value-based design and standardisation. The final section draws the main policy implications of adopting a more upstream approach to technology governance.

Embedding governance in innovation processes

Technological innovation is a major engine of productivity, economic growth and well-being. Its development is shaped by a mix of market, social and political forces. In many parts of the world, people live longer, healthier, and more comfortable lives because of the fruits of innovation. Governments around the globe seek to stimulate innovative activity through the orchestration of innovation systems and the setting of appropriate regulatory frameworks that engage market dynamics as well as the diversity of innovation needs and forms (OECD, 2010a).

While essential for addressing some of society's most pressing challenges, innovation can also have negative consequences for individuals and societies, as witnessed in previous waves of industrial revolution or in current debates around digitization, data privacy, and artificial intelligence. Indeed, the profound and ambiguous societal implications of emerging technologies bring them to the forefront of popular media and political debate. Blockchain technology promises a revolution in business models and transaction transparency, but also calls into question decades' worth of global regulation of financial markets (Berryhill et al., 2018). Autonomous vehicles carry enormous potential, but early experiments also highlight the dangers of their use in real-world environments (ITF, 2015). Digital platforms like Uber or Airbnb have begun to revolutionise entire service sectors, but have also raised concerns about new inequalities, and have occasionally been met with fierce resistance (OECD, 2016a). New developments in bioengineering, including gene editing and do-it-yourself biology kits, have recently triggered a series of global discussions about the future, and a potential ban on CRISPR-Cas9 and other gene-editing technologies (Garden and Winickoff, 2018a). Preventing, correcting or mitigating such potential negative effects while still allowing for entrepreneurial activity to flourish and reaping the benefits of innovation is a key challenge facing policy makers today.

Appropriate governance of emerging technologies is hence the proper task of governments because of the former's capacity to alter – and potentially disrupt – existing social orders, often in uncertain ways. Governing innovation in such ways as to limit potentially negative effects in innovation represents a complementary function of governments in well-functioning innovation systems, in addition to correcting for market, systems, and institutional failures (OECD, 2010; Bozeman, 2002; Smits and Kuhlmann 2004). It balances private sector interests and market dynamics with public good consideration and democratic legitimacy. This task has become more important, yet more difficult, as technology itself has become more complex, pervasive, and convergent. Some have argued that recent development around digital technologies – and their convergence with biological and other material systems – may mark a turning point for reconsidering the role of technology governance (The Economist, 2016; Marchant and Wallach, 2017).

The private sector, too, is increasingly voicing governance concerns. On 10 April 2018, Mark Zuckerberg, CEO of Facebook, the largest social network in the world and one of its most powerful corporations, was questioned before the United States Congress about failures in data protection, the right to privacy and pernicious uses in election meddling. Throughout the hearing, lawmakers raised a wide array of questions on the relationship between innovation and democracy, corporate responsibility in preserving core constitutional values and the disproportionate power of quasi-monopolies in the digital sphere. As Zuckerberg stated in his response, "My position is not that there should be no regulation. [..] I think the real question, as the Internet becomes more important in people's lives, is what is the right regulation, not whether there should be or not" (CBC, 2018). Recently, Microsoft President Bradford Smith has echoed these sentiments for the case of facial recognition software, arguing that "We live in a nation of laws, and the government needs to play an important role in regulating facial recognition technology" (Singer, 2018).

These episodes reflect a broader pattern of unease with the power of technology – and its creators – over our lives. They highlight the seemingly unregulated spaces in which innovative companies like Facebook grow from small start-ups to global giants, as well as the difficulties experienced by policy makers in formulating the right questions – let alone exerting appropriate oversight – in a rapidly changing technological landscape. The perception is growing across the public and private sectors that the future of work, democracy and other aspects of social order will require new forms of governance allowing policy makers to respond to technological change in real time (OECD, 2018).

Box 10.1. Definition of technology governance

Building on previous OECD work, technology governance can be defined as the process of exercising political, economic and administrative authority in the development, diffusion and operation of technology in societies (OECD, 2006; Kaufmann and Kraay, 2007; Carraz, 2012). It can consist of norms (e.g. regulations, standards and customs), but can also be operationalised through physical and virtual architectures that manage risks and benefits. Technology governance pertains to formal government activities, but also to the activities of firms, civil society organisations and communities of practice. In its broadest sense, it represents the sum of the many ways in which individuals and organisations shape technology and how, conversely, technology shapes social order (The Commission on Global Governance, 1995; Greene, 2014).

Several recent trends – some governmental and some market-driven – in the governance of emerging technologies are taking an anticipatory approach. Three instruments in particular for "upstream" innovation governance – participatory agenda-setting, co-creation and test beds, and value-based design and standardisation – show promise as a means of addressing societal goals, concerns and values during the innovation process itself. These instruments tend to emphasise *anticipation*, *inclusiveness* and *directionality* as key ingredients for governance, which can help shape technological designs and trajectories without unduly constraining innovators. The following chapter discusses three promising instruments – participatory agenda-setting, co-creation (e.g. in the form of Test Beds), and value-based design and standardisation – to illustrate how process governance can help augment innovation processes to respond to public and policy concerns.

Reframing governance as integral to the innovation process

The governance of emerging technologies poses a well-known puzzle: the so-called Collingridge dilemma holds that early in the innovation process – when interventions and course corrections might still prove easy and cheap – the full consequences of the technology – and hence the need for change – might not be fully apparent (Collingridge, 1980). Furthermore, early interventions can unduly limit technological options before they are adequately explored.

Conversely, when the need for intervention becomes apparent, changing course may become expensive, difficult and time-consuming. Society and developers may have already made substantial investments in adopting a technology, and set in motion certain path dependencies. Uncertainty and lock-ins are at the heart of many governance debates

(Arthur, 1989; David, 2001), and continue to pose questions about "opening up" and "closing down" development trajectories (Stirling, 2008).

In such conditions of uncertainty, traditional regulatory instruments – e.g. risk assessment, product-based standard-setting, export controls and liability – tend to narrowly focus on immediate or readily quantifiable consequences and their management, or enter only after key decisions about technology design have been made. Yet, many of the issues raised by currently emerging technologies are more fundamental and long-term. For example, current developments in artificial intelligence (AI) research might be subjected to rigid classification, performance standards, estimates of economic gains and losses, and export controls; however, the long-term societal and economic implications for populations, health systems, business and society cannot be predicted with any certainty. Similar patterns can be seen in the field of neurotechnology, where embedded devices and brain-computer interfaces are subjected to existing safety and efficacy regimes, but these regimes may not address long-term ethical questions about human agency and mental privacy (OECD, 2017b; Garden and Winickoff, 2018b).

Several emerging approaches in science policy seek to overcome the Collingridge dilemma by engaging concerns with technology governance "upstream". Process governance shifts the locus from managing the risks of technological products to managing the innovation process itself: who, when, what and how. It aims to anticipate concerns early on, address them through open and inclusive processes, and steer the innovation trajectory in a desirable direction. The key idea is that making the innovation process more anticipatory, inclusive and purposive (Figure 10.1) will inject public good considerations into innovation dynamics and ensure that social goals, values and concerns are integrated as they unfold. By locating governance discussions within the vanguard of innovations, it also ensures that policy makers are not be taken by surprise.

Figure 10.1. Three imperatives of a process-based approach to governance

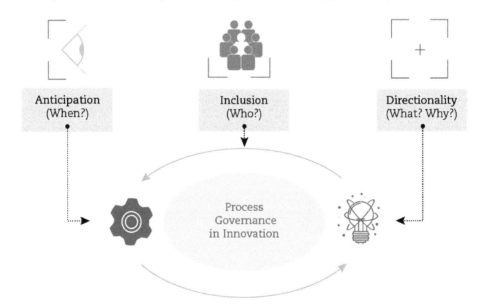

Characteristics of process governance: Anticipatory/upstream

Predicting the path of new technologies is notoriously difficult, whether the context is government regulation, venture capital or academic research. Anticipation – e.g. in the form of structured foresight and informed planning – is a key concern in many policy circles and boardrooms around the globe. From an innovation perspective, end-of-pipe-approaches can be inflexible, inadequate and even stifling (World Economic Forum, 2018). But can anticipation be a systematic component of innovation governance? How can that be done?

Recently, a range of anticipatory and upstream approaches have emerged that may help explore, deliberate and steer the consequences of innovation at an early stage (Box 10.2; Wilsdon and Willis, 2004). They allow for responding to public concerns or changing circumstances along the development trajectory. From an industry perspective, upstream approaches can incorporate public values and concerns, potentially mitigating potential public backlash against technology (see section 9 on design ethics). In OECD countries, frameworks for upstream governance have entered policy debates, e.g. in the context of the "Anticipatory Governance" pillar within the U.S. Nanotechnology Initiative (OECD, 2012). Likewise, under the major EU research-funding programme, Horizon 2020, the Responsible Research and Innovation (RRI) pillar has attempted to mainstream this approach across all research activities, echoed by recent developments in the United States (Box 10.3). Anticipatory governance also mitigates hubristic tendencies in risk management that one can estimate complex risks and guard against failure with authoritative certainty (Jasanoff, 2003, Pfotenhauer et al., 2012).

Box 10.2. In my view: Professor David Guston on "anticipatory governance"

David Guston, Foundation Professor and Director, School for the Future of Innovation in Society, Arizona State University, USA.

The idea of anticipatory governance (Barben et al., 2008; Guston, 2014) is to provide an opportunity to work as productively and pragmatically as possible within the confines of the so-called Collingridge dilemma. To do so, it envisions building three capacities: anticipation, or foresight; integration across disciplines; and public engagement. Building these capacities, both in traditional innovation organisations (like universities and private firms), as well as across society more broadly (in non-governmental organisations and public education), can help create a reflexive approach to innovation that will constantly be re-examining its public purpose and its ability to facilitate responsible changes in society.

Anticipatory governance recognises that at least two changes from current thinking are crucial. One is that governance is not just something that happens in governing institutions like legislatures, courts and regulatory agencies, but that it also happens through the interaction of users with new technologies and through the creative choices that researchers make in laboratories. This "jurisdictional" change means that the bounds of expertise must be expanded from traditional modes, bringing experts in governance into conversation with lab researchers and bringing lay citizens into the conversation altogether.

Two is that anticipation is not about predicting a future state of an innovation, but rather, it is about asking questions about plausible futures so that we may act in the present to help bring about the kind of futures we decide we want. This "temporal" change means that people from many different backgrounds need to work together to imagine futures and begin to build pathways towards them in the present. Neither of these changes resolves the Collingridge dilemma, but together, they give us the best hope of living within it.

Characteristics of process governance: Inclusive/democratic

Publics are normally assigned a passive role in the innovation process, i.e. as end-of-pipe consumers and with a view towards eliciting technology acceptance. This approach has been shown to backfire, e.g. in biotechnology (Irwin, 2001). The benefits of engaging citizens, publics, and systematically excluded actors in policy processes through well-designed exercises, deliberative hearings, panels and comment periods are well-known. Yet, in the domains of science and innovation policy – and particularly in the governance of emerging technologies – these benefits have received much less attention (Jasanoff, 2003; OECD, 2012).

Decades of science and technology studies have shown how value-based choices occur throughout the different phases of the innovation process (Bijker et al., 1987). In shaping what we know, where we go, and how we live and interact, technologies act as a kind of invisible and durable "legislation", as observed by the scholar Langdon Winner (1980) many years ago. If technology designs have "law-like" social consequences, however, then they require some form of democratic accountability. Hence, innovation systems should promote opportunities for public deliberation and participation on the values emerging technologies incorporate, and provide citizens with effective opportunities for appraising and shaping technology pathways (Bozeman and Sarewitz, 2005; Schot and Steinmueller, 2018).

Greater emphasis on public engagement and process inclusivity can therefore help align science and technology with societal goals and needs, a major goal of the Responsible Research and Innovation (RRI) agenda in Europe and elsewhere (Stilgoe et al., 2013; Box 10.3).

Box 10.3. Definitions of RRI in countries of the European Union

- "Responsible innovation means taking care of the future through collective stewardship of science and innovation in the present" (Stilgoe et al., 2013).
- "RRI is a transparent, interactive process by which societal actors and innovators become mutually responsive to each other with a view to the (ethical) acceptability, sustainability and societal desirability of the innovation process and its marketable products (in order to allow proper embedding of scientific and technological advances in our society)" (von Schomberg, 2013).
- "RRI is as much a movement to foster practices and cultures among those engaged in supporting and pursuing innovation, as a concern with appropriate regulatory and governance structures. The engagement of publics in determining what the desirable ends of research are, and how innovation processes can achieve these, is also often seen as a crucial part of responsible practice" (Nuffield Council on Bioethics, 2013).
- "RRI is the ongoing process of aligning research and innovation to the values, needs and expectations of society" (European Commission, 2014).

This emphasis goes beyond the widely acknowledged benefits (and biases) of open or user-led innovation, such as pooling external expert knowledge or collective creativity (von Hippel 2006; Chesbrough 2005). It adds an element of democratic legitimacy to innovation while gauging public concerns and adjusting trajectories accordingly such as to avoid potential backlash (OECD, 2012). By making innovation processes more inclusive and democratic, innovation can provide better opportunities to members of disadvantaged groups, improve the positive impacts of technology for a wider range of actors, and enhance democratic participation in shaping sociotechnical futures.

While the rationale for these engagement mechanisms are increasingly well accepted, their mainstream implementation remains challenging. Who gets to participate how, when, and why in the innovation process? Whose interests predominate? Is the input of expert lead-users more valuable than that of lay citizens? When does public engagement lead to improvement, and when does it begin to hamper innovative activity? Answers to these questions are difficult and highly context-dependent. However, growing experience and literature exist on these questions, and good models can be found across OECD countries (OECD, 2017a, Ch.8). One pathway is to unlock the potential of more open and collaborative forms of innovation through "co-creation" processes, for example in the interaction of disease groups, academic researchers and pharmaceutical companies to develop the next generation of health therapies (Winickoff et al., 2016). This form of inclusion can also enhance the relationship between science and society by building a more scientifically literate, supportive and engaged citizenry.

Characteristics of process governance: Purposive/directional orientation

Commitments to mission-driven versus bottom-up research ebb and flow, and debates about the respective merits and demerits continue apace. In some OECD countries, directionality or "mission orientation" has returned to centre stage (Mazzucato, 2018; OECD, 2016). The challenge of the misalignment between research, commercialisation and societal needs is not new (e.g. in the case of orphan drugs). However, present calls for "directed" and "purposive" transformative innovation display a new level of urgency to better connect innovation to "grand societal challenges" (e.g. the Sustainable Development Goals [SDGs]) (Carraz, 2012; Kuhlmann and Rip, 2014; Schot and Steinmueller, 2016) and respond to the particular needs of emerging economies (Kuhlmann and Ordóñez-Matamoros, 2017).

Mazzucato (2018) suggests that by "harnessing the directionality of innovation, we also harness the power of research and innovation to achieve wider social and policy aims as well as economic goals. Therefore, we can have innovation-led growth that is also more sustainable and equitable." This might point to a stronger role for both the government and the public in defining the goals of innovation and monitoring progress in achieving them. At the same time, mission driven approaches must continue to allow relatively unfettered entrepreneurial activity and provide sufficient market incentives, which points to the challenge of finding the right balance between top-down and bottom-up processes.

Three instruments for process governance in innovation

The three above-mentioned imperatives for an upstream and inclusive approach to technology governance are driving science policies across the public and private sectors, targeting all stages of technology development. A growing number of examples illustrate how innovation should not shy away from societal debates about technological futures: rather, it can actively harness them to improve innovation processes and outcomes.

The following section discusses three instruments of innovation-process governance: 1) participatory agenda-setting for mission-oriented research; 2) co-creation (e.g. in the form of test beds); and 3) design ethics and standardisation phases. All three reflect the dimensions discussed above – anticipation, inclusion and directionality – yet deploy them at different stages and in different ways throughout the innovation process (Table 10.1).

Table 10.1. Process governance in three policy instruments

Imperatives of process governance

	Anticipation	Inclusion	Directionality
Participatory agenda-setting in mission-driven innovation	Anticipate social needs and align innovation by feeding ideas and expectations by the public into new research and development (R&D) initiatives.	Include citizens alongside technical experts, policy makers and companies in bottom-up processes to define R&D priorities.	Clearly articulate the purposes and goals of R&D policies and funding to achieve the desirable sociotechnical outcomes.
Co-creation (e.g. in the form of test beds)	Anticipate potential technical, governance, and public opinion challenges through testing under real-world conditions.	Include users and the other members of the public through open innovation processes at various scales.	Include real-time feedback on desirability and enable small-scale demonstration before broader roll-out in test beds.
Design and standardisation	Design phase interventions to make transparent and promote social values.	Devise multi-stakeholder models to balance expert-driven design.	Articulate social values and goals and integrate them with technology.

Participatory agenda-setting and mission-driven innovation

Science and innovation policy have long wrestled with the question of steerability of technological progress and the role of government in innovation. Traditionally, innovation policy has embraced markets for allocating resources to meet individual and collective demands and a limited role for government interference where market failures or distortions exist. This view has been repeatedly challenged by pushes for mission-driven or sector-specific science and technology policies (Stokes, 1997) – a position reflected in recent discussions on innovation's role in the addressing "grand societal challenges (Kuhlmann and Rip, 2014). This tension can be traced back to Vannevar Bush's post-war science policy manifesto, *Science, the endless frontier*, in which he observed that "science is the proper concern of the government" because it can be mobilised to address important societal challenges, while at the same time warning against overt "government controls" beyond what could be called a hands-off funder-facilitator role (Bush, 1945; Stokes, 1997; Pfotenhauer and Juhl 2017).

Growing concerns about how to best mobilise innovation for the public good and overcome the apparent lack of bold progress have led to calls for a new era of mission-driven research. Scholars like Mariana Mazzucato (2013) have evoked the era of large-scale mission-driven research after the Second World War ("going to the moon") to argue that governments should act "entrepreneurially" and "boldly lead the way with a clear and courageous vision," reaping the benefits of high-risk investments. The more proactive perspective of Mission-driven Innovation 2.0 reflects concerns that science and innovation do not sufficiently meet human needs and public expectations, which in turn affects their public acceptance.

From a governance perspective, then, a key question is who sets the mission, and within what processes? In contrast to previous attempts at mission-driven research, the current wave emphasises anticipatory and inclusive aspects. Today, governments tend to disfavour purely top-down agenda-setting which relies on elected officials, science advisers and other experts. Instead, they are using deliberative processes to better align innovation strategy and societal priorities. For example, the European Commission's Citizen and Multi-Actor Consultation on Horizon 2020 (CIMULACT) has distilled input from EU citizens in 30 countries into a list of 23 distinct research topics for Europe, partly reflected in the European Union's new Horizon 2020 (H2020) research agenda (Box 10.4).

Thus, participatory agenda-setting becomes an idea space and site for upstream governance that allows policy makers to define the very visions and missions driving innovation (OECD, 2017c). It asks what kinds of missions are worth embracing, and how can democratic processes be established to legitimise them? This approach does not consider political and social concerns as external to the innovation process, to be avoided and silenced, but as essential features of any emerging technology, to be explored and incorporated head-on (Pfotenhauer and Juhl, 2017). In this context, controversies can be harnessed as a strategic resource for innovation, enabling discussions about priorities and the distribution of social responsibilities.

> **Box 10.4. Deliberative agenda-setting: Two examples**
>
> In 2015, the EU-funded project CIMULACT engaged more than 1 000 citizens in 30 countries, along with various other actors, in redefining the European Research and Innovation agenda to make it more relevant and accountable to society. The project encouraged participants to formulate their visions for desirable sustainable futures, debate and develop them together with other actors, and transform them into recommendations for future research and innovation policies and topics. The CIMULACT consortium included 29 European members from organisations active in technology assessment, science dissemination, innovation, research and consulting, co-ordinated by the Danish Board of Technology Foundation. Among other things, CIMULACT identified 23 citizen-inspired research topics drawing on 179 "visions" and reflecting 26 distinct social needs, which have since been partly picked up by the European Commission when defining the H2020 research agenda for 2018-20. These citizen-based topics include greater dissemination and access to healthcare innovations; evolving food cultures in growing cities; and mobilising technology to ensure more balanced work-life models in future work models.
>
> In 2014, the Dutch Government began developing a new strategy for science, the National Research Agenda. To maximise support from different social groups, one of the pillars of the development process was public consultation using digital tools, wherein members of the public were invited to "ask a scientist a question". All residents of the Netherlands could submit questions on the website, and access explanations and key words. The questions were analysed and clustered into 248 groups; 3 conferences were organised to add relevant information and aggregate further some of the questions in these groups. A total of 900 people participated in the conferences, which were organised in disciplinary and multidisciplinary discussion groups over several rounds. A panel of experts further reduced the questions to 140. These questions were then linked to the priorities of different national research organisations and also divided into chapters of the final National Research Agenda: 1) Man, the environment and the economy; 2) the Individual and society; 3) Sickness and health; 4) Technology and society; and 5) Fundamentals of existence. The final research agenda described the linkages between the 140 clustered questions and themes from the H2020 programme. By the time the National Research Agenda was released, more than half of those who had submitted a question had received invitations to lectures, public meetings and online fora from a range of organisations.
>
> *Sources*: (OECD, 2017c; CIMULACT, 2017; de Graaf et al., 2017)

Co-creation and test beds

"Co-creation" has emerged as a widely desired key resource in current attempts to enhance innovation processes and outcomes. It is an umbrella term that captures a variety of activities where different innovation actors gather under a joint project to achieve a mutually beneficial outcome. Different disciplines have emphasised different aspects of co-creation, such as social robustness, responsibility, collective creativity, knowledge flows and better alignment of innovation with consumer needs. Co-creation already plays an important role in many current science and innovation strategies of OECD countries, e.g. in Japan's Fifth Science and Technology Basic Plan (Government of Japan, 2016). There, for instance, the Japanese Research Institute of Science and Technology (RISTEX) funds co-creation projects featuring collaborative and prospective technology assessment, and convenes multiple stakeholders around common societal problem formulations.[1]

Why can co-creation help improve the governance of emerging technologies? While innovation was long conceived as happening outside the public eye in secretive corporate R&D departments or created by genius inventors in a garage, the trend in recent years has been a consistent move towards more open, co-creative and responsive forms of innovation. For example, "maker spaces" and "fab labs" have sprung up across universities and municipalities, providing experimental and collaborative workspaces and expertise for young innovators, free of charge or for a small fee. The visible trend towards co-creation offers new resources for steering and governing innovation in the making.

Co-creation facilitates the identification of potential technical flaws and governance challenges through direct feedback from diverse actors, which extends the range of inputs beyond traditional experts or select users. It can also reveal potential public concerns through immediate testing under quasi real-world conditions. For example, if the intention is to build social robots for elderly or patient care in nursing homes or hospitals, then information from patients, relatives, nurses, doctors, insurers and facility managers, alongside scientists and engineers, will likely improve their design. It can be tailored to a specific social environment and enhance the acceptability of the technology.

A number of new co-creation instruments have recently emerged that are particularly promising for questions of technology governance. Prominent examples are test beds and living labs, designated spaces for innovation activity and experimental technology implementation. They aim to test and demonstrate new sociotechnical arrangements in a model environment, under real-world conditions (Box 10.5). Co-creation rationales are also increasingly shifting public procurement practices from a market-based to a governance rationale. With public procurement of innovation, the public sector can act as a co-creator by defining public challenges to be addressed through an innovative solution that is yet to be developed. The novelty is that the government purchases a solution that does not yet exist while simultaneously setting the social, ethical and regulatory conditions under which the innovation should operate. For example, in the European robotics consortium ECHORD++, public procurement of innovation was used to co-develop robotics technology involving firms, universities and municipalities to enhance sewer cleaning and hospital care.

Co-creation still poses challenges for researchers, companies and policy makers, including how to mainstream practices across sectors, regions and scales. The European research consortium Scaling up Co-creation: Avenues and Limits for Integrating Society in Science and Innovation (SCALINGS) is presently exploring ways to expand co-creation in 10 countries and 3 different sectors (robotics, urban energy and autonomous driving).

SCALINGS is both investigating the technical challenges of developing innovative technologies and the social challenges of embedding them in diverse governance regimes.[2]

> **Box 10.5. Test beds: Testing new governance modes for emerging technologies**
>
> Drawing on the popular "grand societal challenges" discourse and the growing insight that adequate policy responses to these challenges will require transformations of both technology and society, test beds (and related initiatives like living labs, real-world laboratories and demonstrators) are sites of collaborative invention, testing and demonstration for future technologies and sociotechnical arrangements in a model environment, under real-world conditions. These increasingly prominent types of co-creation practice are deployed across geographical regions and technical domains to foster innovation. (Engels, Wentland and Pfotenhauer, 2018).
>
> Test beds are particularly prominent in the area of energy transition, smart cities and mobility. For example, in September 2017, Canadian Prime Minister Justin Trudeau announced a partnership between Waterfront Toronto and Sidewalk Labs – a start-up under Google's parent company Alphabet – to turn Toronto's waterfront into "a proving ground for technology-enabled urban environments around the world" (Hook, 2017). The initiative aims to integrate self-driving shuttles, adaptive traffic lights, modular housing and freight-delivering robots, in line with a city commitment to "waive or exempt many existing regulations in areas like building codes, transportation, and energy in order to build the city it envisioned." Elsewhere, test beds for autonomous vehicles are flourishing, affecting rural roads, highways and cities alike. Test-bed projects for smart and sustainable cities, whether in South Korea (Songdo), China (Tianjin) or Abu Dhabi (Masdar City), are experimenting with ways to foster new forms of urbanity and innovation, frequently with the ambition of becoming a model for other cities.
>
> Test beds are providing new opportunities to tackle governance issues in innovation. They offer a glimpse at new sociotechnical arrangements in an "as-if" mode of tentative roll-out, identifying not only glitches in the technology, but also societal responses and governance challenges (Engels, Wentland and Pfotenhauer 2018). Test beds can serve as an instrument to co-develop the very rules and regulations needed to cope with new technologies, and to gauge which existing regulations might be detrimental to adoption. For example, the European Energy Forum in Berlin has re-purposed a historical gas-storage facility into a private research campus that develops and tests new forms of energy, mobility and information technology solutions, blending technology creation-and-use environments (Canzler et al., 2017). Here, building, traffic and infrastructural regulations are being experimented alongside tested technologies, with a view towards scaling them across Berlin and beyond. While public policy has primarily focused on lowering local regulatory barriers in test-bed settings, or blurring boundaries between public and private interests, this experimental approach to governance also provides new opportunities to deliberate new rules and regulations in real time in order to direct innovation towards desirable outcomes. It provides a counterpoint to the widespread notion that regulation is consistently unable to keep pace with innovation (Engels, Wentland and Pfotenhauer, 2018).

Design ethics and standardisation phases

Technology-based standards determine the specific characteristics (size, shape, design or functionality) of a product, process or production method. This form of governance can emanate from both the private sector (e.g. *de-facto* standards in the form of dominant designs) and the public sector (e.g. government regulated vehicle safety standards or mobile phone frequency bands).

Standards are critical for innovation: they define the conditions under which competition takes place, and act as a built-in infrastructure for technology uptake and use within supply chains, markets and society. From an economic perspective, they are desirable as vehicles of efficiency by ensuring interoperability, securing minimum safety and quality, reducing variety, and providing common information and measurement (OECD, 2011). On the other hand, they can also create barriers to entry, distort competition, and be prone to capture. They can also serve as useful vehicles of intellectual property rights (e.g. Blind, 2013), but they also carry the danger of reinforcing monopolistic power and incumbency (Swann, 2000; OECD, 2011).

From a governance perspective, standards are equally important because of their social and ethical implications. Standards "build in" certain norms, values, safeguards and goals into technologies and infrastructures (Bowker and Star, 2000; Busch, 2013; Timmermans and Epstein, 2010). For example, the lack of standardisation for genetic tests (e.g. on cancer risks) may create conflicting diagnoses about an individual's health and required course of action, with downstream effects on who might receive health insurance or be denied coverage because of a pre-existing condition (OECD, 2017b). Emission standards for combustion engines or factories affect public health and the environment, frequently with very unequal distributive effects. The dimensions of airplane seats refer to standardised body measurements, with consequences not only for individuals who do not conform to these measurements, but also for flight safety and economics. Once technological design is standardised – whether in material or code – it shapes human behaviour in a law-like manner and becomes increasingly hard to unseat over time (Lessig, 1999; Winner, 1980). Current technological convergences in production, transportation and energy systems elevate the political stakes of standardisation and integration (OECD, 2017a).

At the same time, careful consideration of product and process standards offers new inroads into the governance of emerging technologies. Recent efforts by technical and policy communities treat standardisation as a point of intervention to incorporate and make explicit certain ethical and political values into the material objects, networks and systems that they are designing.

In nanotechnology, standardisation is seen not just as a means of facilitating commerce through interoperability, but also of promoting health and safety. For example, the "Safety by Design" (SbD) approach seeks to integrate knowledge of potential adverse effects into the process of designing nanomaterials and nanoproducts, and to engineer these undesirable effects out of them (van de Poel and Robaey, 2017; Schwarz-Plaschg et al., 2017). Here, "SbD aims at an integrated and iterative process, where safety information on a certain material, substance or product is integrated from early research and development (R&D) phases onwards" (Gottardo et al. 2017). Drawing on concepts from the construction industry, the approach takes into consideration the projects' life cycle: construction, maintenance, decommissioning and disposal or recycling of waste material (Schulte et al., 2008). As a concept, SbD has been studied extensively in the European projects Nanoreg2[3] and Prosafe.[4]

In AI, concerns about the potential bias of algorithms, the lack of accountability of autonomous systems and potential irreversibility have also sparked debates about design standards. President Emmanuel Macron of France recently called for an anticipatory approach to governance that would "frame" AI appropriately at the design phase *(Thompson, 2018)*:

> *"Because at one point in time, if you don't frame these innovations from the start, a worst-case scenario will force you to deal with this debate down the line. I think privacy has been a hidden debate for a long time…Now, it emerged because of the Facebook issue. Security was also a hidden debate of autonomous driving. Now, because we've had this issue with Uber, it rises to the surface. So if you don't want to block innovation, it is better to frame it by design within ethical and philosophical boundaries."*

This call was later underscored by the Canada-France statement on Artificial Intelligence following the meeting of President Macron with Prime Minister Justin Trudeau of Canada, where both countries "emphasized the need to develop the capacity to anticipate impacts and coordinate efforts in order to encourage trust" (Government of Canada, 2018).

Notwithstanding these calls, questions remain about how and when such framing should take place, and who should undertake it. Numerous stakeholders, including companies such as Google,[5] have issued statements on ethical principles. The OECD, too, is developing recommendations on the ethics of making and using artificial intelligence. The "ethically aligned design" (EAD) Standards for Autonomous and Intelligent Systems, currently being developed by the Institute of Electrical and Electronics Engineers (IEEE), is another potential way forward.[6] EAD comprises more than 100 sets of recommendations (including standards on algorithmic bias; model process for addressing ethical concerns during system design; and transparency of autonomous systems), which can be utilised immediately by technologists, policy makers and academics. However, most of these standards remain a work in progress. Because the AI community aims to be much more inclusive than in typical standard-setting procedures, the working groups at IEEE have operated as fora for public discussion and debate, as much as for technical work.

With design ethics emerging as a potentially powerful tool for translating values into technology, the question arises about how that process is itself governed. A wide array of governance models exists, from purely private standard-setting to mixed public-private fora, like the International Organisation for Standardisation (e.g. Winickoff and Mondou, 2017). Such bodies can be slow and rigid; they also differ widely in how they develop standards and integrate input from diverse stakeholders. Single countries can sometimes dominate standard-setting processes to press technological advantages.

Relevant communities of engineering practice are in a good position to think creatively about finding and standardising technical solutions. However, different technical communities will bring different goals to the task, which may not necessarily align with others within democratic societies. This underscores the importance of inclusiveness and accountability in standard-setting as a key component of innovation: who sets the standards, within what process, and with what claims to legitimacy? In this sense, standard-setting can serve as a stage within the innovation process where more inclusive, purposive and anticipatory forms of governance can be developed.

Policy implications

Recent attention to the governance gaps in digital and other emerging technologies has revealed that traditional end-of-pipe instruments might be ineffective for addressing key issues in a timely manner. In OECD countries, both public and private-sector actors increasingly deploy governance instruments at earlier stages and as an integral part of the innovation process to steer emergent technologies towards better collective outcomes. Anticipation, inclusivity and directionality have emerged as important characteristics for adequate upstream governance of the innovation process. New approaches, such as participatory agenda-setting, co-creation and standardisation, embody these characteristics.

These aspects unfold differently for the three policy instruments discussed above, affecting different stages of the innovation process and shaping outcomes in different ways (Figure 10.2). Participatory agenda-setting draws on structured processes to identify collective needs and concerns, and translates them into research-funding and R&D activities. Co-creation affects R&D practices at various stages and scales, but proves particularly productive in more mature settings (e.g. test beds) where it enables real-time feedback and reveals the governance challenges of emerging technologies. Design ethics scrutinises how ethical and political values are built into technologies; they open up for debate the ways in which emergent technologies will affect society. Note that Figure 10.2 should not be read as a revival of the much-criticized linear model of innovation (Godin 2006, Balconi et al. 2010, Pfotenhauer & Juhl 2017). Rather, it is meant to indicate that process governance can be useful to various types of activity that contribute to innovation, no matter in which order they occur or whatever else might be involved.

Figure 10.2. Upstream governance in the innovation process in three instruments

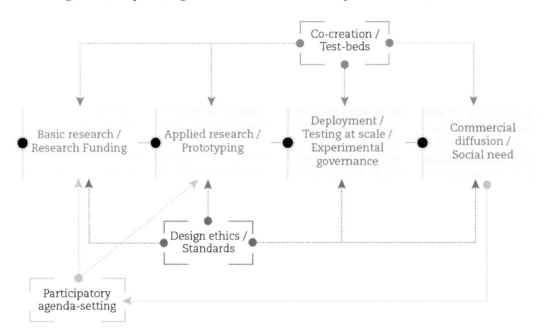

The undiminished pace of technological change suggests that the need for better upstream governance in innovation will continue to grow, partly to enable responsible diffusion downstream for technologies with uncertain consequences. Governments and businesses should seek to enhance their capacities for anticipatory, inclusive and purposive governance throughout the innovation process, and augment their individual capacities through adequate frameworks for transnational governance.

The previous analysis of three policy instruments indicates that governments can build "technology with and for society" in the following ways:

- continue to experiment with, and expand, participatory forms of foresight and agenda-setting, connecting them to funding organisations and national strategy

- bodies; integrate evaluation mechanisms in the design of new governance initiatives from the outset, to improve methods and approaches over time.

- foster opportunities for co-creation among diverse stakeholders for different regions, technologies and scales; exploit opportunities for co-developing new technologies and governance mechanisms, through the responsible use of platforms like test beds.

- use standard-setting to promote the public good and values; support standard-setting processes that function as public fora for democratic deliberation on the governance of emerging technologies, and avoid capture of these fora by narrow interests.

- acknowledge the diversity of innovation practices, needs and rationales across OECD countries, including culturally and politically specific ways of governing emerging technologies; foster co-ordinated international efforts to gather and analyse data and best practices on (upstream) process governance for emerging/converging technologies; build tools and indicators to assess innovation governance against the goals of anticipation, inclusivity and directionality (OECD, 2010).

- develop resources and guidelines for innovation-process governance at an international level; use the capacity for comparison of transnational organisations, such as the OECD or the European Union, to investigate the relative efficacy and context dependency of these process instruments.

Notes

[1] http://ristex.jst.go.jp/hite/en/index.html.

[2] www.scalings.eu.

[3] http://www.nanoreg2.eu.

[4] http://www.h2020-prosafe.eu.

[5] https://ai.google/principles.

[6] The IEEE is a major international association of engineers that produces authoritative technical standards in many fields: https://ethicsinaction.ieee.org.

References

Arthur, W.B. (1989), "Competing Technologies, Increasing Returns, and Lock-In by Historical Events", *The Economic Journal*, Vol. 99/394, pp. 116-131, Wiley, Hoboken, NJ, http://dx.doi.org/10.2307/2234208.

Balconi M., S. Brusoni and L. Orsenigo (2010), In defence of the linear model: An essay. *Research Policy*, Vol. 39/1, pp. 1–13, Elsevier, Amsterdam,. DOI: https://doi.org/10.1016/j.respol.2009.09.013

Barben, D. et al. (2008) "Anticipatory Governance of Nanotechnology: Foresight, Engagement, and Integration," in E. Hackett, O. Amsterdamska, M.E. Lynch, J. Wajcman (eds.), *The Handbook of Science and Technology Studies*, 3rd edition, pp. 979–1000, MIT Press, Cambridge, MA.

Berryhill, J., T. Bourgery and A. Hanson (2018), "Blockchains Unchained", *OECD Working Papers on Public Governance*, OECD Publishing, Paris, https://doi.org/10.1787/3c32c429-en.

Bijker, W., T.P. Hughes and T. Pinch (1987), *The Social Construction of Technological Systems: New Directions in the Sociology and History of Technology*, MIT Press, Cambridge, MA, https://bibliodarq.files.wordpress.com/2015/09/bijker-w-the-social-construction-of-technological-systems.pdf.

Blind, K. (2013), "The Impact of Standardization and Standards on Innovation", working paper, *Compendium of Evidence on Innovation Policy*, Manchester Institute of Innovation Research, Manchester, http://www.innovation-policy.org.uk/compendium/section/Default.aspx?topicid=30.

Bowker, G.C. and S.L. Star (2000), *Sorting Things Out: Classification and Its Consequences*, MIT Press, Cambridge, MA.

Bozeman, B. (2002), "Public value failure: When efficient markets may not do", *Public Administration Review*, Vol. 62/2, pp. 145-161, Wiley, Hoboken, NJ, https://doi.org/10.1111/0033-3352.00165

Bozeman, B. and D. Sarewitz (2005), "Public values and public failure in US science policy", *Science & Public Policy*, Vol. 32/2, pp. 119-136, Oxford University Press, Oxford.

Busch, L. (2013), *Standards: Recipes for Reality*, reprint edition, MIT Press, Cambridge, MA.

Bush, V. (1945), *Science, the endless frontier*, 2nd ed., A Report to the President by Vannevar Bush, Director of the Office of Scientific Research and Development, July 1945, United States Government Printing Office, Washington, DC, https://www.nsf.gov/about/history/nsf50/vbush1945.jsp.

Canzler, W. et al. (2017), "From "living lab" to strategic action field: Bringing together energy, mobility, and Information Technology in Germany", *Energy Research & Social Science*, Vol. 27, pp. 25 35, Elsevier, NY, https://doi.org/10.1016/j.erss.2017.02.003.

Carraz, R. (2012), "Improving Science, Technology and Innovation Governance to Meet Global Challenges", in *Meeting Global Challenges through Better Governance: International Co operation in Science, Technology, and Innovation*, pp. 173-205, OECD Publishing, Paris, http://dx.doi.org/10.1787/9789264178700-en.

CBC (2018), "Zuckerberg sees regulation of social media firms as 'inevitable'", *CBC News*, Toronto, 11 April 2018, https://www.cbc.ca/news/technology/facebook-zuckerberg-users-privacy-data-mining-house-hearings-1.4614174 (accessed 29 June 2018).

Chesbrough, H.W. (2005), Open Innovation: The New Imperative for Creating And Profiting from Technology, Harvard Business Review Press, Cambridge, MA.

CIMULACT (2017), "23 Citizen-Based Topics for Future EU Research", www.cimulact.eu, http://www.cimulact.eu/wp-content/uploads/2018/02/CIMULACT-Booklet-Final-compressed.pdf.

Collingridge, D. (1980), *The social control of technology*, Frances Pinter, London.

The Commission on Global Governance (1995), *Our Global Neighborhood*, Oxford University Press, Oxford.

David, P.A. (2001), "Path dependence, its critics and the quest for 'historical economics'", in P. Garrouste and S. Ioannides (eds.), *Evolution and path dependence in economic ideas: Past and present*, p. 40, Edward Elgar Publishing, Cheltenham, United Kingdom and Northampton, MA.

The Economist (2016), "Frankenstein's Paperclips", *The Economist*, London, https://www.economist.com/special-report/2016/06/25/frankensteins-paperclips.

Engels, F. Wentland and S.M. Pfotenhauer (forthcoming), "Testing the future? Developing a framework for test beds as an emerging innovation policy instrument".

European Commission (2014) *Rome Declaration on Responsible Research and Innovation in Europe*. https://ec.europa.eu/research/swafs/pdf/rome_declaration_RRI_final_21_November.pdf (accessed 1 June 2018), Rome.

Garden, H. and D.E. Winickoff (2018a), "Gene editing for advanced therapies", *OECD Science, Technology and Industry Working Papers*, OECD Publishing, Paris, https://doi.org/10.1787/8d39d84e-en.

Garden, H. and D.E. Winickoff (2018b), "Issues in neurotechnology governance", *OECD Science, Technology and Industry Working Papers*, OECD Publishing, Paris, https://doi.org/10.1787/c3256cc6-en.

Government of Canada (2018), "Canada-France Statement on Artificial Intelligence" (6 June 2018), http://international.gc.ca/world-monde/international_relations-relations_internationales/europe/2018-06-07-france_ai-ia_france.aspx?lang=eng

Government of Japan (2016), "Japan's Fifth Science and Technology Basic Plan" (provisional translation), 22 January 2016, Government of Japan, Tokyo, http://www8.cao.go.jp/cstp/english/basic/5thbasicplan.pdf.

de Graaf, B.D., A.R. Kan and H. Molenaar (eds.) (2017), *The Dutch National Research Agenda in Perspective: A Reflection on Research and Science Policy in Practice*, Amsterdam University Press, Amsterdam.

Green, J.F. (2014), *Rethinking Private Authority: Agents and Entrepreneurs in Global Environmental Governance*, Princeton University Press, Princeton, New Jersey.

Godin, B. (2006), "The Linear Model of Innovation: The Historical Construction of an Analytical Framework," *Science Technology & Human Values*, Vol. 31/6, pp. 639–667, Sage Journals, Thousand Oaks, CA, DOI: https://doi.org/10.1177/0162243906291865.

Gottardo et al. (2017), *NANoREG framework for the safety assessment of nanomaterials*, JRC Science Policy Report EUR 28550 EN, Ispra, doi 10.2760/245972.

Guston, D.H. (2014), "Understanding 'anticipatory governance," *Social Studies of Science*, Vol. 44/2, , pp. 218–42, Sage Journals, Thousand Oaks, CA, DOI: https://doi.org/10.1177/0306312713508669

von Hippel, E. (2006), *Democratizing Innovation*, MIT Press, Cambridge, MA.

Hook, L. (2017), "Alphabet to build futuristic city in Toronto", *Financial Times*, London, 17 October 2018, https://www.ft.com/content/5044ec1a-b35e-11e7-a398-73d59db9e399 (accessed 30 June 2018).

Irwin, A. (2001), "Constructing the Scientific Citizen: Science and Democracy in the Biosciences," *Public Understanding of Science*, Vol. 10/1, pp. 1–18, Sage Journals, Thousand Oaks, CA, DOI: https://doi.org/10.1088/0963-6625/10/1/301.

ITF (2015), "Automated and Autonomous Driving: Regulation under Uncertainty", *International Transport Forum Policy Papers*, No. 7, OECD Publishing, Paris, https://doi.org/10.1787/5jlwvzdfk640-en.

Jasanoff, S. (2003), "Technologies of Humility: Citizen Participation in Governing Science," *Minerva*, Vol. 41/3, pp. 223–244, Dordrecht,. DOI: https://doi.org/10.1023/A:1025557512320

Kaufmann D and Kraay A. (2007), *Governance Indicators: Where Are We, Where Should We Be Going?* The World Bank, Washington DC.

Kuhlmann, S. and G. Ordóñez-Matamoros (2017), "Governance of Innovation in emerging countries: Understanding failures and exploring options", in Kuhlmann, S. and G. Ordoñez-Matamoros (eds.),

Research Handbook on Innovation Governance for Emerging Economies: Towards Better Models, pp. 1-36, Edward Elgar Publishing, Northampton, MA.

Kuhlmann, S. and A. Rip (2014), "The challenge of addressing Grand Challenges... and what universities of technology can do", University of Twente, Enschede, Netherlands, https://ris.utwente.nl/ws/files/13268719/2014_Kuhlmann.pdf.

Lessig, L. (1999), *Code: And Other Laws of Cyberspace*, Basic Books, New York, NY.

Mazzucato, M. (2018), *Mission-Oriented Research & Innovation in the European Union: A problem-solving approach to fuel innovation-led growth*, European Commission, Brussels.

Mazzucato, M. (2013), *The Entrepreneurial State: Debunking Public vs. Private Sector Myths*, Anthem Press, London and New York.

Marchant, G.E. and W. Wallach (2016), *Emerging Technologies: Ethics, Law and Governance*, The Library of Essays on the Ethics of Emerging Technologies, Vol. 3, Ashgate, Farnham, United Kingdom.

Nuffield Council on Bioethics (2013), "Responsible research and innovation", in *Novel Technologies: Intervening in the Brain*, Nuffield Council on Bioethics, London, http://nuffieldbioethics.org/wp-content/uploads/2014/06/Novel_neurotechnologies_Chapter_6_Responsible_research_innovation.pdf (accessed 20 July 2018).

OECD (2018), *Going Digital in a Multilateral World*, OECD Publishing, Paris, https://www.oecd.org/going-digital/C-MIN-2018-6-EN.pdf.

OECD (2017a), *The Next Production Revolution: Implications for Governments and Business*, OECD Publishing, Paris, https://doi.org/10.1787/9789264271036-en.

OECD (2017b), "Neurotechnology and society: Strengthening responsible innovation in brain science", *OECD Science, Technology and Industry Policy Papers*, No. 46, OECD Publishing, Paris, https://doi.org/10.1787/f31e10ab-en.

OECD (2017c), "Open research agenda setting", *OECD Science, Technology and Industry Policy Papers*, No 50, OECD Publishing, Paris. https://doi.org/10.1787/74edb6a8-en.

OECD (2016), "Priorities, strategies and governance of innovation in Sweden", in *OECD Reviews of Innovation Policy: Sweden 2016*, OECD Publishing, Paris, https://doi.org/10.1787/9789264250000-9-en.

OECD (2012), "Planning guide for public engagement and outreach in nanotechnology", OECD, Paris, http://www.oecd.org/sti/emerging-tech/49961768.pdf.

OECD (2011), *Demand-side Innovation Policies*, OECD Publishing, Paris, https://doi.org/10.1787/9789264098886-en.

OECD (2010), *Measuring Innovation: A New Perspective*, OECD Publishing, Paris, https://doi.org/10.1787/9789264059474-en.

OECD (2006), "Applying strategic environmental assessment: Good practice guidance for development co-operation, OECD Publishing, Paris.

Pfotenhauer, S.M. and J. Juhl (2017), "Innovation and the political state: Beyond the myth of technologies and markets", in Godin, B. and D. Vinck (eds.), *Critical Studies of Innovation: Alternative Approaches to the Pro-Innovation Bias*, Edward Elgar Publishers, Cheltenham, United Kindom and Northampton, MA, pp. 68-94.

Pfotenhauer, S.M., et al.(2012), "Learning from Fukushima," Issues in Science and Technology, Vol. 28/3, pp. 79-84, Washington DC, http://issues.org/28-3/pfotenhauer/.

Schot, J. and W.E. Steinmueller (2018), "Three Frames for Innovation Policy: R&D, Systems of Innovation and Transformative Change," *Research Policy*, Vol. 47/9, pp. 1554–67, Elsevier, Amsterdam, DOI: https://doi.org/10.1016/j.respol.2018.08.011.

Schot, J. and E. Steinmueller (2016), "Framing Innovation Policy for Transformative Change: Innovation Policy 3.0.", SPRU Science Policy Research Unit, University of Sussex, Brighton.

Schulte, P.A. et al. (2008), "National Prevention through Design (PtD) Initiative", *Journal of Safety Research*, Vol. 39/2, pp. 115-121, Elsevier and National Safety Council, Itasca, IL, https://doi.org/10.1016/j.jsr.2008.02.021.

Schwarz-Plaschg, C., A. Kallhoff and I. Eisenberger (2017), "Making Nanomaterials Safer by Design?", *NanoEthics*, Springer Netherlands, Dordrecht, Vol. 11/3, pp. 277-281, https://doi.org/10.1007/s11569-017-0307-4.

Singer, N. (2018), "Microsoft Urges Congress to Regulate Use of Facial Recognition", *The New York Times*, New York, 14 July 2018, https://www.nytimes.com/2018/07/13/technology/microsoft-facial-recognition.html (accessed 19 July 2018).

Smits, R. and S. Kuhlmann (2004), "The Rise of Systemic Instruments in Innovation Policy,". *International Journal of Foresight and Innovation Policy*, Vol 1/1, pp. 4–32, Inderscience Publishers, Geneva, DOI: 10.1504/IJFIP.2004.004621.

Stilgoe J, R. Owen and P. Macnaghten (2013), "Developing a Framework for Responsible Innovation," *Research Policy*, Vol. 42/9, pp. 1568–80, Elsevier, Amsterdam, DOI: https://doi.org/10.1016/j.respol.2013.05.008.

Stirling, A (2008), "'Opening Up' and 'Closing Down': Power, Participation, and Pluralism in the Social Appraisal of Technology", *Science, Technology & Human Values*, Vol. 33/2, pp. 262-294, Sage Journals, Thousand Oaks, CA, https://doi.org/10.1177/0162243907311265.

Stokes, D.E. (1997), *Pasteurs Quadrant: Basic Science and Technological Innovation*, Brookings Institution Press, Washington, DC.

Swann, P.G.M. (2000), "The Economics of Standardisation", Final Report for Standards and Technical Regulations Directorate, Department of Trade and Industry, Manchester Business School, University of Manchester.

Thompson, T. (2018), "Emmanuel Macron Talks to Wired about France's AI Strategy", *Wired*, 3 March 2018, https://www.wired.com/story/emmanuel-macron-talks-to-wired-about-frances-ai-strategy/.

Timmermans, S. and S. Epstein (2010), "A World of Standards but not a Standard World: Toward a Sociology of Standards and Standardization", *Annual Review of Sociology*, Vol. 36/1, pp. 69-89, Annual Reviews, Palo Alto, CA, https://doi.org/10.1146/annurev.soc.012809.102629.

van de Poel, I. and Z. Robaey (2017), "Safe-by-Design: from Safety to Responsibility", *NanoEthics*, Vol. 11/3, pp. 297 306, Springer Netherlands, Dordrecht, https://doi.org/10.1007/s11569-017-0301-x.

von Schomberg, R. (2013), "A Vision of Responsible Research and Innovation", in Responsible *Innovation: managing the responsible emergence of science and innovation in society*, Wiley, http://onlinelibrary.wiley.com/book/10.1002/9781118551424.

Wilsdon J and Willis R (2004), *See-through Science: Why Public Engagement Needs to Move Upstream*, Demos, London.

Winickoff DE, Jamal L and Anderson NR (2016), "New Modes of Engagement for Big Data Research", *Journal of Responsible Innovation*, Vol. 3/2, pp. 169-177, Taylor and Francis, Milton Park, UK, https://doi.org/10.1080/23299460.2016.1190443.

Winickoff DE and Mondou M (2017), "The Problems of Epistemic Jurisdiction in Global Governance: the Case of Sustainability Standards for Biofuels", *Social Studies of Science*, Vol. 47/1, pp. 7–32, Sage Journals, Thousand Oaks, CA, https://doi.org/10.1177/0306312716667855.

Winner, L. (1980), "Do Artifacts Have Politics?", Daedalus, Vol. 109/1, pp. 121-136, MIT Press, Cambridge, MA, https://www.jstor.org/stable/20024652.

World Economic Forum (2018), *Agile Governance: Reimagining Policy-making in the Fourth Industrial Revolution*, World Economic Forum, Geneva, https://www.weforum.org/whitepapers/agile-governance-reimagining-policy-making-in-the-fourth-industrial-revolution.

Chapter 11. New approaches in policy design and experimentation

By
Piret Tõnurist

Complexity and uncertainty are core features of most policy making today, and STI policies are no different. This chapter describes and analyses emerging approaches to science, technology and innovation (STI) policy design and implementation. It reviews several new policy tools, such as systems thinking, design thinking, behavioural insights, experimentation, regulatory sandboxes and real-time data analytics, that are transforming STI policy making today. It argues that innovations in policy making should be applied strategically and systemically – they should not be adopted indiscriminately by layering them on top of one another in an ever-expanding 'policy mix'. The chapter concludes by considering the capacities and capabilities required of policymakers in this challenging new environment, and discusses the future outlook for policy design and governance practices.

Introduction

Governments have traditionally played an important role in supporting fundamental science. They have guaranteed scientific autonomy and funding, thereby creating the environment necessary for innovation. At the same time, they are themselves increasingly innovating, experimenting and pushing boundaries in their everyday actions. The literature recognises the quality of public institutions as a powerful driver of economic growth (Rodrik, Subramanian and Trebbi, 2004; Acemoglu and Robinson, 2012). However, it rarely analyses how governments – and the institutions they create – can become "smart". It is therefore important not only to analyse STI policies as separate outcomes, but also to put them in the context of the institutions delivering these policies. This means shining a spotlight on governments' capacity to design and implement effective science, technology and innovation (STI) policies.

Government capacity should not remain static; it needs to adapt to societal and technological changes. New – and often disruptive – technologies, such as the Internet of Things (IoT), blockchain technology and artificial intelligence (AI), are transforming the production and distribution of goods and services, with significant impacts on society (OECD, 2017a). Technological change is also transforming the way government works, operates and interacts with its policy subjects and partners. Increased interconnectivity, platform economies and peer-to-peer production mean that the private and public domains are in flux. The traditional concepts of public value (e.g. transparency, privacy and accountability) connected to both public and private services and products are changing. The uncertainty and risks created by rapid technological change cannot be borne and directed by the private sector alone: governments must evolve and take an active role in the change process. They must harness digital technologies to respond to the impacts of digitalisation and changing citizen demand (OECD, 2014). They must also anticipate, adapt to and mitigate these change processes as part of their STI policy portfolios.

Addressing 21st-century problems with old tools and methods is unlikely to be effective. The speed, scale and complexity of change is ever increasing. Policymakers face an almost impossible task in maintaining stability and confidence in the public system, while rapidly adapting to a new environment characterised by fast-paced change and new demands. Governments must engage in new policy design and implementation and demonstrate dynamic capabilities. They need to understand the impacts of technology, as well as the changing expectations of citizens, companies and innovators, looking deeper into their user experiences in order to experiment and innovate themselves.

Governments are already changing their STI policy design. They are using design thinking and behavioural insights to analyse the changing needs and motivations of researchers, innovators and lead users in order to apply new technological solutions based on users' expertise. They are also seeking to learn from practice and experimentation, creating anticipatory and adaptive ways of working with lead developers and users. These trends are also present in other policy fields, so that STI governance can also learn from innovations in other public-sector domains (e.g. Dutz et al., 2014).

This chapter outlines the promise for improved STI policy making that could arise from design thinking, collective intelligence, behavioural insights, policy experimentation and systems thinking. It highlights the need to build government platforms, anticipate disruptive change, and embrace new skills and capacities for STI policy design. It concludes by discussing the future outlook for this field.

Reaping the benefits of design thinking

Design thinking can enhance the commercialisation of scientific and technological breakthroughs, and has long been linked to STI. Some countries have created specialised organisations to funnel design know-how and talent where it is most needed; examples include the Catapult technology and innovation centres in the United Kingdom (UK Design Council, 2011). Although user-centred methods are often discussed in the context of the technology industry, they are increasingly applied to the delivery of public services (OECD, 2017b). Arguably, policy making and policy implementation are a form of design; however, neither was discussed in design terms until recently. In the last five years, design thinking has taken centre stage in most public-sector innovation toolboxes (Observatory of Public Sector Innovation [OPSI], 2018). In the face of severely declining service satisfaction and trust in government, design thinking stipulates that any policy design – including related to STI – should focus on user or customer needs, rather than on internal organisational needs (Bason, 2016). This approach is rooted in collaborative methods engaging both end-users and service-delivery teams.

Brown (2008) describes design thinking as a discipline using the designer's sensibility and methods to match people's needs with: a) what is technologically feasible; and b) what a viable business strategy can convert into customer value and market opportunity. The increase in design thinking in the public sector has gone hand in hand with digitalisation. Some governments (e.g. Australia, New Zealand and the United Kingdom) have established specific service standards and design toolboxes for digital-service development (Box 11.1). For innovative, user-centred solutions, design thinking presupposes fuzzy front ends[1] that ignore established public-sector silos and operating systems (Table 11.1). This allows them to surpass outdated information systems in government, by prioritising users' needs and experiences. Thanks to its growing popularity, design thinking has become a form of intelligence governments could utilise more systematically, not only to inform more targeted STI policies, but also to initiatives related to digital science and innovation policy (Chapter 12).

Table 11.1. Traditional public-sector context versus design thinking

Problems of "traditional" public sector	Features of design thinking as solutions to public-sector needs
Disjointed incrementalism Cost-driven problem-solving without asking whether the fundamentals are right and user needs are met	*Designing for the fundamental need* Tailored solutions based on user needs
High-risk piloting Large-scale pilots with considerable risk and costs.	*Low-risk prototyping* Small-scale pilots and failing fast and smart
Lack of joined-up thinking Disconnect between analysis of problems, solutions and implementation	*A complete innovation process* Design-led, joined-up innovation process
Lack of citizen engagement No guarantees that citizens' needs are met and limited buy-in to government solutions	*A citizen-centred process* User involvement throughout the innovation process to co-design and test solutions, resulting in higher ownership of solutions
Poor understanding of citizen needs Gap between what people want and what they say want, and ineffective methods to determine the difference	*Direct understanding of citizen needs* Observation of user behaviour to discover unidentified needs
Lack of tangibility Important messages subsumed in information overload	*Dynamic tangibility* Visualisation of relationships and processes through which solutions work
Silo structures Problems with co-ordination and collaboration, both inside and outside of government	*Multidisciplinary teamwork* Ways of assessing the relevancy of actors and devising techniques to help multidisciplinary teams collaborate.
Designing for the average Services and policy are designed for a notional average user in an average situation	*Designing for extremes* Accounting for extremes, ensuring that solutions cover a wide range of users and scenarios

Source: Based on UK Design Council (2013), Design for Public Good.

Box 11.1. Government adoption of design toolkits and standards

Until now, the United Kingdom, the United States, Australia and Canada have mostly led in formalising design thinking, by developing and adopting design toolkits, playbooks and methods. Most of these developments are linked to the digitalisation of government, leading to standardisation through design. In the United Kingdom, the Digital Service Standard is a set of 18 criteria designed to help government create and run optimal digital services (Government of the United Kingdom, n.d. a), and the UK Government Digital Service works with a set of design principles (Government of the United Kingdom, n.d. b). The US Government's digital service agency, 18F, has also developed its own design method tool, "18F Method Cards" and an "Innovation.gov Toolkit: Human-Centred Design. Meanwhile, the Danish Design Centre has led the way in codifying tacit knowledge for the public sector through its Inclusion Toolkit: Designing a User-Centred Living Lab from the Ground Up". Other examples of government-adopted design toolkits include: "Service Design Playbook" from the Government of British Columbia, Canada; and the "User Centred Design Toolkit", Government of South Australia, Australia.

Sources: OPSI (2018), "Toolkit Navigator", https://www.oecd-opsi.org/toolkit-navigator; Government of the United Kingdom (n.d. a), "Digital Service Standard, https://www.gov.uk/service-manual/service-standard; Government of the United Kingdom (n.d. b), "Government Design Principles", https://www.gov.uk/guidance/government-design-principles.

The design-thinking methodology is relatively accessible to government and features seemingly straightforward principles. Yet this is also its main shortcoming: most of its core knowledge is tacit and acquired through practice. What individual designers know, how they implement what they know, how they approach and make sense of their own work, and how they actually perform it are essential to successful design. Little is known about how policymakers identify design problems and design criteria, what professional design expertise they themselves possess, or whether and when they collaborate with outside design professions during the policy-making process (Junginger, 2013). There exists a risk that the approach, when placed in the hands of novice public-sector users, may not live up to its promise.

Numerous innovation toolkits and guides have recently emerged in government. The OPSI at the OECD recently reviewed approximately 230 innovation toolkits. It selected around 150 of these for its Toolkit Navigator (OPSI, 2018), making these approaches more accessible and downplaying the expertise required to apply the methods in practice. Several design organisations and policy labs have emerged that focus on design thinking in the public sector, including the Design Centre and the (now closed) Mindlab in Denmark; the Design Council and Policy Lab in the United Kingdom; Design Driven City in Finland; and the Public Policy Lab in the United States.

While design thinking is sometimes treated as an ideology to rethink complex problems – or even as a panacea for solving most policy problems (UK Design Council, 2013) – it is not a cure for all ills, either in the public or the private sector. One of its core strengths, user centricity, is also its limiting factor. Not all deficiencies in government or in STI policy design come from the front end; many may also be rooted in back-office operations, such as the way governments frame problems. By focusing on user experiences, design thinking may ignore this aspect. Moreover – especially in the field of innovation policy – it may focus disproportionately on the needs and interests of today's user base, ignoring longer-term innovation needs. Thus, design thinking should be coupled with a broader systems-thinking lens and anticipatory governance methods (discussed in more detail in Chapter 10 on technology governance) in order to help identify issues beyond the immediate experiences of researchers and innovators.

Creating collective intelligence

To generate new ideas and innovative solutions, governments have used a variety of tools, including challenges and prizes, such as the US Government's Challenge.gov initiative (Mergel, 2018). Some governments are branching out, co-creating and co-producing innovations and innovative outcomes with citizens. By tapping into various digital crowdsourcing platforms, they have systematically collected ideas, opinions, solutions and data from a wide sample of the general public (Noveck, 2015). Crowdsourcing offers benefits in terms of cost and speed; the potential to find new patterns in large datasets; and the opportunity to conduct near real-time testing and application of new policy approaches (OECD, 2015a). Crowdsourcing can rely on crowd-based resources to design innovative solutions.[2] For example, Mexico City's Mapaton initiative (Box 11.2) uses gamification strategies to encourage citizen involvement. Collective intelligence can also involve more active co-creation of innovations (e.g. through hackathons and living labs) between government and citizens[3] (Almirall and Wareham, 2011; Cardullo and Kitchin, 2017a; Lember, forthcoming). For example, the Agile Islands initiative, spearheaded by Tekes in Finland, uses hackathons for innovation procurement; and the Belgian city of Antwerp is developing its own IoT solution, City of Things, with specific input from local residents in a living-lab format (OECD, forthcoming a). In Canada, the government has launched a Drug Checking Technology Challenge to develop new or improve existing technologies in order to empower the community of people who use drugs to make informed decisions and reduce potential harm (Impact Canada Challenge Platform, n.d.). Longer-term, expert-based collaboration approaches are also emerging as a form of collective intelligence. Some authors (e.g. Mulgan, 2017) are predicting the emergence of a "bigger mind" – human and machine capabilities working together – to solve the great challenges facing the world today.

> **Box 11.2. Mexico City's Mapaton initiative**
>
> Mexico City has one of the largest public transit systems in the world. Its buses provide over 60% of all transit in the city, ferrying about 14 million daily riders on 29 000 buses, covering more than 1 500 routes. However, partly owing to its size and complexity, Mexico City had no bus-related data or maps.
>
> Mexico City's Laboratory for the City (Laboratorio para la Ciudad), an experimental office and creative think-tank reporting to the mayor, partnered with 12 governmental and civil society organisations to develop Mapatón CDMX. This crowdsourcing and gamification experiment maps the city's bus routes through civic collaboration and technology, using smartphones to feed global positioning system data to the authorities. The participants who mapped the most routes and earned the most points won tablets and cash prizes up to MXN 30 000 (Mexican pesos, about USD 1 700 [US dollars]). Because users are concentrated in certain areas of the city, an algorithm was used to assign the most points to neglected routes; the algorithm constantly recalculated the point values of the routes to ensure mapping the maximal number of routes. The city-wide game attracted more than 4 000 participants, who accomplished the main mapping task in two weeks for a total programme cost amounting to less than USD 15 000. Several other cities are considering replicating this platform. The data generated are now available as open government data for others to use and build on, and to guide policy.
>
> *Sources:* OECD (2017c), "Embracing Innovation in Government: Global Trends", *http://oe.cd/eig*; Mendelson (2016), "Mapping Mexico City's vast informal transit system", *www.fastcompany.com/3058475/mapping-mexico-citys-vast-informal-transit-system*.

Collective innovation is also bypassing the public sector altogether. With "civitech", citizens are creating solutions as varied as voter-to-voter communication, opinion matching, watchdogging, online petition sites and hyperlocal news. Some technologies (e.g. blockchain) can facilitate peer-to-peer service delivery (Pazaitis, De Filippi and Kostakis, 2017a). At the city level especially, models of local resilience and self-organisation are emerging, with user-driven innovators generating bottom-up solutions for their communities (von Hippel, 2016). Instead of top-down initiatives co-ordinated by the government or the private sector, collectively produced solutions are being adopted (for example, Wikipedia, or community-owned public taxi services, e.g. in Austin, Texas).

Acknowledging the potential of such bottom-up innovation, governments have sometimes intentionally given control to citizens to decide on initiatives (as with technology co-design workshops).[4] Citizens choose the design and implementation methods, co-create the technologies, and co-ordinate the activities from start to finish (Pazaitis, Kostakis and Bauwens, 2017b). These initiatives, however, can be extremely disruptive to existing public service systems. Governments may need to stay involved and possibly take back control when the risks taken become too large for the system or to guarantee citizens' safety – as when testing privately-led circular-economy solutions in urban settings (OECD, forthcoming a). Governments are also actively creating room for innovators to experiment in public spaces. In 2017, for example, the Estonian Parliament authorised testing self-operated robots in public streets.

Even though collective intelligence for innovation can be a thoroughly positive resource for governments, downsides also exist for digital co-creation and co-production. For example, the increased capabilities for gathering data from everywhere – the IoT – could mean that the scale and reach of co-production grows exponentially. Coupling this with increased data processing capacity, governments can precisely target their collaborations for STI, potentially leading to manipulation, excessive control and "nudging" of researchers and firms actions.

Exploring the promises of behavioural insights

Another major trend in the public sector is the adoption of behavioural insights[5] – "nudges", "budges"[6] and "shoves"[7] (Thaler and Sunstein, 2008) – to influence, rather than direct, the behaviour of policy subjects. Nudges are gentle pushes aiming to change people's behaviour, leaving them the option to choose a route not promoted by government. As they do not specifically regulate people's behaviour, they sometimes extend the governments' scope of action (and the political feasibility of traditional incentives), or make it easier for government to adopt short-term measures that can easily be discontinued after the desired positive change has been achieved. In the field of STI, governments have especially considered nudges to drive technology diffusion – e.g. green innovations (Schubert, 2017). Even when they promote pro-social behaviour, the ethical implications and subversive nature of nudges, which address or exploit cognitive biases, are subject to criticism. This does not mean that behavioural insights should not be used (behavioural biases exist, whether or not they are addressed in traditional policy approaches – behavioural insights help make these choice architectures visible). It does mean that the extent to which they are used to manipulate people, rather than help them make informed choices, should be considered.

The promise of behavioural insights is not new in economics; the concept of behavioural additionality, for example, has been used for some time in evaluations of innovation policy (OECD, 2006). However, STI policy making appears to underutilise behavioural insights and especially rigorous experimentation (e.g. RCTs) that draws on behavioural insights of STI policy subjects. Although approximately 200 institutions worldwide apply them to public policy (OECD, 2017d), OECD member countries mostly apply them to finance, health and safety, and consumer protection, rather than to devise STI policies.

Behavioural insights are generally not used as inputs in agenda-setting and enforcement in the traditional policy cycle; rather, they are most frequently used at a later stage of policy design. Yet they may have great potential for STI policies in the agenda-setting phase – which requires an *inductive* approach, where experiments replace and challenge established assumptions of the "rational" behaviour of people and business. In this way, behavioural insights can inform policy making and implementation with evidence of "actual" behaviours (OECD, 2017d) – especially when those behaviours are changing. It helps understand the complexities and contradictions of human actions, using the derived insights to nudge behaviour. For example, The United Kingdom's Behavioural Insights Team has developed a tool called Predictiv (Box 11.3), which helps governments and other clients run behavioural-insight experiments on a pool of online volunteers, thus scaling and speeding up the process of evidence-informed policy making that STI policy makers could also draw upon. Behavioural insights may be very useful for demand-side STI policies as many barriers to innovation procurement and agile development are actually real or perceived behavioural deficiencies (Georghiou et al., 2014).

> **Box 11.3. Predictiv : Online behavioural experiments platform**
>
> Predictiv is an online platform for running behavioural experiments. It enables governments to run randomised controlled trials with an online population of participants, and to test whether new policies and interventions work before they are deployed in the real world (Figure 11.1). After a short design phase, the tests take one to two weeks to complete, enabling policy makers to obtain responses to questions that would otherwise have taken many months (or years) to answer. As such, it has the potential to profoundly change governments' working methods. While time constraints and political realities sometimes make it hard to run "field trials" on live policy, Predictiv makes experimental methods more accessible.
>
> **Figure 11.1. Predictiv's Approach** 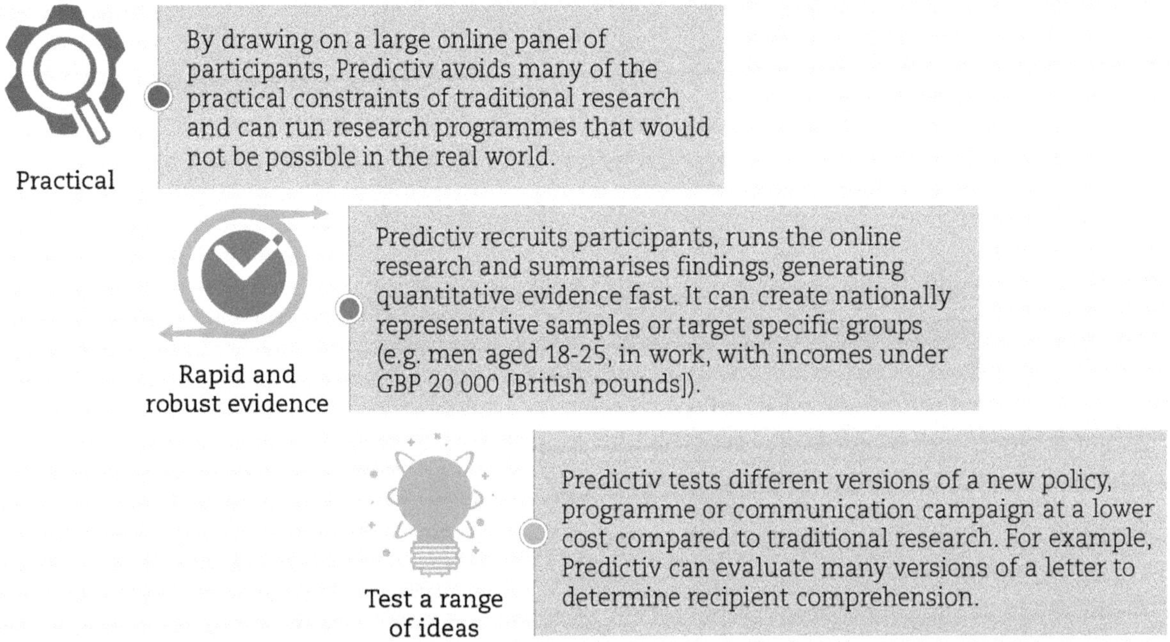
>
>
> Practical — By drawing on a large online panel of participants, Predictiv avoids many of the practical constraints of traditional research and can run research programmes that would not be possible in the real world.
>
>
> Rapid and robust evidence — Predictiv recruits participants, runs the online research and summarises findings, generating quantitative evidence fast. It can create nationally representative samples or target specific groups (e.g. men aged 18-25, in work, with incomes under GBP 20 000 [British pounds]).
>
>
> Test a range of ideas — Predictiv tests different versions of a new policy, programme or communication campaign at a lower cost compared to traditional research. For example, Predictiv can evaluate many versions of a letter to determine recipient comprehension.
>
> *Source:* OECD (2018), "Embracing Innovation in Government: Global Trends 2018", *http://www.oecd.org/gov/innovative-government/embracing-innovation-in-government-2018.pdf*; Predictiv (n.d.), "Predictiv for policymakers & practitioners", *www.predictiv.co.uk/governments.html*.

Experimenting with new STI policy approaches

Policy makers are taking an increasingly active role in creating solutions themselves, rather than facilitating innovation through demand or supply-side policies. As such, they can act as technology makers or innovators in their own right, taking on the uncertainties of innovation through direct policy design, experimentation and implementation activities inside government (Karo and Kattel, 2018). Arguably, governments already support experimentation through the different initiatives and programmes within their STI support portfolios. Some have also started to spur on experimentation directly inside government to devise more innovative services and develop technology. For example, central banks and financial authorities are actively exploring blockchain technologies to support their operations (Berryhill et al., 2018). Among others, NESTA's Innovation Growth Lab and the European Union's Joint Research Centre's Policy Lab have been supporting experimentation in innovation policy for some time.[8] Some commentators have also called for experimental government to meet the policy challenges of today's world (e.g. Breckon, 2015; Mulgan, 2013). They argue that public authorities will need to experiment more and learn iteratively, to gather knowledge and evidence on what works or could work better in a more cost-efficient manner. Many governments are already exploring ways to create "safe spaces" for experimentation inside the public sector, helping civil servants to contend with the uncertainty connected to experimentation processes, and sometimes giving them an explicit licence to fail (OPSI, 2017a). For example, both Canada and Finland have recently adopted formal frameworks to support experimentation within their respective central governments (Box 11.4).

Box 11.4. Central government support for experimentation

Experimentation has been a part of some countries' policy design for some time. For example, What Works Centres in the United Kingdom have been collecting and evaluating evidence from randomised controlled trials since 2013. More recently, central governments, e.g. in Finland and Canada, have begun supporting experimentation in the public sector more explicitly.

Experimentation programme in Finland

In 2015, the Finnish Prime Minister's Office of Finland employed a combined systems and design thinking approach to develop a new policy framework for experimental policy design. As a result, experimentation was incorporated into the strategic government programme and an experimental policy design programme was set up. The new approach to policy design allowed both broader "strategic experiments" (formalised policy trials) – e.g. the universal basic-income experiment – and grassroots experiments, designed to build up an "experimental culture" in the Finnish public sector. In addition to the original six strategic experiments introduced by the Finnish Government, hundreds of experiments and policy pilots have emerged across the country, at both the central and municipal levels. In 2017, the Finnish Government launched its digital platform, Place to Experiment, to promote an experimental culture aiming to develop innovative public policies and services. This work is monitored and supported by the Experimental Finland Team in the Prime Minister's Office.

Experimentation directive in Canada

In 2015, the Prime Minister of Canada issued a mandate to the President of the Treasury Board of Canada Secretariat to support experimentation in government. In late 2016, the mandate was further clarified by a subsequent directive on experimentation, produced by the Treasury Board Secretariat and Privy Council Office. The experimentation directive explicitly linked experimentation to more effective policy making, and called for government departments to allocate part of their programme funding to experimentation.

> *Sources*: OECD (2017e), Systems Approaches to Public Sector Challenges: Working with Change, http://dx.doi.org/10.1787/9789264279865-en; OECD (2017c), "Embracing Innovation in Government: Global Trends", http://oe.cd/eig; Government of Canada (2016), "Experimentation direction for Deputy Heads – December 2016, https://www.canada.ca/en/innovation-hub/services/reports-resources/experimentation-direction-deputy-heads.html.

Building government platforms

Policy making is becoming more data-driven (OECD, 2017f). For countries that are digital frontrunners, the next wave of innovation inside government will rely on new services and solutions, built on linked data, advanced data-processing capabilities, real-time data analytics, and new ways of combining and making sense of information.

Mobile-data collection and advances in real-time data processing will shift policy design from "descriptive" to "predictive", and thereafter to "prescriptive" (Chong and Shi, 2015). Algorithm-based decision-making models are already used in policing and public-space management. Governments are now using them as part of the STI policy portfolio, e.g. to enable better trademark protection in Australia and beyond (Box 11.5). Some are also using text mining, mapping and visualisation tools to monitor innovation, e.g. in the context of the European Commission's Tools for Innovation Monitoring project.[9]

> **Box 11.5. Australian Trade Mark Search**
>
> Over 80% of an average company's value is rooted in its intangible assets, including brands and trademarks, which represent a business's identity. A good trademark identifies a unique product or service in the marketplace. However, the steps necessary to ensure the uniqueness of a company's brand are difficult and time-consuming. IP Australia, the government agency that administers a number of intellectual property (IP) rights in Australia, launched Australian Trade Mark Search to help businesses protect their intangible assets. Powered by industry partner Trademark Vision's image-recognition and AI technology, the solution provides security for businesses by protecting their most important assets, with significant global applicability. The search function was developed as a platform for continuous improvement over time, as user needs, expectations and technical capabilities change. The success of the initiative in enhancing brand identities has led IP Australia to expand the technology to other IP domains. The next public-facing IP Australia search solution will be Australian Design Search, which will allow users to search registered industrial designs using images.
>
> *Source*: OECD (2018), "Embracing Innovation in Government: Global Trends 2018", http://www.oecd.org/gov/innovative-government/embracing-innovation-in-government-2018.pdf.

The biggest trend combining all of the above-mentioned technological functionalities is the increased presence of platforms in both the economy and government (Tõnurist, Lember and Kattel, 2016; Teece, 2018). Platforms facilitate transactions by creating trust and accountability. In the future, innovation through and within government (and, arguably, STI policy implementation) will be influenced by the idea of Government as a Platform (GaaP). New platform-based service designs are already emerging (Box 11.6). In China, for example, the WeChat platform numbers more than 800 million individual and

20 million company users; it combines multiple platforms into one app, with a variety of (private and public) functions and services built into the platform.

Box 11.6. GaaP and the case of eResidency in Estonia

The concept of GaaP envisions that government uses digital technologies "to support the resolution of collective action problems at various levels (city, county, national, regional) through shared software, data and services – and thereby improve the efficiency and effectiveness of government and governance, doing more for less" (Margetts and Naumann, 2017). GaaP rests on the idea that citizens themselves may be involved in delivering digital government through platforms provided by governments, beyond what might be termed the "vending-machine" model of government, where taxes are exchanged for services.

One of the most innovative (albeit contested) GaaP examples is Estonia's eResidency programme. Adopted in 2016, it allows a non-resident of Estonia to apply for (limited) residency, making it possible to use digital public services (such as establishing a company and filing taxes) without being physically located in Estonia. Collaboration is easier among public organisations and private companies that have subscribed and use Estonian e-identity systems. Finland will soon start to use similar data-exchange platforms. Estonia's eResidency can be seen as the first major GaaP proof-of-concept case, allowing new forms of identity and service use beyond governments' traditional remits.

Sources: O'Reilly (2010), "Government as a Platform"; Margetts and Naumann (2017); Lember, Kattel and Tõnurist (2018), "Technological Capacity in the Public Sector: The Case of Estonia".

Anticipating disruptive change

Earlier sections of this chapter have outlined the disruptive nature of existing technologies. Much larger changes are on the way, e.g. autonomous vehicles, drone technologies, blockchain and widespread IoT solutions (see Chapter 2). Governments need to anticipate these changes and consider their implications for public policy. New technologies offer opportunities to improve economic efficiency and quality of life, but they also bring many uncertainties, unintended consequences and risks. Anticipatory governance (see Chapter 10) acts on a variety of inputs to manage emerging knowledge-based technologies and the missions built upon them while such management is still possible (Guston, 2014). It requires government foresight, engagement and reflexivity to facilitate public acceptance of new techno-sciences, while at the same time assessing, discussing and preparing for their (intended and unintended) economic and societal effects. Anticipatory governance considers risk – especially systemic risks – over extended timeframes, and develops the capacity to mitigate uncertainty (e.g. through critical infrastructure and wealth funds).

The benefits and risks of new technologies do not generally befall the same people. Anticipatory governance requires governments to consider which public values should be preserved during the change process, and how technological change – e.g. the adoption of disruptive technologies – affects public values (Box 11.7). Reliance on traditional policy tools is difficult in situations where the future direction of technological innovation cannot be determined. New policy tools – such as normative codes of conduct, regulatory sandboxes and real-time technology assessments – are therefore necessary (Stilgoe, Owen, and Macnaghten, 2013); Australia, Hong Kong, Malaysia, Singapore, the United Arab

Emirates and the United Kingdom, for example, have adopted regulatory sandboxes.[10] This means that government must better operationalise foresight and upstream engagement with technology developers and lead users.

Box 11.7. Regulating the sharing economy: The experience from Canada

In 2014, the transportation network company Uber started operating in Toronto without specific regulatory oversight. The city had to move quickly to implement new regulation and appease the alarmed incumbent industry. To tackle the regulatory challenge and simultaneously preserve the beneficial aspects of a sharing economy, MaRS Solutions Lab served as an independent arbiter, facilitating productive dialogue between the different stakeholders. Utilising systems thinking and design methodologies, the Toronto-based innovation lab proposed a user-centric vision for the regional sharing economy, highlighting the increased public value accompanying the disruptive change. It also helped develop new legislation enabling the city and its citizens to both regulate and benefit from new entrants that disrupt old businesses.

MaRS Solutions Lab developed the "Periodic Table of System Change", a framework for understanding the various elements required to navigate and alter complex systems. The method acknowledges it is not enough to tackle policies and provide solutions for systems to change. For the process to succeed, systems thinking must tackle the capacities of different stakeholders, and how they understand the problems and values connected to them. As part of an anticipatory process, the method also discussed possible future values connected to technological change. To regulate the sharing economy, the City of Toronto needed new competencies, as well as the capacity to understand the newly emerging service models and deal with the unintended consequences. It also had to reframe policies regarding insurance, taxation and market entry. As governments continuously face increasingly complex problems, these insights play a central role in initiating and implementing change.

Sources: OECD (2017e), Systems Approaches to Public Sector Challenges: Working with Change, http://dx.doi.org/10.1787/9789264279865-en; OECD (2017c), "Embracing Innovation in Government: Global Trends", http://oe.cd/eig.

Adopting systems thinking in STI policy making

The design of STI policy is as important as the solutions it seeks to provide, especially in a context of accelerated change. The increase in data analytics alone will force policymakers into real-time decision-making. How should policymakers manage these situations so that they are not locked into reactive policy designs? How can they manage technology upstream and govern innovation in the making, while still demonstrating strategic intent? What do adaptiveness and reflexivity look like in practice? Although many tools and methods exist today to engage in iterative and agile policy making, they should come together more systemically at some stage.

Systems thinking (Box 11.8) is not new to STI (OECD, 2015b). Analysis of "innovation systems" is pervasive, covering national, sectoral and technological perspectives. Yet such perspectives have proven difficult to operationalise in policy settings: they are mostly retrospective and tend not to outline or analyse the real-time choices facing policymakers. An ecosystems-based approach to how government manages innovation both internally and externally is necessary, coupled with the ability to use systems thinking not only as a descriptive, but also as a transformative tool inside government (OECD, 2017e). Some governments and international organisations are building scenarios integrating "socio-technical transitions" to respond to sustainability challenges (Geels, 2004; Geels, McMeekin and Pfluger, 2018). Similar to systems thinking, the concept of socio-technical transitions considers the roles of markets, user practices, policy and culture in the development of new technologies, in addition to the "politics of transitions" (Lawhon and Murphy, 2012). The Swedish Government has used socio-technical roadmaps to determine which large-scale investments it should make in its strategic innovation programmes (Coenen et al., 2017). Austria used them to develop its Industry 4.0 programme.

Box 11.8. Systems thinking and the public sector

Systems thinking is an interdisciplinary approach to understanding how different parts of a system relate to each other, how systems work and evolve over time, and what outcomes they produce. Systems change is an application of that thinking to real-world situations. Systems approaches have developed over the last 75 years to include general systems theory, dynamic systems theory and cybernetics. However, although other sectors have embraced systems change, it is far from established in the public sector, which has only shown interest in applying systems approaches more rigorously in the past decade. In its 2017 report, *Systems Approaches to Public Sector Challenges: Working with Change*, the OECD proposes several tactics for systems change.

Source: OECD (2017e), Systems Approaches to Public Sector Challenges: Working with Change, http://dx.doi.org/10.1787/9789264279865-en.

Lacking system stewardship and clarity, innovation in government will fall back on individual organisations and policymakers. While this may produce pockets of excellence, it will not result in a balanced portfolio of innovative activity inside government. At the OECD, the OPSI is working on these issues as it reviews public-sector innovation.[11] As part of the review process, a new model of how governments innovate internally in policy making is emerging (Figure 11.2). The model involves individual, organisational and systemic elements, and incorporates ways for governments to steward and interact with the system at each level.

Figure 11.2. Determinants of innovation in the public sector

Individual, organisational and system levels

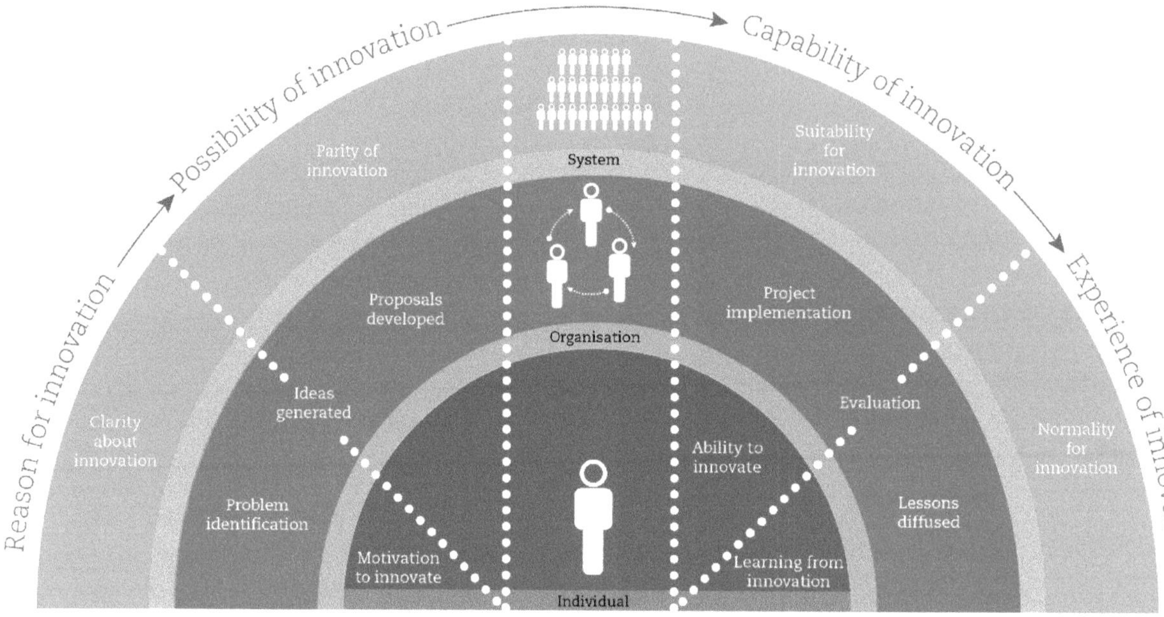

Note: The model is based on a grounded-theory approach as part of the empirical analysis accompanying the review on the Innovation System of the Public Service of Canada. (forthcoming).

Embracing new skills and capacities

To adopt and adapt to the new policy tools and approaches described in this chapter, policymakers need different types and combinations of skills, as well as the organisational capacities to lead and work with change (Lember, Kattel and Tõnurist, 2018). The OPSI has outlined six core skills supporting increased levels of innovation inside the public sector (Figure 11.3). Officials working in a modern 21st-century public service will need to be aware of these core skills in order to support increased innovation in the public sector. However, based on the different types of innovation (e.g. user-centric, mission-oriented or anticipatory), more specific combinations of skills and organisational capacities will be needed. The new tools and methods described in this chapter can help design better policies, but only if they are applied correctly and to the right occasion. For example, while calls have been made for more targeted, challenge-based approaches to STI policy – e.g. the European Commission's new narrative around missions (Mazzucato, 2018) – rapid change significantly raises the risk of lock-in and makes directionality more difficult. This may require different organisational solutions for adaptive and mission-oriented innovation, or different models for balancing the two in practice.

Figure 11.3. Six core skills for public-sector innovation

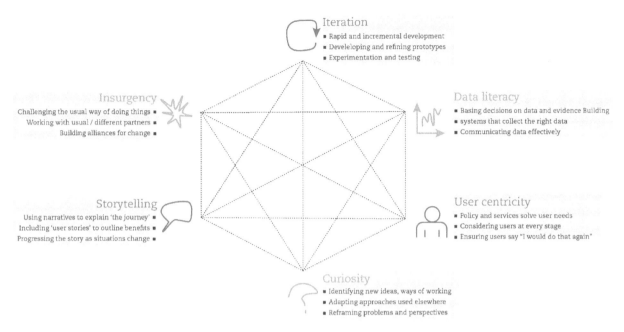

Source: OPSI (2017b), "Core skills for public sector innovation:
A beta model of skills to promote and enable innovation in public sector organisations",
https://www.oecd.org/media/oecdorg/satellitesites/opsi/contents/files/OECD_OPSI-core_skills_for_public_sector_innovation-201704.pdf.

Future outlook for STI policy design

Governments are facing a fundamental sea change, brought about by the increased complexity of socio-technical challenges, globalisation and the digital transformation. Many STI policy challenges are no longer in the hands of single governments. Rather, they are dispersed among networks of governments, innovators, private platforms and users. STI policy will need to tap into new types of demand, new networks and new ways of managing uncertainty. It will have to both provide direction for change (mission-oriented innovation) and adapt to fast-paced changes in technology. If governments do not adapt their operating practices to this new environment, they may become increasingly irrelevant, dysfunctional and disconnected.

With new practical methods of tapping into various dimensions of collective intelligence, STI policy will need to explore more distributed ways of designing and implementing policy. This may mean leaving space for people to experiment and test new solutions by themselves, without direct government involvement or control over the process. The risks connected to this approach should not be ignored. In some cases, government will need to take back control, if the risks become too large for the system or the safety of citizens themselves is at risk. The political dimension of these new policy design tools should also be considered, since they are not value-free (for example, the use of technology to interact with the larger crowd of lead users or technology developers will invariably influence the power of different stakeholders and governments).

Machine-to-human and machine-to-machine interactions are increasingly taking over not only service delivery, but also policy formulation and evaluation. They are creating new types of evidence (personal, real-time, interconnected, etc.) to evaluate policy effectiveness, as well as new ways of implementing policy (e.g. through government platforms). Part of this process is the personalisation of all government services. While this creates the potential for better-quality, timelier evidence to plan STI policies, it also requires the public sector to become more nimble, targeted and adaptive. STI policy is entering an era where real-time change and implementation becomes actual policy making. Yet governments still need to uphold confidence in the overall public system, maintain its stability and manage the long-term risks connected to R&D investments.

Overall, changes in STI policy making will need to be governed systemically, connecting policy intent with the right tools and capacities inside government. This means that the internal design and implementation of STI policies should also adapt to systems thinking. Innovative STI tools should not only be centralised in dedicated units (e.g. innovation labs and agencies), but should also be used by public entities adopting innovation agendas across the board. More sophisticated, diversified innovation strategies and portfolios for STI policy design and implementation will need to emerge inside the public sector. They should be coupled with the right public-sector skills and capabilities, not only to use new policy design tools for maximum impact, but also to steward systemic change itself.

Notes

[1] "The fuzzy front end" is an established term in product design and product development, akin to the writer's well-known "blank page". In design thinking, various strategies for fuzzy front ends have been proposed (e.g. Cagan and Vogel, 2001).

[2] For example, Phones Against Corruption in Papua New Guinea developed user mobile-based reporting to monitor corruption anonymously. In Slovenia, the interactive mobile application "Check the invoice" has been used to reduce the shadow economy, with the help of the public. In some cases, voluntary data repositories have been created, where citizens can donate their personal data (Symons and Bass, 2017) so that they can be used to co-create and co-produce new services; the data are processed, maintained and controlled through various blockchain technologies (Berryhill et al.; 2018). Such repositories include the European Union's DECODE project (European Union, n.d.).

[3] Hackathons are a form of technological co-creation (e.g. government-sponsored, weekend-long prototyping/coding events for citizens, often based on government-provided open data) and a source of new co-creation initiatives (e.g. apps and other technical solutions enabling further co-creation and co-production). Living labs are a bottom-up approach to testing digital technologies with their users in "in-vivo settings". They also aim to solve local issues through community-focused civic hacking, workshops, and engaging with local citizens to co-create digital interventions and apps (Cardullo and Kitchin, 2017b; Schuurman and Tõnurist, 2017).

[4] Technology co-design workshops are a form of participatory design where users and designers express and exchange ideas to develop technology-intensive services (Wherton et al., 2015).

[5] The OECD defines behavioural insights as "an inductive approach to policy making that combines insights from psychology, cognitive science, and social science with empirically-tested results to discover how humans actually make choices" (OECD, n.d).

[6] Behaviourally informed regulatory interventions.

[7] Traditional behaviourally informed bans.

[8] https://blogs.ec.europa.eu/eupolicylab/about-us/.

[9] https://ec.europa.eu/jrc/en/scientific-tool/tools-innovation-monitoring.

[10] The "regulatory sandbox" approach was pioneered by the United Kingdom's Financial Conduct Authority (2015) to address and control the barriers to entry for Fintech firms – small, innovative firms disintermediating incumbent financial services firms with new technology – in the financial landscape. In 2016, the Authority released its "UK sandbox," which allowed innovative FinTech development without requiring a full, strict regulatory testing process. The prerequisite of a sandbox is publicly available criteria that actors need to meet as a prerequisite for entry into the sandbox (meaning that only fulfilling certain criteria they can introduce innovations in the domaign). For further information, see: https://www.fca.org.uk/firms/regulatory-sandbox.

[11] The first OECD review of the public-sector innovation system was carried out for Canada and is forthcoming; the second review will cover Brazil and is preparation.

References

Acemoglu, D. and J. Robinson (2012), *Why Nations Fail: The Origins of Power, Prosperity and Poverty*, Profile Books, London.

Almirall, E. and J. Wareham (2011), "Living Labs: arbiters of mid-and ground-level innovation", *Technology Analysis & Strategic Management*, Vol. 23/1, pp. 87-102, Taylor & Francis Online, London, https://doi.org/10.1080/09537325.2011.537110.

Bason, C. (2016), *Design for policy*, Routledge, London and New York.

Berryhill, J., T. Bourgery and A. Hanson (2018), "Blockchains Unchained: Blockchain Technology and its Use in the Public Sector", *OECD Working Papers on Public Governance*, No. 28, OECD Publishing, Paris, https://doi.org/10.1787/3c32c429-en.

Breckon, J. (2015), *Better Public Services through Experimental Government, Alliance for Useful Evidence*, London, www.nesta.org.uk/sites/default/files/better-services-through-experimental-government.pdf.

Brown, T. (2008), "Design thinking", *Harvard Business Review*, Vol. 1/9, Harvard Business Publishing, Watertown, MA.

Cagan, J. and C. Vogel (2002), *Creating breakthrough products: Innovation from Product Planning to Program Approval*, Prentice Hall, Upper Saddle River, NJ.

Cardullo P. and R. Kitchin (2017a), "Being a 'citizen' in the smart city: Up and down the scaffold of smart citizen participation", *The Programmable City Working Paper*, 30, SocArXiv, Cornell University, Ithaca, NY, https://doi.org/10.31235/osf.io/v24jn.

Cardullo, P. and R. Kitchin (2017b), "Living Labs, vacancy, and gentrification", *The Programmable City Working Paper*, 28, SocArXiv, Cornell University, Ithaca, NY, https://doi.org/10.31235/osf.io/waq2e.

Chong, D. and H. Shi (2015), "Big data analytics: a literature review", *Journal of Management Analytics*, Vol. 2/3, pp.175-201, Taylor & Francis Online, London.

Coenen, L. et al. (2017), "Policy for system innovation-the case of Strategic Innovation Programs in Sweden", *Papers in Innovation Studies*, No. 2017/4, CIRCLE-Center for Innovation, Research and Competences in the Learning Economy, Lund University, Lund, Sweden, http://wp.circle.lu.se/upload/CIRCLE/workingpapers/201704_coenen_et_al.pdf.

Dutz, M. et al. (eds.) (2014), *Making Innovation Policy Work: Learning from Experimentation*, OECD Publishing, Paris, https://doi.org/10.1787/9789264185739-en.

Financial Conduct Authority (2015), "Regulatory sandbox", Financial Conduct Authority, London, www.ifashops.com/wp-content/uploads/2015/11/regulatory-sandbox.pdf.

Geels, F.W. (2004), "From sectoral systems of innovation to socio-technical systems: Insights about dynamics and change from sociology and institutional theory", *Research policy*, Vol. 33/6-7, pp. 897-920, Elsevier, Amsterdam, https://doi.org/10.1016/j.respol.2004.01.015.

Geels, F.W., A. McMeekin and B. Pfluger (2018), "Socio-technical scenarios as a methodological tool to explore social and political feasibility in low-carbon transitions: Bridging computer models and the multi-level perspective in UK electricity generation (2010-2050)", *Technological Forecasting and Social Change*, Elsevier, Amsterdam, https://doi.org/10.1016/j.techfore.2018.04.001.

Georghiou, Luke & Edler, Jakob & Uyarra, Elvira & Yeow, Jillian. (2014). Policy instruments for public procurement of innovation: Choice, design and assessment, *Technological Forecasting and Social Change*, 86. 1–12. 10.1016/j.techfore.2013.09.018.

Government of Canada (2016), "Experimentation direction for Deputy Heads – December 2016", webpage, Government of Canada, Ottawa, https://www.canada.ca/en/innovation-hub/services/reports-resources/experimentation-direction-deputy-heads.html.

Government of the United Kingdom (n.d. a), "Digital Service Standard", webpage, https://www.gov.uk/service-manual/service-standard.

Government of the United Kingdom (n.d. b), "Guidance: Government design policies", webpage, https://www.gov.uk/guidance/government-design-principles.

Guston, D.H. (2014), "Understanding 'anticipatory governance'", *Social Studies of Science*, Vol. 44/2, pp. 218-242, Sage Journals, Thousand Oaks, CA, https://doi.org/10.1177/0306312713508669.

Impact Canada Challenge Platform (n.d.) "Drug Checking Technology Challenge", webpage, https://impact.canada.ca/en/challenges/drug-checking-challenge.

Junginger, S. (2013), "Design and Innovation in the Public Sector: Matters of Design in Policy-making and Policy Implementation", *Annual Review of Policy Design*, Vol. 1/1, pp. 1-11, https://ojs.unbc.ca/index.php/design/article/view/542/475.

Karo, E. and R. Kattel (2018), "Innovation and the State: Towards an evolutionary theory of policy capacity", in *Policy Capacity and Governance*, pp. 123-150, Palgrave Macmillan, Cham, Switzerland.

Lawhon, M. and J.T. Murphy (2012), "Socio-technical regimes and sustainability transitions: Insights from political ecology", *Progress in Human Geography*, Vol. 36, pp. 354-378, Sage Journals, Thousand Oaks, CA, https://doi.org/10.1177/0309132511427960.

Lember, V. (forthcoming), "The role of new technologies in co-production and co-creation", in Brandsen, T. Steen, and B. Verschuere (eds.), *Co-production and co-creation: engaging citizens in public service delivery*, Routledge, London and New York.

Lember, V., R. Kattel and P. Tõnurist (2018), "Technological capacity in the public sector: The case of Estonia", *International Review of Administrative Sciences*, Vol. 84/2, pp. 214-230, Sage Journals, Thousand Oaks, CA, https://doi.org/10.1177/0020852317735164.

Margetts, H. and Naumann, A. (2017), Government as a platform: What can Estonia show the world, *Research paper*, University of Oxford.

Mazzucato, M. (2018), *Mission-oriented research & innovation in the European Union: A problem-solving approach to fuel innovation-led growth*, Publications Office of the European Union, Luxembourg, https://ec.europa.eu/info/sites/info/files/mazzucato_report_2018.pdf.

Mendelson, Z. (2016), "Mapping Mexico City's vast informal transit system", Fast Company, New York, www.fastcompany.com/3058475/mapping-mexico-citys-vast-informal-transit-system.

Mergel, I. (2018), "Open innovation in the public sector: drivers and barriers for the adoption of Challenge.gov", *Public Management Review*, Vol. 20/5, pp. 726-745, Taylor & Francis Online, London, https://doi.org/10.1080/14719037.2017.1320044.

Mulgan, G. (2017), Big Mind: How collective intelligence can change our world, Princeton University Press, Princeton.

Mulgan, G. (2013), "Experimental government", Nesta blog, 8 March 2013, www.nesta.org.uk/blog/experimental-government.

Noveck, B.S. (2015), *Smart Citizens, Smarter State. The Technologies of Expertise and the Future of Governing*, Harvard University Press, Cambridge, MA.

OECD (forthcoming a), *Systems approaches to creating public value in cities*, OECD Publishing, Paris.

OECD (forthcoming b), *The Innovation System of the Public Service of Canada*, OECD Publishing, Paris.

OECD (n.d.), "Behavioural insights", webpage, http://www.oecd.org/gov/regulatory-policy/behavioural-insights.htm (accessed on 15.06.2018).

OECD (2018), *Embracing Innovation in Government: Global Trends 2018*, OECD, Paris, http://www.oecd.org/gov/innovative-government/embracing-innovation-in-government-2018.pdf.

OECD (2017a), *The Next Production Revolution: Implications for Governments and Business*, OECD Publishing.

OECD (2017b), *Fostering Innovation in the Public Sector*, OECD Publishing, Paris, https://doi.org/10.1787/9789264270879-en.

OECD (2017c), "Embracing Innovation in Government: Global Trends", OECD, Paris, http://oe.cd/eig.

OECD (2017d), *Behavioural Insights and Public Policy: Lessons from Around the World*, OECD Publishing, Paris, http://dx.doi.org/10.1787/9789264270480-en.

OECD (2017e), *Systems Approaches to Public Sector Challenges: Working with Change*, OECD Publishing, Paris, http://dx.doi.org/10.1787/9789264279865-en.

OECD (2017f), "Building a data-driven public sector in Norway", in *Digital Government Review of Norway: Boosting the Digital Transformation of the Public Sector*, OECD Publishing, Paris, https://doi.org/10.1787/9789264279742-8-en.

OECD (2015a), *The Innovation Imperative in the Public Sector: Setting an Agenda for Action*, OECD Publishing, Paris, http://dx.doi.org/10.1787/9789264236561-en.

OECD (2015b), *Systems Innovation: Synthesis Report*, OECD, Paris, www.innovationpolicyplatform.org/system-innovation-oecd-project.

OECD (2014), "Recommendation of the Council on Digital Government Strategies, Adopted by the Council on 15 July 2014", OECD, Paris, http://www.oecd.org/gov/digital-government/Recommendation-digital-government-strategies.pdf.

OECD (2006), *Government R&D Funding and Company Behaviour: Measuring Behavioural Additionality*, OECD Publishing, Paris, https://doi.org/10.1787/9789264025851-en.

OPSI (2018), "Toolkit Navigator", Observatory of Public Sector Innovation, website, https://www.oecd-opsi.org/toolkit-navigator/ (accessed 15.07.2018).

OPSI (2017a), "Experimentation is the right to make mistakes", Observatory of Public Sector Innovation, OECD, Paris, https://www.oecd.org/governance/observatory-public-sector-innovation/blog/page/experimentationistherighttomakemistakes.htm.

OPSI (2017b), "Core skills for public sector innovation: A beta model of skills to promote and enable innovation in public sector organisations", Observatory of Public Sector Innovation, OECD, Paris, https://www.oecd.org/media/oecdorg/satellitesites/opsi/contents/files/OECD_OPSI-core_skills_for_public_sector_innovation-201704.pdf.

O'Reilly, T. (2010), "Government as a Platform", in D. Lathrop (ed.), *Open Government: Collaboration, Transparency, and Participation in Practice*, pp. 13-40, O'Reilly Media, Sebastopol, CA.

Pazaitis, A., P. De Filippi and V. Kostakis (2017a), "Blockchain and Value Systems in the Sharing Economy: The Illustrative Case of Backfeed", *Working Papers in Technology Governance and Economic Dynamics*, Vol. 73, Aalborg University, Copenhagen, http://technologygovernance.eu/files/main/2017012509590909.pdf.

Pazaitis, A., V. Kostakis and M. Bauwens (2017b), "Digital economy and the rise of open cooperativism: the case of the Enspiral Network. Transfer", *Transfer: European Review of Labour and Research*, Vol. 23/2, pp. 177-192, Sage Journals, Thousand Oaks, CA, https://doi.org/10.1177/1024258916683865.

Predictiv (n.d.), " Predictiv for policymakers & practitioners", webpage, www.predictiv.co.uk/governments.html.

Rodrik, D., A. Subramanian and F. Trebbi (2004), "Institutions Rule: The Primacy of Institutions over Geography and Integration in Economic Development", *Journal of Economic Growth*, Vol. 9/2, pp. 131-165, Springer Nature.

Schubert, C. (2017), "Green nudges: Do they work? Are they ethical?", *Ecological economics*, Vol. 132, pp. 329-342, Elsevier, Amsterdam, https://doi.org/10.1016/j.ecolecon.2016.11.009.

Schuurman, D. and P. Tõnurist (2017), "Innovation in the public sector: Exploring the characteristics and potential of living labs and innovation labs", *Technology Innovation Management Review*, Vol. 7, pp. 7-14, Talent First Network (Carleton University), Ottawa, https://doi.org/10.22215/timreview/1045.

Stilgoe, J., Owen, R. and Macnaghten, P., 2013. Developing a framework for responsible innovation. *Research Policy*, 42(9), pp.1568-1580.

Symons, T. and T. Bass (2017), "Me, my data and I: The future of the personal data economy", Nesta blog, https://www.nesta.org.uk/report/me-my-data-and-i-the-future-of-the-personal-data-economy.

Teece, D.J. (2018), "Business models and dynamic capabilities", *Long Range Planning*, Vol. 51/1, pp. 40-49, Elsevier, Amsterdam, https://doi.org/10.1016/j.lrp.2017.06.007.

Thaler, R. and C. Sunstein (2008), *Nudge: The gentle power of choice architecture*, Yale University Press, New Haven, CT.

Tõnurist, P., V. Lember and R. Kattel (2016), "Joint data platforms as X factor for efficiency gains in the public sector?", *Working Papers in Technology Governance and Economic Dynamics*, No. 70, Other Canon Foundation and Tallinn University of Technology, Hvasser, Norway, and Tallinn, http://technologygovernance.eu/files/main/2016082501020202.pdf.

UK Design Council (2013), "Design for Public Good: Facts, figures and practical plans for growth", UK Design Council, London, https://www.designcouncil.org.uk/resources/report/design-public-good.

UK Design Council (2011), "Design for innovation", UK Design Council, London, pp. 18-19, https://www.designcouncil.org.uk/resources/report/design-innovation.

Von Hippel, E. (2016), Free Innovation, MIT Press, Cambridge, MA.

Wherton, J. et al. (2015), "Co-production in practice: how people with assisted living needs can help design and evolve technologies and services", *Implementation Science*, Vol. 10/75, pp. 1-10, BioMedCentral, London, https://doi.org/10.1186/s13012-015-0271-8.

Chapter 12. The digitalisation of science and innovation policy

By

Dmitry Plekhanov, Michael Keenan, Fernando Galindo-Rueda and Daniel Ker.

Digitalisation will profoundly affect the public sector and the evidence base on which it formulates, implements, monitors and evaluates public policy. The science, technology and innovation (STI) policy field is no exception. In recent years, many countries have begun to develop digital science and innovation policy (DSIP) initiatives, to help build a picture of the incidence and impact of their science and innovation activities. This chapter provides an introductory overview of DSIP systems in OECD member and partner countries. Drawing on the findings of a recent OECD survey of DSIP initiatives, it first outlines the main characteristics of the DSIP systems currently in use and under development. It then describes the promises and challenges of DSIP systems. It shows that much can be gained from this digital transformation by leveraging the untapped potential of data about STI. However, obstacles and risks also exist. These relate to privacy and confidentiality, interoperability standards, and potential misalignment of incentives between policy objectives and STI actors, including the private sector. If DSIP initiatives are to fulfil their future potential, STI policy needs to address these opportunities and challenges, sometimes at the international level. The chapter concludes by considering the outlook for DSIP systems and possible avenues for policy action.

Introduction

Publicly funded research systems generate, in addition to key research data outputs, considerable amounts of information about the operation of those systems. Policy makers can use this information to monitor the performance and improve the efficiency of their research systems. Emerging digital science and innovation policy (DSIP) initiatives increasingly interconnect various information sources, and apply new technologies and applications that allow policy makers to exploit them more extensively. These systems can help build a picture of the incidence and impact of science and innovation activities, providing potentially valuable tools to facilitate decision-making across the broad spectrum of STI policy and administration. For instance, ministries can use DSIP systems to design, implement, monitor and evaluate policies. Funding agencies can use them to plan, co-ordinate, monitor and evaluate their activities. With the growing wealth of data about research and innovation, DSIP systems could transform the ways in which STI policy is defined and public services are delivered.

Several drivers of change are influencing these developments. First, government itself is undergoing a digital transformation. Digital technologies offer opportunities to increase the access, reach and quality of public services, as well as improve policy making and service design (OECD, 2018, 2014; Ubaldi, 2013). The STI policy field is no exception, although it is not as far along the digitalisation road as other policy areas. Second, the growing adoption of open science (OECD, 2015a; Dai et al., 2018) has created various infrastructures – such as data repositories and interoperability standards – which DSIP systems can readily re-use. Open science has also raised expectations that STI policy should also be open. Third, the emerging interdisciplinary field of science-of-science and innovation policy (Lane, 2010; Husbands Fealing et al., 2011) strongly emphasises developing data and metrics that STI policy makers can apply to their decision-making. Several DSIP initiatives originated in this field, and several others are influenced by it.

This chapter provides an introductory overview of DSIP systems in OECD member and partner countries. Drawing on the findings of a recent OECD survey of DSIP initiatives, it outlines the main characteristics of the DSIP systems currently in use and under development. It then describes the promises and challenges of DSIP systems. Finally, it considers the outlook for DSIP systems and possible avenues for policy action.

The DSIP landscape: A brief overview

"DSIP initiatives" refer to the adoption or implementation by public administrations of new or re-used procedures and infrastructures relying on an intensive use of digital technologies and data resources, to support the formulation and delivery of science and innovation policy. The primary goal of DSIP initiatives is to support certain aspects of the public-policy process, although any actor in the system – including in the private sector – can provide functionalities.

The OECD DSIP project is a first attempt at mapping the landscape of DSIP initiatives in OECD member and partner countries. It addresses the highly specific nature of digital government in the area of science and innovation policy. It includes a survey of 39 DSIP initiatives from 29 OECD and partner countries, which provides much of the evidence used to prepare this chapter.

The results of the survey show that DSIP systems come in many shapes and sizes, making it difficult to classify them neatly. Broadly speaking, one group comprises systems that build on a funding ministry or agency's administrative databases, linking them to other (typically external) data, e.g. to gain insights on funding outputs and impacts. Examples include Argentina's Sistema de Información de Ciencia y Tecnología Argentino (SICYTAR); South Africa's Research Information Management System; Poland's POL-on; and Federal RePORTER in the United States. Another group of DSIP systems consists of analytical solutions (often using machine learning, big data and semantic analysis) that collect and combine data from multiple data sources to provide insights for policy making. Examples include Corpus Viewer in Spain; Arloesiadur in the United Kingdom; SciREX Policymaking Intelligent Assistant System (SPIAS) in Japan; and iFORA in Russia.

While a few DSIP initiatives (e.g. Corpus Viewer) began as part of broader open government/big-data initiatives, most have originated in the STI policy domain. The main operators of DSIP systems captured by the OECD survey are STI ministries and funding bodies. Public research organisations (PROs) that provide governments with strategic policy intelligence services (e.g. evaluation and foresight) also operate DSIP systems in several countries (e.g. Japan and Korea). National statistical offices (NSOs) sometimes play a supporting role, shaped by their core statistical mandate and legislative framework, and the resources available to provide an enhanced range of digital services (Chapter 14).

Figure 12.1 provides a stylised conceptual view of a DSIP initiative and its main components. All of these elements interact in nationally specific ways, reflecting different histories and institutional set-ups. The main elements consist of various input data sources, which feed into a data cycle that is enabled by interoperability standards, including unique, persistent and pervasive identifiers (UPPIs). DSIP systems perform a number of functions and are often used by a mix of users. Box 12.1 highlights several examples of DSIP initiatives.

Figure 12.1. A stylised conceptual view of a DSIP initiative and its possible main components

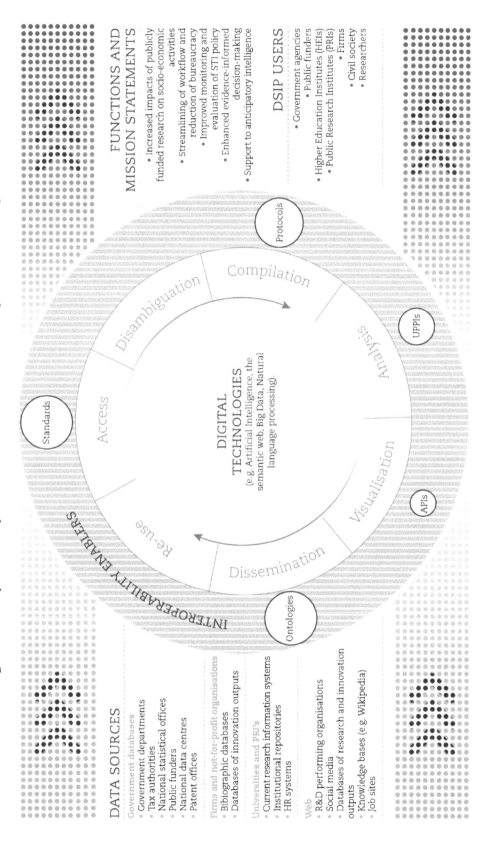

> **Box 12.1. Examples of DSIP systems**
>
> In Argentina, the Ministry of Science, Technology and Productive Innovation uses SICYTAR to evaluate and assess STI policy initiatives, project teams and individual researchers. The system aggregates several databases, covering researchers' curriculum vitae; funded research and development (R&D) projects; information on public and private institutions performing R&D activities in Argentina; and information on large research equipment.
>
> In the Netherlands, the National Academic Research and Collaborations Information System (NARCIS) collects data from multiple sources, including funder databases, current research information systems (CRIS), institutional repositories of research performers and the Internet (Dijk et al., 2006). Data on research outputs, projects, funding, human resources and policy documents collected by NARCIS are used to inform policy makers on research activities undertaken in the Netherlands and to monitor open access. Funders also use the system to identify "white spots" in research to improve resource planning. NARCIS also serves as an important directory of research, providing researchers, journalists, and the domestic and international public with information on the status and outputs of Dutch science.
>
> In Norway, the research-reporting tool Cristin collects information from research institutions, the Norwegian Centre for Research Data and ethics committees. Cristin serves as a foundation for the performance-based funding model of the Ministry of Research and Education. It provides numerous users from government, industry, academia and civil society with verified information on the current status of Norwegian research.
>
> In Japan, the National Graduate Institute for Policy Studies designed the SPIAS system to strengthen national evidence-informed STI policy making. SPIAS uses big data and semantic technologies to process data on research outputs and impacts, funding, R&D-performing organisations and research projects, with a view to mapping the socio-economic impacts of research. SPIAS has been used to analyse leading Japanese scientists' performance before and after receiving grants from the Japan Science and Technology Agency; assess the impact of regenerative medicine research in Japan; and map emerging technologies.
>
> In Spain, Corpus Viewer, developed by the State Secretariat for Information Society and Digital Agenda, processes and analyses large volumes of textual information using natural-language processing techniques. Policy makers use the results of these analyses to monitor and evaluate public programmes, and formulate science and innovation policy initiatives. The system is currently restricted to government officials.

The promises of DSIP

Governments are increasingly launching DSIP initiatives, often with the following objectives:

- *Optimise administrative workflows*: digital tools can help streamline potentially burdensome administrative procedures and deliver significant efficiency gains within agencies. These benefits can also extend to those using public agencies' services, including researchers or organisations applying for (or reporting on) the use of research grants; for example, they can use interoperability identifiers to link their research profiles to grant applications.

- *Support better policy formulation and design*: digitalisation offers new opportunities for more granular and timely data analysis to support STI policy; this should improve the allocation of research and innovation funding. Furthermore, DSIP systems often link data collected by different agencies, providing greater context to policy problems and interventions, and offering possibilities for a more integrated interagency policy design at the research or innovation system level.

- *Support performance monitoring and management*: DSIP systems offer the possibility of collating real-time policy output data. This can allow more agile short-term policy adjustments. It can improve insights into the policy process for accountability and learning in the medium to long term, so that evaluation becomes an open and continuous process. Policy makers and delivery agencies can consider the circumstances that make it possible and meaningful to use other digitally enabled data resources, such as altmetrics of research outputs and impacts (Priem et al., 2010; Sugimoto and Larivière, 2016). They can also rely on other data-collection approaches (e.g. web scraping) to complement and enhance existing approaches to assessing research.

- *Provide anticipatory intelligence*: technologies like big-data analytics can help detect patterns, e.g. emerging research areas, technologies, industries and policy issues. They can support short-term forecasting of policy issues and contribute to strategic policy planning (Peng et al., 2017; Choi et al., 2011; Zhang et al., 2016; Yoo and Won, 2018). For example, DSIP systems could identify job-market demand for specific STI fields and address potential mismatches on the supply side.

- *Help in general information discovery*: DSIP systems often include data on a wide range of inputs, outputs and activities. Policy makers and funders can use these data to identify leading experts in a given field (e.g. identify reviewers for project proposals), as well as centres of excellence (Sateli et al., 2016; Guo et al., 2012). This kind of information also helps researchers and entrepreneurs to identify new partners for collaboration and commercialisation.

- *Promote inclusiveness in science and innovation agenda-setting*: DSIP systems can contribute to the debate with stakeholders on policy options by providing detailed information about the policy problem in an accessible medium, e.g. through interactive data visualisation. The increased transparency provided by DSIP systems can empower citizens by providing them with knowledge about the nature and impacts of ongoing research and innovation. Thus, DSIP may be instrumental in building trust and securing long-term sustainable funding for research and innovation.

Fulfilling these promises will depend on policy makers' readiness to embrace the digital revolution (Box 12.2). It will also depend on meeting several challenges, discussed in the following section.

Box 12.2. In my view: Are science policy makers ready to embrace the digital revolution?

Clinton Watson, Principal Policy Advisor, New Zealand Ministry of Business, Innovation and Employment

Policy makers in science and innovation are charged with designing and overseeing funding mechanisms that funnel billions of dollars of public money into universities, PROs, businesses and not-for-profit entities. Yet despite the huge investments, the science-of-science policy has received almost no funding. Oftentimes, policy makers struggle to demonstrate real societal impacts from investments. Arguably, they have paid more attention to ensuring science systems continue to receive adequate funding and respond to domestic demands. Assessing and demonstrating performance has often played a secondary role to setting high-level objectives and getting money out the door.

Politicians and senior public-sector leaders are increasingly demanding hard evidence of what works and what does not. Science and innovation-related spending is no longer exempt from pressures to provide quantitative evidence of impact. In some countries, the storyline of good science delivering societal outcomes many years down the track is wearing thin. At the same time, hard-to-answer questions on optimal institutional settings, design of funding pots and efficient allocation systems persist. Policy makers need to focus more on supporting monitoring systems, evaluation frameworks and data infrastructures, working with the very researchers and academics they fund.

Advances in information technologies and data-linking techniques are now presenting policy makers with the tools to start answering the hard questions. A handful of countries have developed national research information systems that harvest data from multiple sources. If these systems can be linked to other national data infrastructures (e.g. housing economic, environmental and social data), science policy makers will be in a unique position to demonstrate quantitative relationships between science and innovation, and real-world outcomes. Researchers could also use these linked data infrastructures to prove, for example, that firms collaborating with universities become more productive, or that certain types of research lead to improved environmental outcomes over time. They could also produce useful descriptive statistics, such as the value and growth of spin-out companies.

For several years, I led efforts in New Zealand to improve data holdings on research, science and innovation. Through collaboration between government agencies and key sector bodies, we identified the enduring questions to answer, our data needs, our current data holdings and a high-level roadmap for action. Key challenges were securing trust in data use, developing communication channels within institutions and identifying best practice globally. Implementation has centred on securing sustainable funding, providing detailed communication of benefits, and establishing legal and governance frameworks.

The New Zealand experience and other similar initiatives all point to the social and cultural challenges in building data infrastructures for science and innovation policy. The idea of "social licence", or community acceptance and trust of data use, is in the spotlight. Institutions and researchers need to have assurance that data about their funding, activities and results will be handled appropriately and protected when needed. Many universities and research organisations are also not used to automatic data transfer to a central hub. The funding of national level systems also presents challenges. The optimal cost sharing

> between the central research and innovation ministries, science funders and research providers will differ depending on institutional responsibilities and funding flows.
>
> Science policy cannot afford to be immune to the digital transformation we are witnessing across economies. We need to embrace digitalisation if we are to prove the ongoing worth of science and innovation, and raise the effectiveness of public spending. Policy makers need to support digital tools and their social licence, creating long-term plans for establishing linked data infrastructures, establishing effective governance and funding structures, and building capacity for the science-of-science policy.

Main policy challenges

Realising the potential of DSIP involves overcoming several possible barriers. In their responses to the OECD questionnaire, DSIP administrators identified data quality, interoperability, sustainable funding and data-protection regulations as the biggest challenges facing their initiatives (Figure 12.2). Access to data, the availability of digital skills and trust in digital technologies were somewhat less often cited as challenges.

Figure 12.2. Main challenges facing DSIP initiatives

Percentage of surveyed DSIP systems

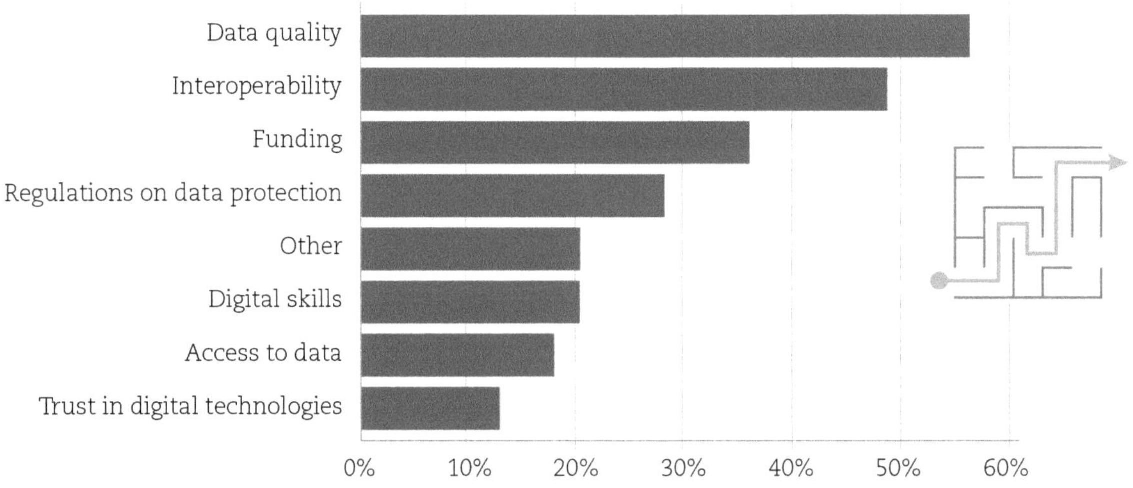

Note: Questionnaire respondents could select more than one challenge facing their DSIP initiatives.
Source: OECD survey of administrators of 39 DSIP systems in OECD member countries and partner economies.

StatLink https://doi.org/10.1787/888933858335

Policy makers wishing to promote DSIP in their countries face further systemic challenges, including overseeing fragmented DSIP efforts and multiple (often weakly co-ordinated) initiatives; ensuring the responsible use of data generated for other purposes; and balancing the benefits and risks of private-sector involvement in providing DSIP data, components and services. Figure 12.3 summarises and organises the main challenges in implementing or using DSIP systems. The section below elaborates on each challenge.

Figure 12.3. Challenges in implementing and using DSIP systems

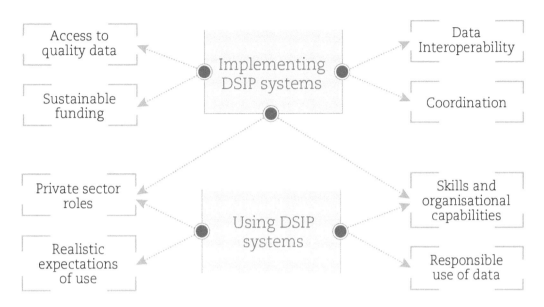

Access to quality data

Most DSIP systems draw upon different data sources to provide new insights that cannot be obtained through working with each data source separately. For example, they link data on inputs and outputs to provide insights on the impacts and efficiency of public research funding. Most of the DSIP systems surveyed incorporate data on research outputs (typically academic publications), research organisations, research funding (i.e. project and grant awards), research personnel and research projects (Figure 12.4). Some DSIP systems include data on research equipment and facilities, as well as research impacts (including citations and media mentions).

Figure 12.4. Types of information harnessed for DSIP systems

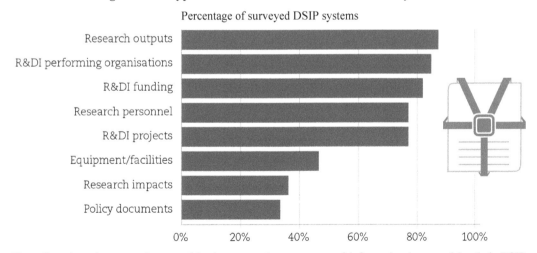

Note: Questionnaire respondents could select more than one type of information harnessed by their DSIP initiatives.
Source: OECD survey of administrators of 39 DSIP systems in OECD member countries and partner economies.

StatLink ⟶ https://doi.org/10.1787/888933858354

While data reusability is a major source of efficiency promised by DSIP systems ("enter once, re-use often"), respondents to the OECD survey of DSIP administrators cited data quality as a major challenge (Figure 12.2). Data quality is a multi-dimensional concept, encompassing relevance, accuracy, credibility, timeliness, accessibility, interpretability, coherence and cost efficiency (OECD, 2011). It ultimately defines whether data serve a given purpose. Data used in DSIP systems may have been generated for different or related purposes, meaning that users must assess quality factors for each intended application. Data are predominantly sourced from a mix of funding agencies (typically their administrative data, e.g. databases of grant awards) and research performers (e.g. university CRIS), as well as proprietary bibliometric and patent databases. However, available data may not capture precisely what is needed for the DSIP system (need for relevance/interpretability); alternatively, they may be presented in an unstructured format that is complicated to process (need for accessibility/coherence). Fixing this may require further complementary resources, including additional metadata; algorithms for data processing; and secure digital infrastructures for (shared) data storage, processing and access. The costs involved may discourage more widespread data sharing, particularly when its benefits are not always obvious to those providing the data (OECD, 2017).

Other potential barriers exist to open-data sharing. These include bureaucratic competition and conflicting interests among government organisations and individual departments, and notions that any value to be extracted from administrative data should be initially – and primarily – the preserve of the data owners. Several systems provide tiered access to their data, whereby policy officials inside the host organisation can access more granular data. A lack of trust in the manner in which shared data will be used may also hinder sharing. For example, organisations may be legitimately concerned that their data will be misused or poorly interpreted by users with an inadequate understanding of its meaning and limitations. As semi-autonomous agents, organisations may also fear the unwanted scrutiny of their operations that open administrative data might invite. Privacy and confidentiality are also major concerns when re-using data collected for other purposes (Lane et al., 2015.

Data interoperability

The databases used in DSIP systems have often been locally designed, without adherence to common standards. Hence, many of the data relevant to STI policy are stored in inaccessible silos, complicating data re-use. Ensuring data compatibility is not only potentially beneficial to policy makers and other stakeholders managing national research and innovation systems, it can yield considerable benefits for individuals and organisations doing (or reporting on) research. If an individual data item is made interoperable, it can be re-used across multiple systems, meaning it can be provided to authorities only once. Interoperability also allows diffusing updates across systems more easily and automatically comparing information from multiple sources (e.g. checking the consistency of project-funding reports submitted by researchers and employers). An integrated and interoperable system leads to a considerable reduction in the reporting and compliance burden, freeing up more time and money for research itself.

Research organisations, funders and non-profit organisations have started designing standards, vocabularies and protocols that connect and disambiguate research data and metadata to improve interoperability between silos. Some DSIP systems use existing national identifications (IDs) – e.g. business registration and social security numbers – as well country-specific IDs for researchers (Figure 12.5). In recent years, attempts have been made to establish international standards and vocabularies to improve the international interoperability of DSIP infrastructures. These include UPPIs, which assign a standardised code that is unique to each research entity, persistent over time and pervasive across various datasets. One example is Open Researcher and Contributor ID (ORCID), which aims to resolve name ambiguity in scientific research by developing unique identifiers for individual researchers. Figure 12.1 sets out several prominent examples.

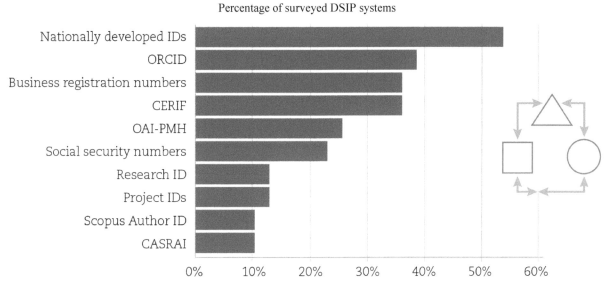

Figure 12.5. Use of interoperability enablers in DSIP systems

Note: Questionnaire respondents could select more than one type of interoperability enabler used in their DSIP initiatives.
Source: OECD survey of administrators of 39 DSIP systems in OECD member countries and partner economies

StatLink https://doi.org/10.1787/888933858373

Table 12.1. Examples of interoperability enablers in DSIP and related systems

Type	Examples
UPPIs for STI actors	ORCID
	Digital Object Identifier (DOI)
	Global Research Identifier Database (GRID)
	International Standard Name Identifier (ISNI)
	Ringgold ID
Author IDs generated by publishers/indexers	Researcher ID
	Scopus Author ID
Management standards for data about STI	Common European Research Information Format (CERIF)
	Consortia Advancing Standards in Research Administration Information (CASRAI) Dictionary
	VIVO ontology
Protocols	Open Archives Initiative Protocol for Metadata Harvesting (OAI-PMH)

In addition to the reduced administrative burden, interoperability allows quicker, cheaper and more accurate data matching, making existing analyses less costly and more robust, and facilitating new analyses. Interoperability can produce more timely and detailed insights, enabling more responsive and tailored policy design. Furthermore, the gradual emergence of internationally recognised UPPIs makes it easier to track the impacts of research and innovation activities across borders, and map international partnerships.

As Figure 12.5 shows, several DSIP systems also use the CERIF data-management standard to promote uniform management and exchange of research information.[1] A smaller number use the standard dictionary of research administration developed by CASRAI. Semantic ontologies can also help improve interoperability among DSIP systems. For example, the VIVO project has developed an ontology for research information that enables a federated search of organisations, researchers and activities, and their relationships. However, because few DSIP systems currently use semantic technologies, none of the systems included in the OECD survey have deployed it yet.

Thus, although identifiers, standards and protocols have proliferated, interoperability remains a major challenge (Figure 12.2). In the absence of national strategies to ensure the architectural coherence and interoperability of public databases, there is a risk that data sources will become fragmented, undermining the functionality of DSIP systems and raising the costs of data integration.

Domestic and international co-ordination

As individual ministries and agencies in the policy system increasingly rely on digital tools to exploit their administrative data, DSIP initiatives are proliferating. Some are ambitious in scope and seek to become national data hubs; many are tentative experiments, limited in scope and with little cross-departmental co-ordination. They lack the mandate (and resources) to expand, and prefer to remain manageable and modest.

This fragmented landscape presents likely drawbacks, including inefficient, overlapping efforts; missed opportunities to establish interoperability standards that would improve data exchange; and greater funding uncertainties. However, simultaneously running multiple small-scale experiments may present benefits, by providing more space for innovation and promoting a more agile DSIP ecosystem. Moreover, if these distributed initiatives adopted common data-management frameworks (including interoperability standards), then a modular landscape of interconnected DSIP initiatives could emerge. This interconnectedness could have some advantages over a single dominant system, by achieving greater service scalability and flexibility, as well as more seamless integration with external digital infrastructures. It could also support data integration from a wider range of ministries and agencies funding R&D, and thus provide a more complete picture of national policy and funding for research and innovation.

For these benefits to be realised, however, there needs to be co-ordination of data-management frameworks. Most countries already have ministries and agencies that formulate high-level national digital strategies, establish technological architectures and promote good practices in public-data management. However, although these measures can provide the necessary conditions for DSIP to flourish, they are insufficient on their own. For instance, specificities to the STI domain around data sources, interoperability standards and the intended (and unintended) uses of DSIP data require further – but still underdeveloped – co-ordination and support at the STI policy level. Most countries still lack dedicated plans for DSIP, and only a few (e.g. Norway and New Zealand) have appointed lead agencies to formulate and co-ordinate common frameworks for STI policy-related data management. In the absence of national co-ordination mechanisms, the wide adoption of international (or private-sector) interoperability standards could provide an "invisible hand" for co-ordination, but DSIP owners would still need to prepare and agree to share their own administrative data. Thus, beyond finding technical solutions for system interoperability, DSIP is ruled by political considerations and compromise.

Sustainable funding

Most of the DSIP systems surveyed are funded by their host organisations' operating budgets, which could be a positive sign of their long-term survival. However, many DSIP systems are relatively new, and it is difficult to estimate their sustainability. More than one-third of the surveyed DSIP administrators pointed to funding as a challenge for their systems, the third highest ranked challenge after data quality and interoperability (Figure 12.2). The individuals charged with building and maintaining DSIP systems often underestimate the magnitude of the task, particularly with respect to data access, disambiguation and linking; this can lead to significant project delays and cost overruns. As with any infrastructure, maintenance and use costs may also be higher than the initial investment costs.

Skills and organisational capabilities

A distinction should be made between the skills and organisational capabilities needed to use DSIP systems in policy making and analysis, and those needed to build and maintain digital infrastructures, which DSIP administrators ranked as a low-level challenge (Figure 12.2) – perhaps because many systems use well-established digital tools and techniques easily mastered by existing digital teams. When faced with more challenging problems, the individuals implementing DSIP infrastructures can readily buy the necessary technical expertise on the market.

A few DSIP initiatives are experimenting with more advanced digital tools, such as semantic technologies to link datasets, algorithms to support big-data analytics, and interactive visualisations and dashboards to promote data use in the policy process. For example, Spain's Corpus Viewer uses natural-language processing techniques to process and analyse large volumes of textual data on Spanish research funding (Box 13.1). In the United Kingdom, the Arloesiadur project[2] – a partnership between the Welsh government and NESTA, with inputs from a company specialising in data visualisation – combines traditional indicators with data from social networks, company websites and collaboration platforms, to provide interactive visualisations of research and innovation networks in Wales (Mateos-Garcia et al., 2017). Together with IBM, the long-established Flanders Research Information Space is exploring ways to use web scraping to capture Flemish research outputs scattered across the web.

To date, the public sector has rarely used advanced digital tools in its DSIP initiatives. This reluctance may stem from the costs of hiring digital-technology professionals with expertise in big data, machine learning and natural-language processing, which can be prohibitive, given competition from the private sector for these skills. It may also reflect policy data needs, which remain quite straightforward (c.f. many policy makers eschew advanced econometric studies in favour of simpler indicators). Approaches such as semantic analysis tend to be quite technical; interpreting the information they provide requires certain skills.

Considering the skills and organisational capabilities needed to utilise the data and functions of DSIP systems, the STI policy-making community is increasingly attracting quantitatively literate officials with backgrounds in various analytical disciplines (Chapter 14 on next-generation data and indicators). A striking number of DSIP initiatives seem to target this specialised audience of analysts: many DSIP users are evaluators and analysts who act as intermediaries, processing the data before feeding it to decision-makers. This situation could change in the future thanks to advances in visual analytics, which could open up DSIP systems to a wider range of non-analyst users, both in government and beyond.

At the level of the policy organisation, a mix of capabilities is required, including technical staff with specialised skills in data curation and stewardship, to manage the use of necessary standards and metadata. Policy analysts and decision-makers would find it useful to possess statistical skills, i.e. knowledge of key concepts and statistical software. Existing staff can accumulate some of these capabilities gradually, by upskilling through massive open online courses; this is a more cost-effective option than hiring expensive data scientists. In this way, DSIP initiatives could benefit from a process of cumulative organisational learning and deploy increasingly ambitious technologies.

The private sector plays an increasingly important role in DSIP systems. For example, various academic publishers, web service companies and data-management systems provide access to proprietary databases, digital analytical tools and unique identifiers. Beyond the simple provision of services, these relations encompass different levels of public-private co-operation, such as joint development of methods and tools to analyse research impact, and collaboration on the design and implementation of digital platforms for policy-making purposes.

Three companies with long-standing ties with the academic research community – Elsevier (the world's largest academic publisher and owner of the Scopus index), Holtzbrinck Publishing Group (owner of Springer Nature and Digital Science) and Clarivate Analytics (formerly part of Thomson Reuters and owner of the Web of Science index) – are developing digital solutions for workflow management and research analytics on top of in-house databases of research outputs. By acquiring and developing digital tools that complement their product portfolios, and building interoperability linkages between in-house and external solutions, they are creating digital platforms of interconnected digital products with similar functionalities to publicly owned DSIP systems. They are using machine learning, natural-language processing and big-data analytics to exploit in-house databases. They are also designing new add-on analytical services to monitor and assess research and innovation activities.

Although they do not formally provide DSIP solutions to governments, large information and communication technology firms provide some of the building blocks for DSIP. Some, like Google Scholar, Microsoft Academic, Baidu Scholar and Naver Academic, have already transformed scientific and technical discovery with their search engines. DSIP systems only minimally rely on these solutions, but this could change as they become more sophisticated (e.g. by deploying artificial intelligence and semantic-search tools) and provide wider coverage of research outputs.

Private-sector involvement in DSIP initiatives offers several benefits. Private firms can often provide off-the-shelf, well-developed solutions and building blocks for DSIP. These can be implemented quickly and at an agreed cost, sparing the public sector the need to develop the necessary in-house skills beforehand. As highlighted earlier, private companies can also promote interoperability through their standards and products; moreover, the largest firms operate across national borders, and can therefore promote international interoperability. This can expand the scope and scale of data within a DSIP system utilising these products and standards; for example, policy makers can compare the features of their own research systems with others. At the same time, governments often expect their open public data to spur innovation (e.g. new products and services) in the private sector.

Potential risks also exist when the public sector relies on the private sector for DSIP systems and components. For example, outsourcing data-management activities to the private sector may result in a loss of control over the future development trajectory of DSIP systems; reliance on proprietary products and services may lead to discriminatory access to data, even if these concern research activities funded by the public sector; and the public sector's adoption of commercial standards for metrics may drive the emergence of private platforms exhibiting network effects that are difficult to contest. Furthermore, while methods and algorithms are sources of competitive advantages, the secrecy surrounding them can undermine trust in such systems, particularly when they are used to assess research performance.

Realistic expectations of use

By re-using and combining data from a variety of sources, DSIP can provide policy makers with a broader view of the research and innovation landscape, and consequently furnish evidence to help them allocate funding. However, expectations around the uses of DSIP should avoid a "naïve rationalism" that ignores the inherent messiness of policy making. DSIP can inform policy judgement, but it cannot and should not provide a "technical fix" to what are ultimately political judgements, shaped by competing values and uncertainty. If they were "open by design", DSIP systems could promote inclusiveness in science and innovation agenda-setting, making it less technocratic and more democratic. Whatever the policy setting, an embedded and routine use of DSIP will depend not just on digital technologies, but also on favourable social and administrative conditions promoting their adoption.

Responsible use of data

Private and confidential data make up a considerable portion of the data processed by the public sector, and can be potentially useful in DSIP systems. However, these data must be used responsibly (OECD, 2013). This often means anonymising or aggregating them – e.g. when the identity of individual companies would become apparent in more granular data. More than one-quarter of the DSIP administrators surveyed highlighted data-protection regulations as a challenge (Figure 12.2).

More than half of the DSIP systems surveyed play a role in research assessment. Some, like the Cristin system in Norway, the Lattes Platform in Brazil, and the METIS system in the Netherlands, are the primary sources of data for national research assessments. However, some evidence exists that non-policy actors are also using DSIP data – e.g. to assess the performance of individuals – raising concerns over the responsible use of linked open data generated for other purposes. DSIP could reinforce some existing misuses of data (e.g. reliance on journal-impact factors in various types of assessments), which could further distort the incentives and behaviour of individuals and organisations (Edwards and Siddhartha, 2017; Hicks et al., 2015).

Over-reliance on data is dangerous when its interpretation is problematic – hence the need to improve our understanding of STI processes, to make sense of the data. DSIP systems offer a great opportunity to develop such an understanding, as the data can be made available to a broad community of researchers, who could further develop the emerging field of science-of-science and innovation policy (Lane, 2010; Husbands Fealing et al., 2011). AI-based tools can also be mobilised to promote such an understanding. Barring that, DSIP systems will simply result in even more data being interpreted – and often misinterpreted – in many ways.

Although few surveyed DSIP administrators reported a lack of trust in digital technologies as a major challenge (Figure 12.2), this could change with the introduction of newer and more advanced technologies and processes, e.g. machine learning and big-data analytics. These technologies rely on notoriously opaque algorithms, which could undermine trust in DSIP-based solutions. The use of data sources with questionable provenance – e.g. data derived through web scraping of company websites – is another potential source of mistrust, which should be treated with care.

Future outlook

Digital content and processes will play an important role in the future policy design, operational delivery and governance arrangements of research and innovation. Governments cannot continue to work in analogue mode when society and the economy are increasingly working in digital mode. The rapid and broad uptake of digital technologies and data across the public sector will place increasing pressure on governments to rethink the management of core policy processes and activities, including with regard to STI policy.

The digital transformation of STI policy and its evidence base is still in its early stages. As digitalisation becomes increasingly pervasive, uncertainties remain as to what it will cover, who will take the lead, and what roles existing actors (including NSOs, as data clearinghouses for statistical purposes) will play. The consequences on the relations (including governance arrangements) between STI actors are also uncertain. Moreover, international co-operation could take different forms and perform different functions in the future DSIP landscape.

STI policy makers could assume a relatively passive stance in the face of these developments: their activities – including the evidence base they use to inform their decision-making – will inevitably become increasingly digitalised. Alternatively, they could adopt a more active stance, shaping the DSIP ecosystems to fit their needs. This will require strategic co-operation, through significant interagency co-ordination and sharing of resources (such as standard digital identifiers), and a coherent policy framework for data sharing and re-use in the public sector. Since several government ministries and agencies formulate science and innovation policy, DSIP ecosystems should be founded on the principles of co-design, co-creation and co-governance.

In a desirable future scenario, DSIP infrastructures will provide multiple actors in STI systems with up-to-date linked microdata to help inform their decision-making. They will erode information asymmetries, and empower a broad group of stakeholders to participate more actively in the formulation and delivery of science and innovation policy. Policy frameworks will have resolved privacy and security concerns, and national and international co-operation on metadata standards will have addressed interoperability issues. Best practices in the responsible use of DSIP systems will have taken hold, informed by widely accepted norms of acceptable use. While the private sector will provide supporting infrastructures and services, the public sector will own its data, ensuring they remain outside of "walled gardens", for others to readily access and re-use.

Considerable scope also exists for international mutual learning and co-operation in developing digital data infrastructures for STI policy. Given the global nature of science and innovation activities, there could be particular benefits in establishing further international standards – including strengthening existing OECD legal and informal guidance instruments – to take full stock of the potential and challenges of DSIP.

Notes

[1] Administrative standards relate closely to, but do not fully align with, OECD statistical standards (e.g. Frascati Manual [OECD, 2015b]) or definitions contained in OECD legal instruments. As noted in Chapter 14 (on next generation data and indicators), it is important to ensure closer correspondence between these different international standards and the standards used by countries and supranational organisations, such as the European Union.

[2] "Innovation Directory" in Welsh.

References

Choi, S. et al. (2011), "SAO network analysis of patents for technology trends identification: A case study of polymer electrolyte membrane technology in proton exchange membrane fuel cells", *Scientometrics*, Vol. 88/3, pp. 863-883, Springer International Publishing, Cham, Switzerland, https://doi.org/10.1007/s11192-011-0420-z.

Dai, Q., E. Shin and C. Smith (2018), "Open and inclusive collaboration in science: A framework", *OECD Science, Technology and Industry Working Papers*, No. 2018/07, OECD Publishing, Paris, https://doi.org/10.1787/2dbff737-en.

Dijk, E. et al. (2006), "NARCIS: The Gateway to Dutch Scientific Information", in *Digital Spectrum: Integrating Technology and Culture – Supplement to the Proceedings of the 10th International Conference on Electronic Publishing*, pp. 49-58, ELPUB, Bansko, Bulgaria, https://elpub.architexturez.net/doc/oai-elpub-id-233-elpub2006.

Edwards, M. and R. Siddhartha (2017), "Academic Research in the 21st Century: Maintaining Scientific Integrity in a Climate of Perverse Incentives and Hypercompetition", *Environmental Engineering Science*, Vol. 34/1, pp. 51-61, Mary Ann Liebert, Inc. Publishers, New Rochelle, NY, http://online.liebertpub.com/doi/pdf/10.1089/ees.2016.0223.

Guo, Y et al. (2012), "Text mining of information resources to inform forecasting innovation pathways", *Technology Analysis & Strategic Management*, Vol. 24/8, pp. 843-861, Routledge, London, https://doi.org/10.1080/09537325.2012.715491.

Hicks, D. et al (2015), "Bibliometrics: The Leiden Manifesto for research metrics", *Nature*, Vol. 520/7548, pp. 429-31, Macmillan Publishers, London, https://doi.org/10.1038/520429a.

Husbands-Fealing, K. et al. (eds.) (2011), *The Science of Science Policy: A Handbook*, Stanford Business Books, Stanford University Press, Stanford, CA, http://www.sup.org/books/title/?id=18746.

Lane, J. et al. (2015), "New Linked Data on Research Investments: Scientific Workforce, Productivity, and Public Value", *NBER Working Paper*, No. 20683, National Bureau of Economic Research, Cambridge, MA, http://www.nber.org/papers/w20683.

Lane, J. (2010), "Let's make science metrics more scientific", *Nature*, Vol. 464, pp. 488-489, Macmillan Publishers, London, https://doi.org/doi:10.1038/464488a.

Mateos-Garcia, J., K. Stathoulopoulos and S. Bashir Mohamed (2017), "An (increasingly) visible college: Mapping and strengthening research and innovation networks with open data", *SocArXiv Papers*, University of Maryland, College Park, MD, https://doi.org/10.17605/OSF.IO/3CU67.

OECD (2018), "Going Digital in a Multilateral World", Interim Report of the OECD Going Digital Project, Meeting of the OECD Council at Ministerial Level, Paris, 30-31 May 2018, OECD, Paris, http://www.oecd.org/going-digital/C-MIN-2018-6-EN.pdf.

OECD (2017), "Key Issues for Digital Transformation in the G20, report prepared for a joint G20 German Presidency/OECD conference, Berlin, 12 January 2017, OECD, Paris.

OECD (2015a), "Making Open Science a Reality", *Science, Technology and Industry Policy Papers*, No. 25, OECD Publishing, Paris, http://dx.doi.org/10.1787/5jrs2f963zs1-en.

OECD (2015b), *Frascati Manual 2015: Guidelines for Collecting and Reporting Data on Research and Experimental Development*, The Measurement of Scientific, Technological and Innovation Activities, OECD Publishing, Paris, https://doi.org/10.1787/9789264239012-en.

OECD (2014), "Recommendation of the Council on Digital Government Strategies", Adopted by the OECD Council on 15 July 2014, OECD, Paris, http://www.oecd.org/gov/digital-government/Recommendation-digital-government-strategies.pdf.

OECD (2013), *The OECD Privacy Framework*, OECD, Paris, http://www.oecd.org/sti/ieconomy/oecd_privacy_framework.pdf.

OECD (2011), "Quality Dimensions, Core Values for OECD Statistics and Procedures for Planning and Evaluating Statistical Activities", OECD, Paris, http://www.oecd.org/sdd/21687665.pdf.

Peng, H. et al. (2017), "Forecasting potential sensor applications of triboelectric nanogenerators through tech mining", *Nano energy*, Vol. 35, pp. 358-369, Elsevier, Amsterdam, https://doi.org/10.1016/j.nanoen.2017.04.006.

Priem, J. et al. (2010), Altmetrics: A manifesto, 26 October 2010, http://altmetrics.org/manifesto (accessed 5 February 2017).

Sateli, B et al. (2016), "Semantic User Profiles: Learning Scholars' Competences by Analyzing Their Publications", in González-Beltrán A., F. Osborne and S. Peroni (eds.), *Semantics, Analytics, Visualization. Enhancing Scholarly Data*, SAVE-SD 2016, Lecture Notes in Computer Science, Vol. 9792, Springer, Cham, Switzerland, https://doi.org/10.1007/978-3-319-53637-8_12.

Sugimoto, C. and V. Larivière (2016), "Social media indicators as indicators of broader impact", Presentation given at OECD Blue Sky Forum, Ghent, www.slideshare.net/innovationoecd/sugimoto-social-media-metrics-as-indicators-of-broader-impact.

Ubaldi, B. (2013), "Open Government Data: Towards Empirical Analysis of Open Government Data Initiatives", *OECD Working Papers on Public Governance*, No. 22, OECD Publishing, Paris, https://doi.org/10.1787/5k46bj4f03s7-en.

Yoo, S.H. and D. Won (2018), "Simulation of Weak Signals of Nanotechnology Innovation in Complex System", *Sustainability*, Vol. 10/2/486, MDPI, Basel, https://doi.org/10.3390/su10020486.

Zhang, Y. et al. (2016), "Technology roadmapping for competitive technical intelligence", *Technological Forecasting and Social Change*, Vol. 110, pp. 175-186, Elsevier, Amsterdam, https://doi.org/10.1016/j.techfore.2015.11.029.

Chapter 13. Mixing experimentation and targeting: innovative entrepreneurship policy in a digitised world

By

Carlo Menon

Innovative entrepreneurship plays a key role in our societies, being an engine of job creation, innovation, and inclusiveness. In light of that, policy makers are aware of the importance of creating a fertile entrepreneurial ecosystem. However, only a tiny minority of new firms eventually becomes a successful innovative business. Improving the prediction of this success ex ante would allow governments to target their support to start-ups, and could alter the balance between targeted and non-targeted (e.g. reducing entry and exit barriers) policy approaches. This chapter first presents the main arguments in favour of public support to innovative entrepreneurship. It then discusses how newly available big data and machine learning techniques could help policy makers design more effective policies through more precise targeting of firms with the highest growth potential. The chapter then focuses on venture capital, which, besides filling the equity gap, also aims to target these innovative firms. It concludes with a discussion of the factors that will affect the balance between targeted and non-targeted policies in the future.

Balancing targeting and experimentation: new developments in policy support to innovative entrepreneurship and venture capital

This chapter critically overviews the policy debate about high-growth and innovative entrepreneurship, addressing the most crucial and timely policy questions. Although they represent only a tiny sub-sample of all new firms, high growth firms are crucial to create new employment, foster innovation, and increase productivity in the long run, taking advantage of the ongoing digital transformation and other disruptive innovations. However, a number of scholars argue that entrepreneurship is following a "secular" declining trend, especially in the United States, although recent evidence show signs of recovery across OECD countries (Decker et al., 2016). At the same time, the range of policy instruments that are available to support start-ups and the associated volume of finance and technical expertise provided to support these firms has probably never been so high. Public intervention in venture capital – one of the most popular instruments to identify and financially support innovative firms – has also become increasingly popular over the last years. A central question is whether these policy interventions are effective, and whether they are targeting the "right" start-ups and entrepreneurs.

These questions – concerning the type of interventions that best support the emergence and development of high-growth and innovative firms – have attracted increasing attention in academic and policy circles. This chapter critically reviews this debate. Considering that only a tiny minority of new firms contribute to economic growth, the effectiveness of untargeted entrepreneurship policy has been questioned by some scholars, who argue that public resources should be concentrated only on firms with the highest growth potential. This in turn poses the related question of whether it may be possible to identify high potential firms ex ante, possibly leveraging on new opportunities provided by big data and innovative predictive analytics techniques (e.g. machine learning). Alternatively, if success proves to be unpredictable ex-ante, ultimately this would lead to an important role of market experimentation to detect successful ventures. In such a case, policy makers should therefore opt for a "let 100 flowers bloom" approach, in which entry and exit barriers are reduced, and entrepreneurs can test their business ideas in the market, growing fast if they prove to be successful, and exiting smoothly and rapidly if they are not.

Of course, this is a stylised framework; in practice, innovative entrepreneurship policies typically attempt to strike the right balance between these two extreme approaches. An illustrative "real world" example is public investments in venture capital (VC), which represents the main instrument for public intervention in innovative entrepreneurship across many OECD economies. This policy typically targets the most successful start-ups, given that VC investments tend to involve less than 1% of new firms. At the same time, the need to complement the private VC market with public investments is often motivated by filling in an equity financing gap in technological areas that require more experimentation and risk-taking. These are often also areas where market success is less predictable and private businesses may contribute to wider social objectives (e.g. climate change mitigation, public health, etc.), thus providing an additional rationale for public policy. But is government VC an effective policy instrument to identify and support innovative and high growth start-ups? While the necessary evidence base to answer this question is still incomplete, this chapter overviews the available evidence. A number of policy trade-offs that should be taken into consideration in the implementation of this policy emerge. The chapter also stresses the need to widen the available evidence base to better inform policy making.

How has innovative and high-growth entrepreneurship evolved in recent years?

A number of recent empirical contributions based on administrative data have shown that entrepreneurship as a whole has been declining before and in the aftermath of the 2007-9 financial crisis. This is particularly evident in the United States, where a "secular" decline in business dynamism and new firm creation since the 1970s has been observed (Decker et al., 2016). Since 2000, the decline in dynamism and entrepreneurship in the US has been accompanied by a decline in high-growth young firms. Although similarly long time-series are not generally available for other economies, the decline in entrepreneurship has also been observed over the 2000s' in other OECD countries (Blanchenay et al., 2016). However, in many OECD economies the number of new firms created appears to have recovered in the aftermath of the financial crisis, reaching in many cases pre-crisis highs.

Furthermore, more recent data on venture capital (VC) and related funding [1] reflecting the dynamism of the tiny share of new entrants with high-growth potential – suggest that VC-funded entrepreneurship has been booming over the last couple of years. The total amount of VC funding granted across OECD countries in 2016 is substantially higher than before the 2007-9 international crisis. This is mainly explained by a steep increase of VC investments in the United States, which more than doubled over the same period (Figure 13.1).

Therefore, the secular decline in entrepreneurship rate and in dynamism that has recently spurred concerns among policy makers may be a rather composite phenomenon, with different indicators pointing to contrasting trends. The mixed evidence is also a consequence of the very different phenomena that are generally referred to as "entrepreneurship". While a slowdown in so-called "subsistence" entrepreneurship may not necessarily be bad news, a decline in high-growth start-ups and in innovative entrepreneurship may have serious long-lasting negative effects, given the role these firms play for aggregate growth and job creation. This chapter finds that a more fine-grained understanding of the heterogeneity of entrepreneurship is also crucial for policy.

Figure 13.1. Venture capital investments over time in Europe and in the United States

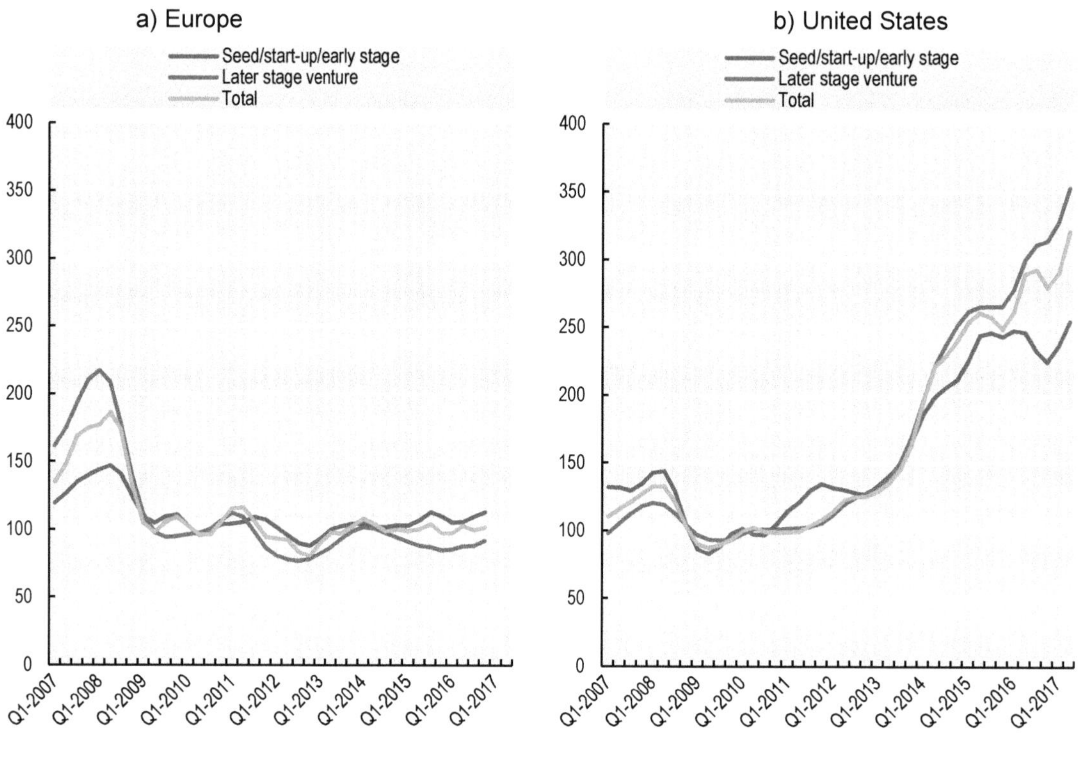

a) Europe

b) United States

Note: Trend-cycle, 2010 = 100.
Source: OECD (2017), Entrepreneurship at a Glance 2017, OECD Publishing, Paris http://dx.doi.org/10.1787/entrepreneur_aag-2017-en based on Invest Europe Yearbook 2016 and National Venture Capital Association/PitchBook Report, 2017Q2

StatLink ⟶ https://doi.org/10.1787/888933858392

Why and how should innovative and high-growth entrepreneurship be publicly supported?

Deploying effective entrepreneurship policies is a priority across many OECD economies, in the light of the evidence that new and young firms contribute disproportionally to job creation across OECD countries. The 2017 edition of the EC/OECD STI policy survey identified 167 different initiatives in participating countries related to targeted support to young innovative enterprises (EC/OECD, 2017). Several OECD countries are also developing comprehensive and organic policy frameworks to support innovative entrepreneurship. The need for effective policy interventions also rests on the important role that innovative start-ups can play in meeting broader environmental and social objectives. Start-ups are supposed to be more effective in introducing disruptive and breakthrough innovations (e.g. Egli, Johnstone and Menon (2015)) that provide new solutions to long-standing problems, because they do not suffer from the "organisational inertia" that may instead hamper the development of radical innovations by established incumbents (Henderson and Clark, 1990). Innovative entrepreneurship can promote inclusiveness, which is also high in the policy agenda given growing concerns that economic inequality undermines social cohesion.[2]

The fact that innovative entrepreneurship plays a crucial role for economic growth and wellbeing does not alone grant the need for policy intervention. Rather, the motivation for policy intervention to support innovative start-ups arises out of the widespread belief that there are three general types of market failures that may hamper their development:

- Capital market failures which arise from information asymmetries that affect new or small firms in general;

- Knowledge market failures that make it difficult for innovative firms and their shareholders to capture the full value of their innovation efforts; and

- Positive externalities that are not priced and therefore imply that the social value of entrepreneurship is higher than private returns.

These three sets of market failures may reinforce one another, making it particularly difficult for innovative start-ups to attract the necessary inputs to grow. In principle, this justifies the need for policy intervention. However, the fact that there is a *need* for policy intervention does not necessarily imply that any type of policy would *work*. There are many areas in which there is a consensus on the existence of market failures, but very little understanding of which policy levers should be activated to fix them.

Even more than the type of support and the design of policy instruments essential to the effectiveness of public policies aiming to support high-growth entrepreneurship is the issue of which entrepreneurs and start-ups should be supported. The fact that only a tiny proportion – typically less than 5% – of start-ups eventually grows and innovates (e.g. Calvino, Criscuolo and Menon (2015)) is typically overlooked in the public debate on entrepreneurship. In this context, Shane (2009) argues that encouraging more people to become entrepreneurs is "bad public policy". Rather, "policy makers should stop subsidizing the formation of the typical start-up and focus on the subset of businesses with growth potential".

This narrative, however, should be counterbalanced by the evidence that experimentation is a crucial ingredient of successful entrepreneurship. According to this view, the policymaker should not attempt at "picking winners" and should rather let the market select. A viable alternative approach could thus be "let one hundred flowers bloom". Within this framework, potentially successful entrepreneurs should be enabled to experiment with various innovative strategies and technologies while having the ability to scale up or down, in the event of productivity shocks. There is indeed considerable empirical evidence that suggests that firms need space to experiment with various innovative ideas. Instances of failing in the past are common amongst successful start-ups (Eggers and Song, 2015). Importantly, making room for experimentation may be particularly beneficial for start-ups with a wider societal impact, whose success is not strictly measured in terms of jobs created or profit generated.

The role of the policy maker, in this context, would be to streamline both the entry and the exit of businesses, also by designing an insolvency regime that is not perceived as too "punitive". In practice, however, this entails a number of policy trade-offs which are not easy to solve. For instance, insolvency procedures that are "pro-entrepreneur" and allow for a "fresh-start" would facilitate exit on the one hand, but, on the other hand, they would also increase risk for lenders, thus restricting access to financial resources for prospective entrants. Moreover, because start-ups are small and relatively less organised in comparison to incumbents, it may be difficult for them to communicate their needs directly to policy makers. Although the reality of policy support to innovative start-ups is of course a mix of both, one can distinguish analytically between a targeted and non-targeted approach (Figure 13.2).

Policy bottlenecks that are generally detrimental for all businesses can be particularly harmful for small start-ups (Calvino, Criscuolo and Menon, 2016). In some circumstances, the policy environment may have been implicitly designed with the needs and conditions of incumbents in mind, meaning that horizontal structural reforms that are particularly helpful to start-ups are delayed or not implemented. This may also depend on regulation being tailored to the prevailing technology adopted by incumbents, rather than to the innovative technology used by start-ups. Across OECD countries, a number of advocacy groups have been established to help facilitate dialogue between start-ups and government officials. This is commendable, as the policy debate across several OECD economies appears to focus more on saving distressed firms, rather than favouring the birth of new ones. The critical importance of framework conditions, such as civil justice efficiency, open science, fair taxation, free movement of talents across borders, and a dynamic labour market in fostering innovative entrepreneurship, need to be further emphasised.

Figure 13.2. The two approaches to encourage innovative start-ups

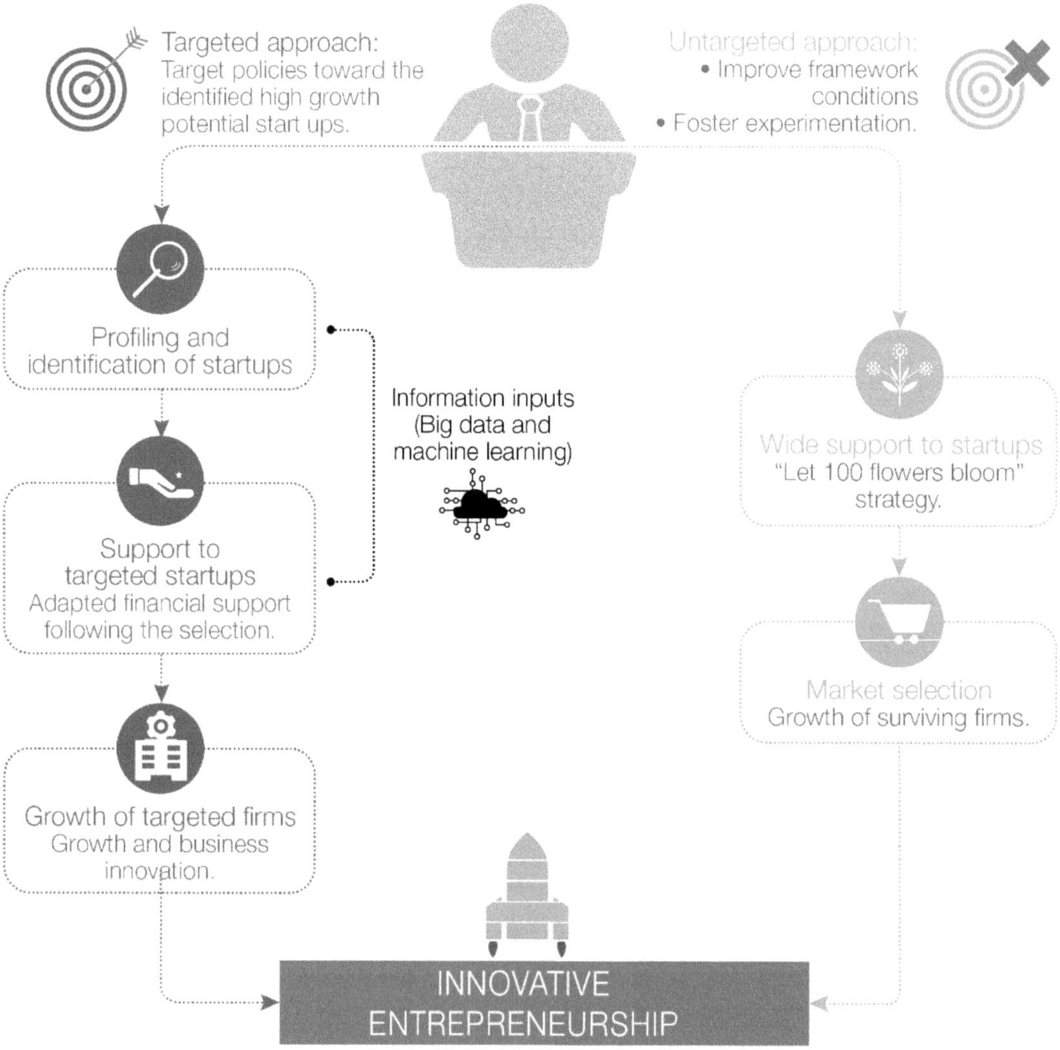

Do big data and machine learning applications open new avenues to identify high growth potential start-ups?

Targeting policy interventions only on start-ups with high growth potential poses the crucial policy question of whether policy makers can identify high-potential start-ups. The issue of whether start-ups' growth can somehow be reliably predicted based on observable characteristics is highly debated in the economic literature, particularly following the increased availability of firm-level data (Geroski, 1999; Coad, 2009; Guzman and Stern, 2016). Despite many efforts from econometricians, there has been limited success in identifying firm (or entrepreneur) characteristics predicting subsequent growth dynamics. The combined explanatory power of independent variables is usually low, typically less than 10% (Coad, 2009). A number of academics have argued that the systematic components of growth and performance are by far overshadowed by its randomness (Geroski, 1999; Coad, 2009). Some scholars even suggest that the factors that are expected to explain firm growth path so far are quite erratic and not very meaningful (Coad et al., 2013). Therefore, these scholars maintain that the actual firm level determinants of growth are still somewhat unknown.

One of the difficulties in identifying successful entrants is the lack of detailed data on the characteristics of firms and entrepreneurs "ex-ante", i.e. at the moment in which they create the new company. As many of these firms are very small entities, limited public information is available from administrative sources. In addition, comprehensive measures of "success" are also not readily available from traditional sources, especially when innovation is deemed to be an important component. However, advances in communication technology have opened up an era of big data, making information on both firm characteristics at entry and subsequent performance more accessible. At the same time, advances in information processing hardware and software make it easier for machine learning tools to analyse the growing accumulation of data. This enables the identification of complex relationships and clusters of similar firms, which may be used to more effectively identify successful high growing entrants (Box 13.1).

As a consequence of these improvements on the data side, a rising number of scholars have begun to challenge the idea that growth is random and unidentifiable (Guzman and Stern, 2015; Guzman and Stern, 2016; Åstebro and Tåg, 2017). Guzman and Stern (2015), for example, state that while random factors and unobservable characteristics influence the success of entrepreneurs, the divergence in performance and the effects on various entrants can be explained by observable differences in ex-ante firm characteristics. The authors employ data on entrepreneurs at a similar stage of their entrepreneurial career to design measures of firm characteristics linked to entrepreneur quality. Quality measures include if the founder names the start-up after him/herself, if the start-up purchased or carried out attempts to protect intellectual property (such as a registered trademark or patent) and if the firm has a legal form oriented toward equity financing (i.e. undergoing incorporation or locating in Delaware). Using these measures, they estimate the relationship between growth outcomes (firms that achieve an IPO or high value acquisition within six years of entry) and initial start-up characteristics, and find that a few characteristics allow one to construct a predictive model that determines entrepreneurial quality.

Similarly, Ng and Stuart (2016) apply machine learning algorithms to datasets containing a large set of socio-demographic variables for hundreds of thousands of entrepreneurs in the US tech sector. The authors show that the group of individuals possibly classifiable as "entrepreneurs" actually comprises two distinct clusters (which they name "hobos" and "highflyers") that have diametrically opposed characteristics, namely in terms of social positions and career pathways. Transitions across the two groups are also extremely rare.[3] The authors argue that these findings support a different way to define and study "entrepreneurship", which could take into account the two distinct phenomena that are usually covered by the same term. However, the study also supports the idea that big data and machine learning algorithms could be informative to predict successful entrepreneurship, and possibly also to improve the targeting of entrepreneurship policy.

Box 13.1. Machine learning, econometrics, and economics

Machine learning techniques are becoming increasingly popular in economic analysis as an alternative or a complement to traditional regression analysis to solve practical estimation problems. As Athey (2018) states, "machine learning will have a dramatic impact on the field of economics within a short time frame".

Regression analysis is used to precisely estimate a number of coefficients on a limited number of variables selected a priori based on a model derived from theoretical hypotheses. The estimated coefficients allow the calculation of marginal effects and elasticities, e.g. it is possible to state that start-ups with at least one female founder receive on average 70% less funding, everything else being equal. However, this approach is not informative on whether the linear combination of selected variables is doing a good job in explaining the phenomenon under scrutiny, compared to possible alternative variable combinations and selections.

Supervised machine learning techniques are designed instead, in general, to solve a prediction problem – given all the available information on start-ups, the objective is to identify the variables that are the best predictors of the outcome of interest (e.g. probability of being acquired). The question is not how the variable Y changes with respect to a change of the variable(s) Xs, but rather how the variable Y can be predicted out of a sample based on a wide set of Xs, and possibly which of those are the most important ones. While traditional econometrics has also being dealing with prediction (for instance to estimate economic forecasts), machine learning algorithms perform substantially better. Given that they require large datasets to provide precise predictions, their application suits particularly well "big data" sources that are becoming increasingly available.

The other big difference, compared to traditional econometrics, is that no theoretical model is needed: the analyst can simply "let the data speak". The algorithm is typically fed with the largest available set of variables, and the result is a list of variables ranked by importance, i.e. by their explanatory power in the prediction exercise. The major advantage of machine learning techniques lies in their flexibility. Because no functional form is imposed on the data a priori, these techniques are capable of finding appropriate models for data with varied structure and complexity.

The biggest danger with machine learning algorithms is using them to attach a causal interpretation to the result or to test general economic theories, which can be highly misleading. While the distinction between causality and correlation is prominent in the econometric literature, it is seldom mentioned in machine applications. However, the fact that a set of variables can predict very precisely an outcome does not imply that the same variables are affecting that outcome, as omitted variables or reverse causality could play a role. The classic example of Athey (2018) is the prediction of hotel occupancy rates using room prices. While low prices predict low occupancy rates, lowering the prices further would not lead to lower occupancy rates.

The findings that "highflyers" are systematically different from "hobos" does not necessarily imply, however, that the policymaker should ignore the latter. For instance, this kind of "subsistence" entrepreneurship may still be important to achieve inclusiveness, especially if it creates employment opportunities for individuals who are discriminated against or "red-lined" in the regular labour market. For instance, anecdotal evidence point to Uber having provided a foothold in the job market for thousands of undereducated youngsters of immigrant descent living in French *banlieues*.[4] The lack of transitions across the two groups can also be symptomatic of a lack of opportunities for the "hobos", due to obstacles to social mobility that could possibly be removed. The normative implication of these findings would point to tackling the barriers and the obstacles that hamper social mobility and limit the opportunities for a large set of the population. This is related to a recent US study (Bell et al., 2017) showing that children from high-income (top 1%) families are ten times as likely to become patent inventors as those from below-median income families, while differences in innate ability, as measured by test scores in early childhood, explain relatively little of these gaps.

How to target policies toward the identified high growth potential start-ups?

Once the high growth start-ups have been identified, or at least the scope of potential firms has been narrowed down, policy instruments must be designed to specifically target these firms. In other words, the *analytical* identification of these firms must be translated into administrative and policy *actionable* terms. This points to the challenging issue of how to maximise policy impact by refining policy targeting and the eligibility rules – a topic that has been examined only tangentially by the relevant literature. Even policies with an identical "average treatment effect" on supported firms would have a radically different aggregate effect depending on the eligibility rules, if the way in which firms react to the policy is highly heterogeneous.

Indeed, one important factor driving the success of public interventions is whether the target population reacts to the policy incentives as expected, i.e. in whether it complies with the policy objectives (Andini et al., 2017). Everything else being equal, a policy may fall short of the desired objectives because the eligibility rules are not fully effective in filtering the target population. Policy effectiveness can therefore increase by better targeting the sub-sample that is more likely to take-up and benefit from the policy, i.e. the policy "compliers".

Identifying the observable characteristics of the compliers is a policy prediction problem that suits well machine learning algorithms, which are designed to minimize the out-of-sample prediction error by exploiting all available information (Box 13.1). Machine learning can therefore be used ex-ante to inform the policy design phase, complementing traditional policy impact assessment approaches.

At the same, some factors of complexity should be taken into account. First, the relevant question for the policy maker is not only "*which are the high-growth start-ups*", but also "*which are the high-potential start-ups that do not grow because of the existence of the market failures that the policy is seeking to correct*". The start-ups to be targeted are not those that would grow in any case, but only those for which the policy makes a difference. In principle, answering the latter question requires performing a predictive exercise both in the presence and absence of the market failures, in order to be able to compare the two scenarios. The effectiveness of the policy would be improved if there is a part of the population that is excluded, but that share the same characteristics of the treated sample for which the policy has a positive impact. These considerations are partially related to the emerging literature on the intersection of machine learning with causal impact evaluation in econometrics (Athey, 2018; Chernozhukov et al., 2018).

Despite these challenges, there is huge potential for machine learning techniques to help policymakers to design more effective policies through more precise targeting. Researchers and policymakers should work together to assess the actual feasibility of this approach.

Is government venture capital an effective instrument to select and support high growth start-ups?

Equity financing for innovative start-ups is not only a tool to provide them with the financial resources and expertise needed for their early-stage development; it is also a mechanism to screen and identify high growth innovative start-ups. Governments, especially in Europe, have increasingly used this mechanism to complement their intervention portfolio and influence the type of investments toward those start-ups that most need it from a public mission perspective. This section provides an assessment of the state of the art of knowledge on the effectiveness of public VC and early-stage financing as a policy tool.

Governments are active VC investors in many OECD economies. According to the EC/OECD (2017), equity financing is the most popular instrument to support access to finance for innovative firms across OECD countries. In countries like Canada or Korea, more than 50% of VC-backed start-ups have received some form of public equity support (Brander, Du and Hellmann, 2015). Different degrees of government involvement exist. On the one hand, some programmes entail financial support to existing private VC funds with no direct control over management of the funds (e.g. "funds of funds"). On the other hand, some schemes involve direct government ownership of VC funds.

Government intervention in the VC market is justified by the existence of market failures of the private VC market. Indeed, innovations introduced by VC-backed start-ups may bring about important social benefits, often exceeding private ones. Given the public good nature of innovations, start-ups are likely to be underfunded compared to the welfare-maximizing level of funding. This is particularly true for young firms developing innovations that take longer to get to market, or those that generate further social benefits (e.g. inclusive start-ups, start-ups developing green technologies, start-ups in the health sector). Additionally, government venture capital initiatives can target companies for which they have informational comparative advantage (e.g. in the sectors of health and defence) and signal start-up quality to traditional investors (Lerner, 2002).

There are also important risks associated with government VC investments. For instance, government VC may displace private investments if they are targeting the same kinds of start-ups (Brander, Du and Hellmann, 2015). While the evidence on the effectiveness of public VC capital is equivocal, the majority of studies suggest that public VC do not crowd-out private investments. Brander, Du, and Hellman (2015) and Leleux and Surlemont (2003) show that government VC funding seems to cause greater amounts of money to be invested as a whole, both at the industry and firm level. The evidence on the impact of government VC on firm performance is also quite limited, and conclusions are mixed. On average, private VC-backed companies appear to perform better than public VC-backed companies in terms of total investments and successful exits (Brander, Du and Hellmann, 2015), innovation output (Bertoni and Tykvová, 2015), sales and employee growth (Grilli and Murtinu, 2014), although there are also several success stories. However, the form of investment that is associated with the best performance of companies consists of heterogeneous syndicates involving both public and private investors. Given the policy relevance of the topic and the huge amount of public resources invested, additional empirical evidence would be extremely valuable. Innovative sources of data, e.g. Crunchbase, provide new opportunities in this context, despite some limitations (Box 13.2).

> **Box 13.2. Using Crunchbase for economic and managerial research**
>
> Crunchbase (www.crunchbase.com) is a commercial database on innovative companies maintained by Crunchbase Inc. The database was created in 2007 within the Techcrunch network, but its scope and coverage has increased significantly over the past few years. As reported by the Kaufmann Foundation, the database is increasingly used by the venture capital industry as "the premier data asset on the tech/startup world".[5] Dalle, den Besten, & Menon (2017) present a detailed discussion of the database and its potential for economic, managerial, and policy-oriented research.
>
> Crunchbase raw data are obtained through two main channels: a large investor network and community contributors. These data are processed by the Crunchbase analyst team with the support of artificial intelligence algorithms, in order to ensure accuracy and scan for anomalies. Additionally, algorithms continuously search the web and thousands of news publications for information to enrich profiles.
>
> Compared to commercial databases covering similar information and frequently used for economic research, Crunchbase has major advantages: it contains cross-linked information on companies, their funders, and their staff; it is partially crowd-sourced, which adds to the comprehensiveness and timeliness of the database; it is updated on a daily basis; and it is structured in an accessible way. The comprehensive information on the profile of founders and the timeliness of the data are two of the characteristics of Crunchbase that make it particularly valuable for policy analysis. The VC industry evolves very rapidly – for instance, China went from having almost no investments in artificial intelligence in 2015 to being the second biggest global player after the US in 2017 – therefore more traditional sources of data may fail to cover the main trend early enough.
>
> Breschi, Lassébie, and Menon (2018) discuss the coverage and representativeness of the database, compared to some benchmark data sources that are more commonly used in the literature. The general message of the benchmarking exercise is that Crunchbase has a better coverage of VC deals and start-ups than comparable data sources. The country-year comparison with aggregated sources on VC investments also suggests that the coverage of Crunchbase is sufficiently exhaustive across OECD member countries and four large emerging economies (i.e. Brazil, China, India, and Russia), with few exceptions.
>
> The database also suffers from a number of limitations. First, the historical dimension of the database is mainly limited to the snapshot of companies that have been active recently. Start-ups that failed and ceased operations are likely not to have left any trace in the database. Therefore, spurious ascending growth trends of deal number or investments may appear in the data. This calls for caution in the examination of trends over time. Second, the amount invested in the VC deal is not disclosed in around 20% of cases. Third, the classification of investors into different groups (e.g. corporate, government, etc.) is not always accurate, and requires additional refinement and cross-validation work. The database has been refined and expanded by OECD in a number of different dimensions. A particularly important addition consists in the matching of patent data from PATSTAT, for

> both companies (patent applicants) and people (patent inventors) listed in the database (Tarasconi and Menon, 2017).

Even if direct positive effects fail to materialise on the "average" start-up, there is reason to believe that under specific circumstances government VC may be an effective complementary policy – at least in some sectors and for a period of time (Breschi et al., forthcoming). For example, government venture capital initiatives can target companies for which they have informational comparative advantage. This would be the case for sectors which are heavily regulated, and thus for which governments have preferential access to information on future market conditions. It would also be true for sectors for which government is a significant source of demand (e.g. in the sectors of health and defence).

Another motivation often put forward to support the role of government VC is its role in kick-starting emerging technologies that are deemed too risky by private investors. If this were generally true, a relatively higher share of government VC should be observed at the beginning of an exponential growth of aggregate investments in a sector. In order to test this hypothesis, Breschi et al. (forthcoming) calculated the total number of deals (upper panel) and investments (lower panel) by quarter in the following rapidly emerging technologies: drones; virtual reality; artificial intelligence; apps; 3D printing; blockchain; and cloud computing (Figure 13.3).[6] The data show that these sectors have experienced an exponential growth in both the number of deals and the total amount of investments. However, private investments appear to anticipate – rather than follow – the inflow of public support, with the latter, however, reaching almost the same level of private investments in 2016. This descriptive evidence therefore seems to suggest that the actual kick-starting of the technology is done by private investors, with public money playing an important role in expanding the market during the consolidation of the technology.

Figure 13.3. Investments in rapidly expanding technologies: public-mixed and private VC

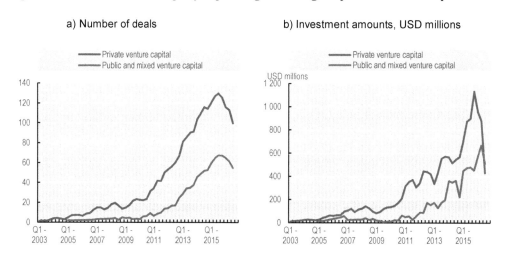

Note: Data are preliminary. The rapidly expanding technologies are the following ones: drones; virtual reality; artificial intelligence; apps; 3D printing; blockchain; and cloud computing. The identification of the companies is based on keyword search on the company short descriptions. Excludes rounds where the type of investor is unknown.
Source: Breschi et al. (forthcoming), based on Crunchbase data (http://www.crunchbase.com).

StatLink https://doi.org/10.1787/888933858411

Overall, a complex policy trade-off is emerging. On the one hand, the model that appears to provide more "value for money" is the mixed one, in which public investments follow *pari passu* their private counterparts and mimic their investment strategies. However, inevitably this form of investment is less effective in achieving the "public good" objectives that are mentioned above. It is also the strategy with the highest risk of crowding out private actors. On the other hand, fully public investments – which in theory may be much more effective in filling the equity gap and in providing finance to areas that require more risk-taking and experimentation – appear to underperform their private counterparts, according to the (limited) available evidence. However, private investments may not be the best benchmark in this context, and it is extremely complicated to measure all outcomes of interest beyond market returns. Further research is needed to properly inform this important area of policy making.

Regardless of the model, investments in government VC in contexts in which the market is immature or almost non-existent are unlikely to succeed, as the exiguity of the private VC market is likely to be the symptom rather than the disease. This does not necessarily imply that there is a shortage of innovative entrepreneurial ideas; rather, the main problem appears to be the difficulty of transferring the ideas "from the lab to the market". This may be due to many different factors, e.g. the lack of managerial skills, the difficulty for innovative firms to attract resources (labour and capital) due to market rigidities, a scarce demand of innovative good and services in the local market, policy failures that impose an extra cost on risk, etc. In such a situation, the policymaker should assess carefully the bottlenecks that hamper the development of a dynamic entrepreneurial ecosystem, taking the whole start-up life cycle into account, as the lack of private equity investments may actually be the consequence of very weak growth prospects for start-ups at later stages of their development (Box 13.3).

Box 13.3. In my view: When innovation waterworks are clogged, we should remove obstructions downstream

Marco Cantamessa, Department of Management and Production Engineering, Politecnico di Torino

I sometimes have the feeling that economists and policy-makers tend to overestimate the role of inputs to the innovation process and the supply side of markets for innovation. The number of patents, of startups, VC investments, are key statistics that are constantly monitored and eventually incentivized, often by pouring public money in the system, "*since private investors will not*". The underlying assumption is that – if the innovation waterworks are clogged – building upstream pressure will solve the problem. As an engineer, I beg to differ, since I would rather try to remove obstructions from downstream.

Metaphors aside, I see a significant risk of neglecting the complexity of innovation systems. Innovation systems do not only comprise multiple tiers of venture capitalists and budding entrepreneurs bearing innovative ideas, but also experienced entrepreneurs, advisers, professional services firms, a talented workforce, suppliers, companies that might acquire startups, open minded regulators and – most important of all – customers, which includes consumers, other businesses, or government entities.

Lacking customers, all the rest loses significance. Even more, in the case of 'corporatist' economies where markets are *de facto* not contestable, unassailable incumbents cannot be competed against by entrant startups, nor will they consider them either as suppliers or as

> acquisition targets. Now, lacking customers and/or real opportunities for startups to enter markets and grow, why should a dearth of VC investors be considered as a market failure, and not as the proof that markets work very well?
>
> I understand policymakers' willingness to funnel money into VC funds, or in Funds of Funds. It is a direct way to declare they have "thrown money at the problem", but I am sometimes afraid that this may serve as an alibi for sidestepping the much more difficult task of freeing startups from the shackles that hinder their growth. Moreover, this approach tends to support the myth that Venture Capital is simply about money, while it is much more. Allocation of capital to startups is a complex exploratory process, carried out on assets characterized by huge Knightian uncertainty and information asymmetries. The only way to make it work is to ensure a liquid market with multiple VC funds interacting with many startups, combined with a competitive ecosystem able to progressively select the best firms. Moreover, capital allocation is just the first step of the VC process, while the second and most important one is the active support provided by General Partners to founders in growing their firms, recruiting managers, making deals, and structuring the next financing round or exit. In other words, the VC industry will work properly if it is based on a large number of funds, staffed by experienced General Partners. So, while it is likely that a thriving startup ecosystem may nurture an adequate number of such professionals and generate a robust VC industry, I have some doubts that a VC industry may arise out of financial endowments alone, and lead to a thriving startup ecosystem.
>
> Leading scholars like Edmund Phelps and Amar Bhide have argued that the main input to the innovation process is culture: a culture where change and competition are viewed as values *per se* and not as something to be 'managed'. A culture where it is customary to take clear and rapid decisions, without ambiguity and endless deferrals. A culture where not only failure, but also glaring success, are not frowned upon. It is not an easy task, but a worthwhile endeavor indeed.

Future outlook: toward new balances between targeted and non-targeted policies?

Both policy approaches based on targeting or providing wide support to high growth potential firms have their own challenges, costs – actual costs or opportunity costs – and benefits. In practice, policy makers set out to strike the right balance between targeting and fostering experimentation. A number of attributes, including formal and informal institutional set-ups, greatly influence this balance and make it country-specific. This balance also evolves over time with policy learning and changes in preferences.

New empirical evidence, leveraging on big data and machine learning techniques, could influence this balance in the near future. Their potential to fruitfully inform the refinement of eligibility rules of entrepreneurship policy – with a view to focusing on the specific group of start-ups that see their growth potential hampered by market failures – is growing. They therefore hold great potential for better targeting of entrepreneurship policy, which could significantly improve its effectiveness. At the same time, in rapidly changing markets, idiosyncratic and unobservable factors will always play an important role, thus start-up success will preserve a degree of unpredictability. Therefore, direct and targeted policy interventions will certainly always have to be complemented with horizontal reforms in order to ensure an overall business environment conducive to entrepreneurship and experimentation. Start-ups will have to be able to attract resources and to scale-up if successful, and to exit smoothly if unsuccessful.

The balance between targeted and non-targeted approaches will also be increasingly influenced by the need to address the "grand challenges" of our time – from climate change to food security and aging. While the radical innovations brought to life by visionary entrepreneurs are essential to face some of these challenges, these are by nature surrounded by strong ("radical") uncertainties. As shown by studies in economics and sociology of science and innovation, periods of disruptive changes do not lend themselves well to policies aiming to pick the "best" alternatives. Most innovations in these turbulent periods are systemic, emerging from trials and errors of various combinations of technological and social innovations. This makes any attempt to identify ex ante firms with the greatest potential more challenging or, even worse, detrimental to the process of change as it limits experimentation. In such context, a subset of firms with higher growth potential are not "revealed" to the world; their potential for growth emerges and increases through interactions with their environment, allowing faster learning and larger investments in some of them.

Growing inequalities, another mounting concern, might affect the balance in the same direction of a more open and experimental approach for supporting innovative entrepreneurship. The need for inclusiveness will call for policies making high-growth entrepreneurship more accessible to talented "outsiders" in order to foster social mobility.

While big data and machine learning will without doubt improve the capacity of policy makers to identify the sub-sample of start-ups with high growth potential that could critically benefit from targeted policy support, new societal challenges will require more experimentation in years to come. No one can predict the result of this "moving target" process on the innovative entrepreneurship policy in different national contexts. However, it is clear that countries will have to both invest in and monitor progress in techniques targeting firms, and continue to reform their economic system to make the process of experimentation more efficient.

Notes

[1] In this chapter, the term "venture capital" is used with a very general meaning, referring to all forms of equity funding for high-growth and high-risk entrepreneurial venture. The term therefore also encompasses forms of financing that are not properly venture capital, such as angel investments and seed and early stage funding.

[2] For instance, there is evidence that innovative entrepreneurship fosters social mobility in the United States (Aghion et al., 2015), while minority communities, particularly those of South/East Asian origin, have played increasingly important roles in US science and technology sectors (Chellaraj, Maskus and Mattoo, 2008).

[3] In passing, this finding may have important normative implications for the debate on inclusive entrepreneurship, which are not discussed by the authors.

[4] See *the Financial Times*, "Uber: a route out of the French banlieues", 3/3/2016, by Anne-Sylvaine Chassany; available online at https://www.ft.com/content/bf3d0444-e129-11e5-9217-6ae3733a2cd1 (retrieved on 16th July 2018)

[5] http://www.kauffman.org/microsites/state-of-the-field/topics/finance/equity/venture-capital accessed on September 11th, 2017.

[6] The choice of these fields, as well as the procedure to tag the related companies in Crunchbase, follows closely the work done by Crunchbase experts on "buzzword" technologies, available at

https://news.crunchbase.com/news/ahead-buzzword-curve-finding-investors-front-top-tech-trends/ accessed on 12th June, 2018.

References

Aghion, P. et al. (2015), *Innovation and Top Income Inequality*, National Bureau of Economic Research, Cambridge, MA, http://dx.doi.org/10.3386/w21247.

Andini, M. et al. (2017), "Targeting Policy-Compliers with Machine Learning: An Application to a Tax Rebate Programme in Italy", *SSRN Electronic Journal*, http://dx.doi.org/10.2139/ssrn.3084031.

Åstebro, T. and J. Tåg (2017), "Gross, net, and new job creation by entrepreneurs", *Journal of Business Venturing Insights*, Vol. 8, pp. 64-70, http://dx.doi.org/10.1016/j.jbvi.2017.06.001.

Athey, S. (2018), The Impact of Machine Learning on Economics, Working paper.

Bell, A. et al. (2017), *Who Becomes an Inventor in America? The Importance of Exposure to Innovation*, National Bureau of Economic Research, Cambridge, MA, http://dx.doi.org/10.3386/w24062.

Bertoni, F. and T. Tykvová (2015), "Does governmental venture capital spur invention and innovation? Evidence from young European biotech companies", *Research Policy*, Vol. 44/4, pp. 925-935, http://dx.doi.org/10.1016/j.respol.2015.02.002.

Blanchenay, P. et al. (2016), "Cross-country Evidence on Business Dynamics over the Last Decade: From Boom to Gloom?", Vol. Unpublished working paper.

Brander, J., Q. Du and T. Hellmann (2015), "The Effects of Government-Sponsored Venture Capital: International Evidence", *Review of Finance*, Vol. 19/2, pp. 571-618, http://dx.doi.org/10.1093/rof/rfu009.

Breschi, S., N. Johnstone, J. Lassebie, C. Menon (Forthcoming), "Never walk alone? A cross-country analysis of governments venture capital investments", *OECD Science, Technology and Industry Policy Papers*, Forthcoming.

Breschi, S., J. Lassébie and C. Menon (2018), "A portrait of innovative start-ups across countries", *OECD Science, Technology and Industry Working Papers*, No. 2018/2, OECD Publishing, Paris, http://dx.doi.org/10.1787/f9ff02f4-en.

Calvino, F., C. Criscuolo and C. Menon (2016), "No Country for Young Firms?: Start-up Dynamics and National Policies", *OECD Science, Technology and Industry Policy Papers*, No. 29, OECD Publishing, Paris, http://dx.doi.org/10.1787/5jm22p40c8mw-en.

Calvino, F., C. Criscuolo and C. Menon (2015), "Cross-country evidence on start-up dynamics", *OECD Science, Technology and Industry Working Papers*, No. 2015/6, OECD Publishing, Paris, http://dx.doi.org/10.1787/5jrxtkb9mxtb-en.

Chellaraj, G., K. Maskus and A. Mattoo (2008), "The Contribution of International Graduate Students to US Innovation", *Review of International Economics*, Vol. 16/3, pp. 444-462, http://dx.doi.org/10.1111/j.1467-9396.2007.00714.x.

Chernozhukov, V. et al. (2018), Generic Machine Learning Inference on Heterogenous Treatment Effects in Randomized Experiments, National Bureau of Economic Research, Cambridge, MA, http://dx.doi.org/10.3386/w24678.

Coad, A. et al. (2013), "Growth paths and survival chances: An application of Gambler's Ruin theory", *Journal of Business Venturing*, Vol. 28/5, pp. 615-632, http://dx.doi.org/10.1016/j.jbusvent.2012.06.002.

Coad, A. (ed.) (2009), *The growth of firms: A survey of theories and empirical evidence*, Edward Elgar Publishing.

Dalle, J., M. den Besten and C. Menon (2017), "Using Crunchbase for economic and managerial research", *OECD Science, Technology and Industry Working Papers*, No. 2017/08, OECD Publishing, Paris, http://dx.doi.org/10.1787/6c418d60-en.

Decker, R. et al. (2016), "Where has all the skewness gone? The decline in high-growth (young) firms in the U.S.", *European Economic Review*, Vol. 86, pp. 4-23, http://dx.doi.org/10.1016/j.euroecorev.2015.12.013.

EC/OECD (2017), *STIP Compass: International Science, Technology and Innovation Policy (STIP)* (Database), April 2018 version, https://stip.oecd.org.

Eggers, J. and L. Song (2015), "Dealing with Failure: Serial Entrepreneurs and the Costs of Changing Industries Between Ventures", *Academy of Management Journal*, Vol. 58/6, pp. 1785-1803, http://dx.doi.org/10.5465/amj.2014.0050.

Egli, F., N. Johnstone and C. Menon (2015), "Identifying and inducing breakthrough inventions: An application related to climate change mitigation", *OECD Science, Technology and Industry Working Papers*, No. 2015/4, OECD Publishing, Paris, http://dx.doi.org/10.1787/5js03zd40n37-en.

Geroski, P. (1999), "The growth of firms in theory and in practice", CEPR Discussion Papers, Vol. 2092.

Grilli, L. and S. Murtinu (2014), "Government, venture capital and the growth of European high-tech entrepreneurial firms", *Research Policy*, Vol. 43/9, pp. 1523-1543, http://dx.doi.org/10.1016/j.respol.2014.04.002.

Guzman, J. and S. Stern (2016), The State of American Entrepreneurship: New Estimates of the Quantity and Quality of Entrepreneurship for 15 US States, 1988-2014, National Bureau of Economic Research , Cambridge, MA, http://dx.doi.org/10.3386/w22095.

Guzman, J. and S. Stern (2015), Nowcasting and Placecasting Entrepreneurial Quality and Performance, National Bureau of Economic Research, Cambridge, MA, http://dx.doi.org/10.3386/w20954.

Haskel, J. and S. Westlake (2017), *Capitalism without Capital: The Rise of the Intangible Economy*, Princeton University Press.

Henderson, R. and K. Clark (1990), "Architectural Innovation: The Reconfiguration of Existing Product Technologies and the Failure of Established Firms", *Administrative Science Quarterly*, Vol. 35/1, p. 9, http://dx.doi.org/10.2307/2393549.

Leleux, B. and B. Surlemont (2003), "Public versus private venture capital: seeding or crowding out? A pan-European analysis", *Journal of Business Venturing*, Vol. 18/1, pp. 81-104, http://dx.doi.org/10.1016/s0883-9026(01)00078-7.

Lerner, J. (2002), "When bureaucrats meet entrepreneurs: the design of effective `public venture capital' programmes", *The Economic Journal*, Vol. 112/477, pp. F73-F84, http://dx.doi.org/10.1111/1468-0297.00684.

Ng, W. and T. Stuart (2016), "Of Hobos and Highfliers: Disentangling the Classes and Careers of Technology-Based Entrepreneurs", Vol. Working paper.

OECD (2017), *Entrepreneurship at a Glance 2017*, OECD Publishing, Paris, http://dx.doi.org/10.1787/entrepreneur_aag-2017-en.

Shane, S. (2009), "Why encouraging more people to become entrepreneurs is bad public policy", *Small Business Economics*, Vol. 33/2, pp. 141-149, http://dx.doi.org/10.1007/s11187-009-9215-5.

Tarasconi, G. and C. Menon (2017), "Matching Crunchbase with patent data", *OECD Science, Technology and Industry Working Papers*, No. 2017/07, OECD Publishing, Paris, http://dx.doi.org/10.1787/15f967fa-en.

Chapter 14. Blue Sky perspectives towards the next generation of data and indicators on science and innovation

By

Fernando Galindo-Rueda

Just like its subject matter, the world of data, indicators and analysis on science, technology and innovation (STI) is experiencing profound transformations that call for co-ordinated action by users and producers of STI data and statistics. This chapter brings together discussions and perspectives shared at the OECD Blue Sky Forum 2016. It presents several key trends presenting both challenges and opportunities. It first considers major developments, starting with the latest trends in policy uses of STI data and statistics. It then assesses how digitalisation has transformed the production of STI data, and examines the major issues facing traditional and new producers of STI data and statistics. It concludes by discussing the future outlook and key governance perspectives in STI measurement and analysis, including the potential role of the OECD.

I often say that when you can measure what you are speaking about, and express it in numbers, you know something about it; but when you cannot measure it, when you cannot express it in numbers, your knowledge is of a meagre and unsatisfactory kind; it may be the beginning of knowledge, but you have scarcely, in your thoughts, advanced to the stage of science, whatever the matter may be

William Thompson, *Lord Kelvin (1883)*

Introduction

Just like the field of science, technology and innovation (STI) itself, the data, statistics, indicators and empirical analysis on the structure and dynamics of science and innovation systems have changed significantly in recent years. Many of these trends are expected to intensify over the next decade. This chapter examines expert views on the way digitalisation and the aftermath of the global financial crisis will shape the production and use of STI data and indicators. It builds on the discussions and outcomes of the OECD Blue Sky Forum 2016, a major global conference organised by the OECD on the future of STI data and indicators in Ghent, Belgium (Box 14.1). It also examines work carried out since. Science ministers of the OECD countries endorsed this initiative as a vehicle to "continue improving statistics and measurement systems to better capture the key features of science, technology and innovation" (OECD, 2015). This chapter reviews key messages from the Blue Sky Forum 2016 and considers the outlook for STI statistics production and policy use in the coming years.

Box 14.1. The OECD Blue Sky Forum on STI data and indicators

Every ten years, the OECD convenes and engages the policy community, data users and data providers in an open dialogue to review and develop the OECD long-term agenda on STI data and indicators. This event is part of the OECD Committee for Scientific and Technological Policy's Programme of Work; its organisation is entrusted to its Working Party of National Experts on Science and Technology Indicators (NESTI). Known as the "OECD Blue Sky Forum", the title reflects the intention to provide a long-term, unconstrained discussion on evidence gaps in STI, as well as on initiatives the international community can take to identify and address related data needs. Previous editions held in Paris (1996) and Ottawa (2006) have been influential in setting the path for a series of national programmes supporting research aiming to advance the scientific basis for science and innovation policy (Marburger, 2007, 2011). They have also informed the 2010 OECD Innovation Strategy (OECD, 2010a) and its measurement agenda (OECD, 2010b).

Among its objectives, the Blue Sky Forum 2016 set out to:

- review the main conceptual underpinnings and use of current frameworks for STI indicators and data infrastructure initiatives
- explore the role of digital infrastructures in creating new opportunities for measurement and analysis, as well as the challenges inherent in existing collection standards and STI indicator quality
- provide new opportunities for collaboration and strengthen the dialogue between policy makers, data users and providers

> - lead to a forward-looking and policy-relevant roadmap on STI measurement the OECD could implement in collaboration with its membership, other international organisations and experts.
>
> More detailed information on the 2016 OECD Blue Sky Forum, including papers, presentations and videos, can be found at: http://oe.cd/blue-sky.

In defining the scope of this chapter, it bears noting that the STI field possesses unique features as a domain for policy analysis that have to do with the nature of knowledge and how it is created and diffused, leaving a limited number of traces. These features make it particularly difficult not only to measure STI concepts, but also often to demonstrate how the different elements are connected through cause and effect. The empirical study of innovation (defined in Box 14.2) and innovation policy faces the challenge of seeking to measure how activities that are themselves difficult to measure affect other outcomes that are also difficult to measure (Krugman, 2013). Besides, there are few instances in which experiment-like conditions arise naturally or can be reproduced in the empirical analysis of STI. The scope for experimentation in innovation policy has increased in recent years (Nesta, 2016), but remains limited.

Understanding the nature, outcomes and eventual impacts of science and innovation activities requires the ability to observe and understand action at multiple levels of analysis across the entire system (Lane et al., 2015). Over the least decades, the main objective of key actors in the world of STI data and statistics has been to ensure that the frameworks and tools used in this area align with the broadly espoused view of innovation as a highly interconnected and dynamic system (Soete, 2016).

> **Box 14.2. Defining and measuring innovation**
>
> Although firmly grounded in knowledge development, innovation requires implementation: innovations derive from the creation and application of ideas. As defined in the 2018 edition of the Oslo Manual (OECD/Eurostat, 2018): "An innovation is a new or improved product or process, or combination thereof, that differs significantly from the unit's previous products or processes and that has been made available to potential users (product) or brought into use by the unit (process)."
>
> This neutral, measurement-oriented definition can be applied to the economy and society as a whole. Innovations may succeed or fail in meeting their objectives. Over time, a diverse range of features and unforeseen effects may appear. From a policy perspective, understanding innovation is critical to understanding how policies can harness it to drive growth and well-being, while managing the potential downsides of individual innovations and broad innovation systems.

While these substantive conceptual and practical challenges may have resulted in lagging evidence for empirical analysis of STI compared to other data-rich areas (Bakshi and Mateos-García, 2016), they have not prevented producer and user communities from creating and applying conceptual frameworks, data infrastructures, indicators and analytical efforts to shed light on the functioning of STI systems. This chapter considers major developments, starting with the latest trends in policy uses of STI data and statistics. It then assesses how digitalisation has transformed the production of STI data, and examines the major issues facing traditional and new producers of STI data and statistics.

It concludes by discussing the future outlook and key governance perspectives in STI measurement and analysis, including the potential role of the OECD.

New perspectives on the policy use of STI data, statistics and analysis

The STI measurement infrastructure has developed over time to serve the needs of science users and innovation policy makers for statistics and related analysis, often for advocacy purposes. As indicated in Chapter 6 on the digitalisation of science and innovation policy (DSIP), policy demand for data is broader, covering not only inputs to policy definition, but also the operational delivery and management of public services. In addition to policy makers, businesses and institutional managers increasingly rely on science and innovation data to support both their day-to-day and strategic decisions, e.g. when assessing the intellectual-property landscape before deciding to enter a market. However, published statistics and related research are generally too broad for management purposes; hence, their overall relevance is largely determined by how they influence public policy.

Changes in policy user interests following the crisis

In the immediate response to the 2008 global financial and economic crisis, most countries recognised the importance of sustaining economy-wide research and innovation capabilities. However, the crisis had a profound impact on the STI policy questions that guide decisions on data collection and analysis. Its very roots challenged the implicit assumption that innovation is inherently good and should be promoted, generating a heightened sense of responsibility about its outcomes and impacts on society. The ensuing wave of government financial austerity measures led to a tighter and more competitive budgetary environment, triggering requests for evidence that would allow STI policy makers to present the best possible case for economy and finance ministers to provide public support for STI. The raised bar for quality evidence across all competing policy areas entails demonstrating the economic return of investment in STI compared to other government investments.

The debates about the origins of innovation and the government's role in relation to companies have intensified (Mazzucato, 2015). The generalised productivity slowdown has also renewed mainstream policy analysts' interest in measuring the link between innovation and growth. They are questioning the tools used to measure both (Brynjolfsson and MacAfee, 2011; Gordon, 2016; Coyle, 2017) and ascertaining whether research productivity has been declining (Bloom et al., 2017).

In this context, conceptual models need empirical evidence to allow testing hypotheses about the operation of STI systems and the role of policies. Data development requires new models and concepts. Dosi (2016) humorously described the limits of theories and value measurement in STI as searching for "the marginal impact of an extra gram of butter to the taste of the cake". However, major policy discussions still focus on explicit or implicit value and credit attribution. As econometric analysis and value-for-money assessments have entered the debate on science and innovation policy, and results are used to justify policies, a very significant change in the data and analysis landscape is taking place. Some question whether this could be the start of an "arms race" in reported impacts, and whether studies that report massive impacts could be generating unrealistic expectations of returns from STI investment. If true, this may well affect STI programmes with modest returns, as well as programmes that are harder to evaluate.

The struggle to communicate and act upon STI data and indicators

Communicating complex messages for policy making presents significant challenges. Policy makers demand simple indicators to monitor and benchmark STI systems that directly relate and can inform their key decisions. Experts meeting at Blue Sky described the role of STI policy makers as managing uncertainty. This requires an "options-thinking approach" – i.e. considering a wide range of small seed investments, which open opportunities for more decisive investments in specific areas that are chosen when uncertainty is resolved. Both quantitative data and narratives ("story telling") are necessary to address uncertainty, and communicate the results from research on science and innovation to policy makers (Feldman, 2016).

Ministers participating in the Blue Sky Forum 2016 welcomed the opportunity to reflect on the data they use, and how they use them (Box 14.3). Some even reported being struck by the extent to which science and innovation can be driven by targets and composite indicators. They recognised the need to identify insightful indicators on science and innovation actors and their linkages, and examine the enabling infrastructure.

Box 14.3. Selected senior policy makers' perspectives from the Blue Sky Forum 2016

There is a need to open and enrich the production of indicators towards a diversified set of skills, together with theoretical advancements regarding the use of indicators and the navigation through data sets. The next generation of data should guide new forms of international cooperation giving priority to science, education and mobility. Data is also needed to foster the collective action of governments, public institutions and the private sector to promote the diversification of education and research systems leading to technological change, as well as a participatory approach to science and innovation.

Manuel Heitor, Minister for Science, Technology and Higher Education, Portugal

The quality of our measurement needs to keep pace with our societies. The ability to compare ourselves internationally helps us immensely in our domestic policy decision-making. We must recognise the interaction between fundamental knowledge and innovation. We need to collaborate internationally to harmonise open data standards and use evidence to increase transparency, accountability and citizen engagement.

Kirsty Duncan, Minister of Science, Canada

When I became minister, I was struck by how much science and innovation policy is driven by indicators and targets. Researchers and universities change their behaviour in response to them. We need to keep under review how indicators are produced and used. Indicators can become less useful as they become more widely used. Good evidence is more important for policy makers at times of budgetary constraints.

Elke Sleurs, State Secretary with responsibility for Science Policy, Belgium

To achieve a real impact on our society, over the next ten years we should strive to maximise the value of the massive quantity of data available to us […] I firmly believe the next ten years will not be about producing data, as much as it will be about understanding data. […] If we are serious about the growth enhancing and job creating role of research and innovation, then we need to be able to demonstrate and prove those.

Carlos Moedas, European Commissioner for Research, Science and Innovation

Source: OECD (2016), Blue Sky Forum 2016, http://oe.cd/blue-sky.

Problems with indicator-driven science and innovation policy

The purpose of innovation data and indicators is to be used by policy makers – without that, the whole enterprise would be a failure. But could there be too much of a good thing? The potential misuse of indicators – which are commonly used to set targets – is one concern, and experts argue that indicator-driven policy should not be viewed as equivalent to evidence-based policy (Polt, 2016).

Targets without data are a common problem. Several domestic and international initiatives focusing on quantitative targets keep on being launched. In many cases it has soon transpired that suitable data for monitoring target fulfilment were not available for many of these initiatives, and no significant resources were devoted to addressing these gaps. In other instances, setting unrealistic targets without adequate analysis may undermine interest in measurement. It may also divert resources from activities that do not directly respond to the indicator – e.g. innovation efforts beyond research and development (R&D) that may be just as important to the objective. When R&D investment targets are difficult to meet, incentives may exist to blame the measurement effort and "shift the goalposts" (Bakshi et al., 2017), rather than question the choice of indicator.

A key concern is the potential abuse and misuse of STI indicators that oversimplify reality on the sole basis of what can easily be measured, obfuscating their interpretation and generating some complacency about what is and is not important (Martin, 2016). A majority of Blue Sky participants intensely criticised the use of composite science and innovation indexes, which combine multiple widely available indicators and rank countries' or regions' performance –implying that a higher value or rank is preferable from a societal perspective. Although they recognised the value of simplicity in communicating a high-level message, they saw considerable conceptual and practical problems in current practice.

Changing demands for STI data from a more sophisticated user base

The STI policy-making community has been increasingly attracting officials with backgrounds in various analytical disciplines and a good understanding of empirical evaluation tools and the importance of good data and hypothesis building to support them. Such an expert community, embedded within policy ministries and agencies, can ensure that data are considered at the appropriate stages of policy planning and evaluation. As part of building a culture for evaluating everything, policy makers are asked to impose the integration of data and evaluation requirements into programmes supporting science and innovation (Jaffe, 2016). In several instances – such as the disbursement of public funds – public interest and accountability for policy actions may override privacy concerns.

A greater focus on the human perspective

The policy community appears to agree on the need to place individuals and communities at the centre of science and innovation-related policy design and analysis. Citizen-based analytics are considered as enabling better targeting of public services (Gluckman, 2017). At the Blue Sky Forum 2016, ministers and other policy makers and experts strongly advocated for "human-centred policy design", and called for systematic collection of data about individuals (Heitor, 2016). This represents a shift in the collection and use of evidence on STI systems from an organisation-based perspective towards considering the "human factor", such as scientists' decisions to return to their home country or work in industry, or their motivation for developing a new solution to a particular community problem. Information on career decisions can contribute to STI supply-side policies, but the dimension of individual and collective demand for innovation should also be considered. Policy questions and measurement efforts need to be framed in terms of society's engagement in innovation systems and preparedness for the changes brought about by innovations.

The policy community is also underscoring the need to better characterise participatory R&D processes and agenda-setting on innovation policy. The goal is to help engage scientific institutions and actors with civil society – highlighting the need for collaboration with scientists, engineers and users, to understand the process and impact of knowledge production. The development of data toolkits can help explore choices in research landscapes and spur citizen participation in decision-making (Rafols et al., 2016), as well as raise popular buy-in for science and innovation.

Demand for more accessible, granular and linked data

Growing requests from general users of statistical data for more complex solutions, tailored to their needs and relatively simple to use, are another major trend (Peltola et al., 2017). Research and policy users of STI data and statistics with more advanced analytical capabilities expect data to be micro-based, with the finest possible degree of granularity, to support aggregate statements not only about a country or sector, but also about relatively small geographical areas, organisations, teams and even individuals. They demand infrastructures that allow accurately linking and using data for a range of statistical and research purposes (Hicks, 2010). Data linking has been significantly facilitated by improvements in information and communication technology – including advances in machine learning – but remains a significant challenge in the absence of universally adopted identifiers.

A move to more granular data raises potential issues: measurement can have a direct impact on individuals' incentives and behaviour, potentially undermining its ultimate utility. Aggregate indicators serve as a guide on the functioning of entire systems and inform the policy direction. But what happens when disaggregated data are used to base more targeted action? A more sophisticated discussion needs to take place about the relationship between data, indicators and policy uses. Questions are raised about whether the statistical use of certain indicators at an aggregate level (as at the OECD and other international organisations) represents an implicit endorsement of the same indicators' use at a far more disaggregated level, for non-statistical purposes. This is a common concern regarding specific bibliometric indicators, which are also often used for research assessment; this practice is criticised by signatories to the San Francisco Declaration on Research Assessment (DORA, 2012) and the Leiden Manifesto on Research Metrics (Hicks et al., 2015). It can be difficult to explain that the same indicator may be appropriate for one purpose, and not for the other. However, the consensus is that the underlying microdata need to be available, to facilitate the production of more complex and nuanced aggregate indicators and analyses, which in turn help break down and comprehend the various components of headline indicators.

Data beyond jurisdiction boundaries: Towards truly global STI statistics

Previous trends pointing to the increased globalisation of STI systems have continued – and possibly accelerated – over the past decade. Without the ability to map the creation and circulation of knowledge and related financial flows across countries, it is impossible to characterise science and innovation, or measure their drivers and impacts. However, generating STI statistics still relies on a largely national approach. Privileging the nation as the natural scale of analysis is a built-in bias of statistics, which years of economic change have progressively eroded (Davies, 2017).

What is preventing the adoption of truly global standards and measurement practices? Path dependencies imply substantial adjustment costs to measure STI phenomena similarly across countries; rendering jurisdiction-specific data truly interoperable and global is still a long way off. Born-global data require a more significant co-ordination effort, including reaching agreements on data sharing and standards. Global companies are partly occupying this space, because their cross-border activities are less constrained than those of official organisations. Policy makers should ask themselves whether commercial databases represent an appropriate basis for sustainable data infrastructures – at least to meet their global statistical needs.

Thanks to their global nature and the replication of best practices, science and research activities are becoming increasingly similar, implying that measurement methods should also converge. This may, however, not be the case for the innovation culture, which retains a strong local component (Bauer and Suerdem, 2016). Even in the age of global brands and hyper-connected individuals, innovation critically depends on social, spatial and historical contexts that are essentially local. Hence, integrated local approaches to measurement are needed to support common policy learning at the global level.

Innovation for the SDGs

One key recent driver of global statistical data on STI has been the world leaders' definition and endorsement of the 17 Sustainable Development Goals (SDGs) in the context of the 2030 Sustainable Development Agenda. Improving the data is key to attaining these goals. As part of Target 9.5, countries have pledged to "enhance scientific research [and] upgrade the technological capabilities of industrial sectors in all countries" (United Nations, 2015). A key issue for STI policy makers is the ability to monitor the link between science and innovation on the one hand, and the entire range of global sustainability objectives – from poverty and hunger eradication, to equality and climate action – on the other. Those links are not easily traced or exposed solely by using indicators. Given the multidimensional nature of the SDGs, monitoring the overall role of science and innovation requires accumulating the findings from empirical studies of policy experiments around the world. Enhanced use of meta-analysis in this area should be further encouraged.

Timelier data

At the Blue Sky Forum 2016, EU Commissioner for Research, Science and Innovation Carlos Moedas worried that most of the data he used to take policy decisions was outdated (Moedas, 2016). The timing of STI evidence shapes the types of decisions it can support. Most STI indicators are fundamentally structural and exhibit limited variation over small periods; they can help inform the development of long-term plans to attain strategic goals. However, in times of rapidly disruptive change, timelier and frequent data become more critical, owing to the risks of basing decisions on information that is no longer relevant. Timeliness is also critical when measuring possibly short-lived processes, such as entrepreneurship and business dynamics.

In this context, many organisations consider applying nowcasting methods using complementary sources, instead of relying solely on models. Today, many surveys include questions about respondents' intentions to invest in R&D or innovation over the current and coming year; this has become part of OECD statistical guidance (OECD, 2015;

OECD/Eurostat, 2018). Data from quarterly and annual reports to investors and regulators informing about investment and product launches as well as hiring campaigns with new job descriptions can be obtained from online sources. This more frequent information creates the need to filter higher levels of "noise", and identify the optimal balance between structural and conjunctural data. More active nowcasting also requires more tolerance from policy makers regarding revised statistical data, as commonly occurs when measuring key macroeconomic indicator (e.g. gross domestic product).

Digitalisation: The expanding frontier for STI data and statistics

Exploiting the digital trace of science and innovation

Digital technologies and data are transforming business processes, the economy and society. Digitalisation represents a major force for change in the generation and use of STI data and statistics. STI systems have become remarkably data-rich: information on innovation inputs and outputs that was only recorded in highly scattered paper-based sources is now much easier to retrieve, process and analyse. The use of digital tools by researchers and administrators leaves digital traces that can be used to develop new databases and applied to indicators and analysis. The digitalisation of the patent application and scientific publication processes has already provided rich and widely used data resources for statistical analysis. Digitalisation is rapidly extending to other types of administrative and corporate data, e.g. transactions (billing and payroll data); website content and use metadata; and generic and specialised social media, in which STI actors interact with their peers and society.

Data practitioners view these new "big data" as "uncomfortable data", i.e. datasets that are too large to be handled by conventional tools and techniques (Alexander et al., 2015). The fuzzy boundary between qualitative and quantitative data is a striking example. Methodologies (e.g. user testing or interviews) traditionally considered as qualitative can now be conducted on a very large scale and quantified – text, images, sound and video can all be "read" by machines. Natural language-processing tools automate the processing of text data from thousands of survey responses or social media posts into quantifiable data. These techniques can help alleviate some of the common challenges facing STI statistics (e.g. survey fatigue and unfit-for-purpose classification systems applied differently by human coders) and generate adaptable indicators. Effective application of these methods relies on fit-for-purpose, high-quality systems to collect qualitative information consistently and avoid potential manipulation. Administrative database managers become important gatekeepers of data quality, but the information providers still need incentives.

Big data implies risks in exploiting datasets with possible defects and biases not recognised by the researchers; difficulties in evaluating big-data techniques and analysis, especially using conventional criteria (such as falsifiability); and complexities in explaining these techniques – and their value as evidence for policy evaluation – to decision-makers and the public. One case in point is altmetrics – where "alt" stands not only for "alternative", but also "article-level" (Priem et al., 2011) – which offer great promise and attract plentiful attention. However, altmetric indicators are essentially citation-based and do not reflect actual use or impact; more concerningly, they can be easily gamed, e.g. by bots. Research shows that altmetric indicators do not broaden the geographic or cross-disciplinary dissemination of science, but they do help heighten the profile of authors whose work, for good or bad reasons, generates attention. Sugimoto et al. (2016) conclude that altmetrics have not been the expected panacea, and recommended: a) providing disaggregated (rather than composite) altmetric indicators; b) accounting for outliers and gaming; c) expanding both evidence and scholarship sources; and d) reframing the conversation around the meaning of "broader impacts".

Moving progressively away from fixed scales of analysis (such as the nation) towards variable categories and dealing with vast new databases requires a different way of searching for patterns, trends, correlations and emergent discourses. Visualisations "mapping" interdependencies between individuals, teams, institutions and research domains can support the interactive exploration of large amounts of abstract data (Börner, 2010). These map-like representations – comprising nodes representing objects like researchers and their work with various labels, their positions and the highlighted edges connecting them – approximate a complex reality. However, their effective interpretation requires a qualitative narrative that is consistent with the underlying data, as well as visualisations of counterfactual scenarios. A rich research agenda is assessing how users assimilate statistical data, using various forms of experimentation and tests.

Surveys in the era of big data

Surveys are the cornerstone of statistical data – especially official data – on STI. Compared to data arising "organically" from administrative or commercial processes, survey data are "designed". Two elements constrain their potential: asking questions of individuals and organisations relies both on respondents' memories and formal records, and on their willingness to collaborate and provide truthful answers. The increased complexity of measurement constructs can sometimes exceed respondents' attention span and reporting capacity; they may also lack incentives to keep records of the information that surveys (and the policy makers promoting them) aim to retrieve.

Some question whether the shift to big data is the precursor to the demise of surveys, while others, paraphrasing Mark Twain, will argue that reports of the death of surveys "are greatly exaggerated". The manner in which surveys are carried out has indeed changed, as online surveys have largely displaced more expensive non-digital methods. Surveys are also more targeted towards areas where other data sources are less effective (Callegaro and Yang, 2018). The ease with which surveys can be conducted using electronic tools (including do-it-yourself survey platforms) has resulted in an explosion of surveys both in general and in the area of STI studies. A downside of this apparent "democratisation" is that these surveys often fail to meet basic statistical quality requirements, including for safeguarding privacy and confidentiality. This surge represents a growing source of fatigue for respondents; it results in lower expected response rates to non-compulsory (and compulsory, but difficult-to-enforce) surveys and may undermine trust. STI policy makers should co-ordinate and apply standards to their sponsored survey efforts.

Some of the transformational power of new STI data sources stems from their multidimensionality and possibilities to interconnect the different types of subjects and objects that are covered in them. The strengths of these organic data sources are hard to reproduce in surveys, which are traditionally conceived to identify key actors and the presence of pre-defined types of interactions, rather than trace those linkages. Digital solutions applied to survey tools can help address this gap. They are viewed as key components of the move towards "rich data" and are crucial to validating and augmenting the quality of big-data sources (Callegaro and Yang, 2018). Rather than competing with alternative sources, surveys will increasingly focus on the crucial information that cannot be obtained otherwise. Recent experience shows that trust and credibility will be the most crucial factors determining the success of survey efforts in the digital era.

A digitally enabled decentralisation of value measurement?

Building on the themes of trust and the new possibilities deriving from digital technologies, the STI community should think beyond current tools and data sources. It could explore, for example, the transformative potential of distributed ledger technology like blockchain in science and innovation systems (Soete, 2016). These technologies present alternatives to trusted third-party intermediary models aiming to assert "quality" or simply "truth" (a key function for government agencies), by shifting to tamper-proof models where that responsibility becomes distributed. It has been argued that countries or communities with weaker "rule-of-law" enforcement systems may find this particularly relevant. While the use of blockchain in STI and STI measurement remain more of an aspiration than a reality, it is worth reflecting on the potential role of collective intelligence when asserting the veracity and importance of the growing information contained in statistical data or indicators. STI indicators need to move beyond just counting items (i.e. papers, patents or even citations) so that they are able to distinguish contributions of different value.

Perspectives for producers of STI data and statistics

Data scientists and STI statistics

The great achievement of statisticians has been to reduce the complexity and fluidity of national populations into manageable and comprehensible facts and figures (Davies, 2017). Today, statistical skills are in high demand. Google's chief economist, Hal Varian, foresaw this trend when he argued in 2009 that the "sexy job in the next 10 years" would be "statistician" (Varian, 2009). This prediction has come true for data scientists, a category of statisticians at the junction of software programming and decision sciences, who are equipped to marshal the current data-driven boom in artificial intelligence capabilities. Data scientists thrive on access to data and cumulative algorithmic knowledge. Much of the basic expertise in this area is accessible to anyone by tapping into online courses and other instructional material, but true expertise requires using data towards a defined goal. Data scientists are powering major developments across the entire STI analysis community, handling extremely challenging tasks such as data extraction, disambiguation, data linking and topic analysis. Many now work for companies that are changing the landscape of STI data and statistics, offering a combination of services to meet the needs of scientists, administrator, firms and policy makers. It has become increasingly common for policy organisations to tap into this community's expertise, not only through conventional contracts, but also by organising problem-oriented prizes and challenges, like the disambiguation and visualisation challenges (e.g. the 2016 Cancer Moonshot Challenge) organised by the United States Patent and Trademark Office.

Official STI statistics at a crossroads

Following up on the pioneering efforts by academics and research organisations, national statistical organisations (NSOs) are now the backbone of many STI statistics. Working within a legal framework and applying minimum professional standards, their independence and objectivity has allowed them to attain official statistics status. Preserving privacy and confidentiality has been a longstanding concern and driver of official STI statistics. The highly concentrated nature of many STI activities is one of the main causes of the lack of granularity in published statistics as data gets aggregate to avoid disclosing information about individual organisations to preserve confidentiality.

A measure of the success of STI statistics is that economic statistics now incorporate several key dimensions of the "knowledge economy" in the System of National Accounts. A key lesson is that economic statistics can reflect STI issues, as long as tested data sources are available to address potential gaps. Accounting for innovation in the digital and knowledge-based economy is part of the new frontier.

However, NSOs have been experiencing major disruptions to their business models. Competition from alternative data sources is one factor: for example, many users seem willing to forego the limitations of publicly disclosed business data on R&D or other activities. NSOs struggle to hire young employees with high qualifications in data and statistics, as they are lured towards better-paying, cutting-edge companies. With few exceptions, governmental fiscal austerity has reduced the resources to maintain – let alone develop – new statistical data infrastructures. Process and product innovation within NSOs is constrained by day-to-day operational requirements, which absorb resources and leave little room for adapting business models to changing demands and new opportunities (Rozkrut, 2016). Too often, the role of NSOs is restricted to investigating their activities at the national level, making it difficult to address inherently global phenomena. NSOs need to exchange information and co-ordinate their activities to pursue their mission. They may not always consider customising national data for international comparison as a top priority, leading to many gaps in international statistics (such as those produced by OECD). A misguided drive for efficiency may leave little room for triangulating between different data sources that examine issues from different yet complementary perspectives (Bean, 2016).

As they struggle to capture more fluid identities, attitudes and economic pathways, traditional forms of statistical definition and classification are under strain. For example, what is a firm, an employee or a research field in the digital era? Efforts to represent socio-economic changes in terms of simple, well-recognised indicators risk losing legitimacy if users do not recognise themselves in them (Davies, 2017). As STI measurement moves increasingly into capturing individual behaviours and attitudes, respondents will also demand to define themselves in their own terms.

Nevertheless, the role of NSOs is largely considered critical to STI policy evidence. No other existing entities are endowed with the adequate formal authority and responsibilities to become viable alternatives to NSOs. NSOs are uniquely placed to objectively assess the reliability of new data sources and methods, and conduct representative statistical surveys. They can provide the clearing houses requested by researchers, combining information sources in ways other organisations cannot (National Research Council, 2014). It is therefore relevant for research-funding organisations supporting STI analysis to consider the role of NSOs, facilitating their work by providing the basic data infrastructure, and making it linkable and usable. While this represents a slight change from the traditional model, entirely centred on releasing aggregate statistics, many NSOs use the full potential of microdata infrastructures and partnerships to serve their core business needs, e.g. assess their own data quality. Lack of access lowers users' interest in promising initiatives; it may lead to terminating otherwise promising developments requiring further national and global consolidation. For example, innovation surveys partly owe their survival and diffusion to conscious efforts to make microdata available for analysis. Researchers need to understand better how NSOs operate; they need to present their research proposals so as to deliver operational benefits to NSOs, which may otherwise struggle to support access infrastructures. Many participants at the Blue Sky Forum 2016 demonstrated how such partnerships can be built over time.

Engaging STI researchers and administrators in data development

Since its emergence as a recognisable form in the late 1950s, practised by a handful of people interested in the subject, science policy research is now an established discipline practised by thousands of researchers (Martin, 2016). A common question asked by the more data-oriented group within this community – a major source of ideas for developing and using data sources – is what it takes to scale up to the national level and compel NSOs to adopt these ideas in their statistical enquiries, at home and internationally. While the absorptive capacity of NSOs is limited, many academics have succeeded in putting forward persuasive cases for NSOs to experiment with new questions, often by demonstrating their feasibility and policy relevance.

Empirical researchers are often faced with either buying or making their own data. They find the increasingly commercial control over many STI sources challenging, since they have to secure resources to pay for licences covering proprietary data. Commercial control is also a potential hindrance for public-sector organisations, which often end up paying to access information about research they themselves funded, or that data companies have secured from public registers over the years. Researchers are also actively creating data, using different approaches and gaining more recognition for their efforts. Around the world, initiatives such as the Science of Science and Innovation Policy in the United States; Japan's Science for RE-designing Science, Technology and Innovation Policy; and the EU Framework Programmes involve researchers in collaborative undertakings aiming to link the available data sources on STI and make them broadly accessible. The long-term sustainability of these databases after an initial wave of papers have been written is, however, still an open question, which may require deeper interaction with NSOs.

Administrative requirements and procedures are powerful enablers of statistical data in STI. Current research information systems (CRIS) contain a growing trove of valuable data that serve several purposes, including allowing universities to complete R&D surveys. Meanwhile, librarian collection-management needs have contributed to the emergence of bibliometrics. The social and policy drive for publicly funded institutions to manage "open science", as well as ensure appropriate management of knowledge resources, positions CRIS managers as critical providers of research metadata – i.e. data about research and innovation (Chapter 6). Concurrently, institutional librarians, repository managers and administrators have been particularly active in developing and promoting common standards for CRIS (Chapter 12).

Infrastructures and tracing and interoperability standards have increasingly emerged not from traditional standard-development organisations, but from ad-hoc organisations and consortia formed for specific purposes. This broad movement stems from dissatisfaction with data that are inconsistently specified, country-specific, and prevent researchers and policy makers from sharing or linking cleaned datasets. Access to these data has been limited by a patchwork of laws, regulations and practices, which are unevenly applied and interpreted (Haak et al., 2012). The creation of open digital object identifiers for individuals, organisations, STI outputs (e.g. documents) and relational information (e.g. ownership and citation) is a key element of this new infrastructure. These efforts need to address strong incentives among actors to keep key information proprietary and attain market power. While outside the business sector database developments have been in most cases initially propped up by government (such as the platforms *VIVO* in the United States and *Lattes* in Brazil or the OpenAire or RISIS infrastructure projects), some initiatives are supported by philanthropic sources which have been increasingly promoting the creation of data and metadata commons that can help map and understand STI systems. Examples include the *ORCID* persistent researcher identifier (Chapter 12), *opencorporates.com*, a database to retrieve and map information on companies and their complex ownership-based interconnections (Tett, 2018) or lens.org, a tool for combined patent and scholarly search. Although generating statistical data and analysis is not the primary driver for these initiatives, they present considerable potential for applied statistical work. However, they are yet to be fully tested, and statistical representativeness is a key issue to consider for such purpose.

These new initiatives should aim for greater consistency with existing and new standards related to STI statistics. Statisticians should in turn take greater notice of them and major changes that are transforming what – and how – individuals and organisations can report on their STI activities. A starting point for convergence could involve NSOs testing these new resources within their regular internal business processes to assess their comprehensiveness and consistency, e.g. in terms of classifications. NSOs have been using all sorts of available registers and information sources for decades to keep their own sampling frames up to date.

Policy and governance perspectives for STI measurement

A call for action for the STI indicator and policy analysis community

The STI indicator community exists to create data and metrics to gain shared understanding, evaluate policy alternatives and identify gaps (Stern, 2016). Like certain crafts, evidence-based policy making is a mix of art and science (Harayama, 2016). While digitalisation brings closer the utopia of integrating evidence processes into decision-making, it also brings about significant changes in information markets and consumer behaviours. Today, digital platforms build relations, establish truth or falsity, and prioritise events (Baldacci and Pelagalli, 2017). In this new world, data are captured first and research questions come later, often leaving little room for high quality statistics and experts. Without good governance and data interpretation skills at all levels, this may in turn reduce statistics to the role of attention-grabbing "clickbait".

The idea of a common public good is also worth defending. Like many other strands of statistics, innovation statistics were created as tools through which policy makers could view society, gradually developing into something in which academics, businesses and civil society also have a stake. As the business of innovation analytics grows, secrecy surrounding data methods and sources can be a competitive advantage, not to be given up voluntarily. In what Davies (2017) described as a move to a post-statistical society, policy makers need to consider that statistical "facts" may become privatised or diluted in the surrounding "noise".

Long-term perspectives to move beyond what is easy to measure

The nature of data, statistical and quantitative analysis in the relatively young and multidisciplinary domain of innovation is such that significant time lags exist between the formulation of a new user need and the provision of a solution. In the meantime, priorities may shift to other subtopics, resulting in a misalignment of statistical evidence with user demands. It is therefore important not only to anticipate future user needs in order to develop new data sources, but also to secure and fully utilise the available data sources and infrastructures, in order to deal with unexpected and time-sensitive questions as they arise. This agenda requires considerable transparency and accountability in public policy (García,

2016), highlighting once more the interdependence between good data and statistics, and good policy.

Considering the entire value chain of STI data generation and use

Arguably, it is time to stop focusing on single indicators and consider instead the entire STI data value chain. The objective should be to consider both interdependencies that span the full data cycle and data reusability in different settings, possibly for initially unintended purposes and applications. Data play different roles, from feeding into agenda-setting and policy design, to supporting implementation and policy evaluation. The business case for data can be more easily articulated if all their uses and implications, including confidentiality and privacy, can be examined holistically. The cost of developing new sources to answer specific questions may be prohibitive, but linking different data sources can provide insights that could not be derived from working separately with the different components. Thus, several policy questions can be addressed not by collecting new data, but by meaningfully connecting existing sources.

Reaping the opportunities of digitalisation

The transformational potential of digitalisation is being felt in all dimensions of data production and use. The data revolution that facilitated the emergence of new STI evidence communities and actors has brought considerable dynamism and change to the field. However (as was already noted at the Blue Sky Forum 2006), much work is still being carried out in silos, which are difficult to connect. New tools and data sources need to be viewed as toolkits, rather than as silver bullets for policy makers. Looking ahead, the measurement community will need to utilise all types of data sources and methods to meet their objectives, and engage in partnerships to this end. Since Blue Sky 2016, Germany and several other countries have embarked on their own "Blue Sky initiatives" (expanding into measurement areas inspired by the experiences shared in Ghent), and Australia is currently embarking on a major review of its STI data and indicators.

The STI evidence community needs to address the persistent and significant disconnect between users and producers of STI data, statistics and analysis. Building capabilities and encouraging co-ordination among different actors will be necessary to allow new data infrastructures to emerge. Most solutions aiming to build infrastructures that transform evidence capabilities rely on social change rather than technology, underpinned by community engagement and trust building. Driving progress requires identifying the major obstacles to developing infrastructures – often starting within public administrations, where data are fragmented and synergies foregone. Lack of policy awareness can block improvements to the legal framework for data exchange and re-use. It can also hinder the implementation of sustainable "business models" for data, which consider the intended statistical use by policy makers.

Conclusions and future outlook

Progress in STI measurement is expected to continue to be incremental, based on refinement of existing tools and experimentation. Indeed, most of the issues and solutions identified at the Blue Sky Forum 2006 (see OECD, 2007; Gault, 2013) are still relevant more than ten years later, and will likely remain relevant in the future. As in many other areas of research on socio-economic systems characterised by reflexivity, it is difficult to dispel the perception that STI policy questions formulated several decades ago still lack conclusive evidence, even as new questions continue to emerge. However, the landscape in which data and statistics on STI are produced and used has undergone major transformations, creating new opportunities to match the growing challenges. Policy awareness and understanding of specific areas will be transformed by the availability of new data sources, and new opportunities to combine them with existing data. These data will often come from unanticipated sources. However, traditional organisations – such as NSOs – will continue to play major roles, albeit informed by new practices and methods.

By reflecting on its own achievements and shortcomings, including the role of the OECD (Box 14.4), the STI community is reaching growing consensus on the need to challenge assumptions and move beyond sterile debates (including debates opposing traditional data sources to new data sources; economic-impact measures to social-impact measures; and narratives to hard numbers).

In a desirable future scenario, an appropriate mix of instruments and disciplines will be used to address specific evidence needs and develop solutions that can globally scaled up to achieve international comparability and synergies in highly interconnected STI systems. Defining such needs requires a higher level of policy engagement with data producers, as well as a higher degree of literacy in empirical methods. A sign of engagement is that STI policy makers are increasingly able to use quantitative arguments in their discussions with peers from other policy areas.

Box 14.4. Blue Sky messages for future OECD work on STI data and indicators

Discussions at the Blue Sky Forum 2016 offered a series of reflections on the historical contribution and future role of the OECD, highlighting its leading role in promoting evidence-based STI policy both domestically and internationally. Some argued this could be achieved by reaching out to new actors presenting the interest and potential to transform STI data and statistics. The OECD could also contribute to national efforts to develop an evidence culture by: 1) empowering NSOs to access and use relevant commercial and administrative data; 2) enhancing the availability and interoperability of administrative data on science and research funding, not only to benefit statistical evidence, but also to enhance the governance of science and innovation systems; and 3) providing more hands-on guidance to practitioners. The OECD DSIP project (Chapter 12) brings together policy and data perspectives as a direct outcome of these Blue Sky discussions.

Blue Sky 2016 participants also recommended that the OECD consolidate and extend its work on defining standards, compiling statistical information, building a global infrastructure and instructing data users worldwide. They advised the OECD to focus on areas where it is uniquely placed to do so, and to prioritise global policy relevance and international comparability. In particular, they recommended that the OECD:

- prioritise collecting evidence on the role of individuals in STI systems
- secure statistical information directly from key STI actors worldwide, to identify key emerging challenges and possible responses in more timely fashion
- extend the framework for measuring innovation beyond business
- promote secure international infrastructures and institutional agreements that make it easier to link and analyse microdata sources, making intensified use of projects based on distributed analysis across countries when common infrastructures are not possible

> - map public efforts to support research and innovation geared towards a range of societal objectives and challenges, and identify global funding gaps
> - provide evidence on the incidence and impact of known and hidden forms of public support for innovation
> - integrate research and innovation elements in economic statistics, and develop frameworks that help account for the contribution of investment in knowledge and its diffusion to economic performance within and across countries
> - ensure that STI statistics capture globalisation and digitalisation phenomena, and demonstrate the vast interconnectedness and major interdependencies of global STI systems.

Realising a vision in which data and statistics on science and innovation become part of the mainstream requires concerted action to make this future possible and define its trajectory. The pressure will increase to displace, control or appropriate valuable public information to serve private interests. As public information goods, statistics will be used to assert or dispute facts about how innovation occurs and changes our societies. Aligning private and public interests in data and evidence will be a major test for the future governance of STI data and statistics. The Blue Sky process represents an extremely valuable vehicle for the international community to engage more broadly and sustainably in the run-up to the next Blue Sky Forum in 2026.

References

Alexander, J., M. Blackburn and D. Legan (2015), "Digitalization in R&D management: Big Data. Research", *Research-Technology Management*, Vol. 58/6, pp. 45-49, Innovation Research Interchange and Routledge, Arlington, VA and Abingdon, United Kingdom.

Baldacci, E. and F. Pelagalli (2017), "Communication of statistics in post-truth society: The good, the bad and the ugly", *Statistical Working Paper*, Eurostat, Publications Office of the European Union, Luxembourg, http://ec.europa.eu/eurostat/documents/3888793/8223142/KS-TC-17-005-EN-N.pdf/deae4e95-c2b0-43bd-b963-a362002db02c.

Bauer, M. and A. Suerdem (2016), "Relating 'Science Culture' and Innovation", paper submitted to the OECD Blue Sky meeting on Science and Innovation Indicators – Theme: "Trust, culture, and citizens' engagement in science and innovation", Ghent, https://www.oecd.org/sti/097%20-%20OECD%20Paper%20attitudes%20and%20innovation_v4.0_MB.pdf.

Bakshi, H. et al. (2017), "The new UK R&D target: Why it's now time to move the measurement goalposts", Nesta blog, London, https://www.nesta.org.uk/blog/the-new-uk-rd-target-why-its-now-time-to-move-the-measurement-goalposts/.

Bakshi, H. and J. Mateos-García (2016), "New Data for Innovation Policy", paper submitted to the OECD Blue Sky Forum 2016, Ghent, http://www.oecd.org/sti/106%20-%20Bakhshi%20and%20Mateos-Garcia%202016%20-%20New%20Data%20for%20Innovation%20Policy.pdf.

Bean, C. (2016), "Independent Review of UK Economic Statistics: Final report", UK Government, London, www.gov.uk/government/publications/independent-review-of-uk-economic-statistics-final-report.

Bloom, N. et al. (2017), "Are Ideas Getting Harder to Find?", *NBER Working Paper*, No. 23782, National Bureau of Economic Research, Cambridge, MA, https://doi.org/10.3386/w23782.

Börner, K. (2010), *Atlas of science. Visualizing what we know*, MIT Press, Cambridge, MA.

Brynjolfsson, E. and A. McAfee (2011), *Race against the Machine – How the Digital Revolution Is Accelerating Innovation, Driving Productivity, and Irreversibly Transforming Employment and the Economy*, Digital Frontier Press, Lexington, MA.

Callegaro, M. and Y. Yang (2018). "The Role of Surveys in the Era of 'Big Data'", in Vannette, D. and J. Krosnick (eds.), *The Palgrave Handbook of Survey Research*, Palgrave Macmillan, Cham, https://doi.org/10.1007/978-3-319-54395-6_23.

Coyle, D. (2017), "Do-It-Youself digital: The Production Boundary and the Productivity Puzzle", *ESCoE Discussion Paper*, Vol. 2017/1, 13 June 2017, Economic Statistics Centre of Excellence, London, http://dx.doi.org/10.2139/ssrn.2986725.

Davies, W. (2017), "How statistics lost their power – and why we should fear what comes next. The long read", *The Guardian*, London, https://www.theguardian.com/politics/2017/jan/19/crisis-of-statistics-big-data-democracy.

DORA (2012), San Francisco Declaration on Research Assessment. Web version accessed from https://sfdora.org/read/.

Dosi, G. (2016), contribution to Blue Sky panel on "New models and tools for measuring science and innovation impacts", OECD Blue Sky Forum, Ghent, http://ocde.streamakaci.com/blueskyagenda/.

Feldman, M. (2016), keynote address at Blue Sky panel on "Science and Innovation policy-making today: what big questions are begging for an answer?", OECD Blue Sky Forum, Ghent, http://ocde.streamakaci.com/blueskyagenda/.

García, C. (2016), contribution to Blue Sky panel on "Looking forward: what data infrastructures and partnerships?", OECD Blue Sky Forum, Ghent, http://ocde.streamakaci.com/blueskyagenda/.

Gault, F. (2013), Innovation indicators and measurement challenges, in: Fred Gault, (ed.), *Handbook of Innovation Indicators and Measurement*, Edward Elgar, Cheltenham, UK and Northampton, MA, USA, 441-464.

Gluckman, P. (2017), "Using evidence to inform social policy: the role of citizen-based analytics: A discussion paper", Government of New Zealand and European Commission, Auckland and Brussels, www.pmcsa.org.nz/wp-content/uploads/17-06-19-Citizen-based-analytics.pdf.

Gordon, R.J. (2016), *The Rise and Fall of American Growth: The U.S. Standard of Living since the Civil War*, Princeton University Press, Princeton.

Haak, L. et al. (2012), "Standards and infrastructure for innovation data exchange", *Science*, Vol. 338/6104, pp. 196-197, American Association for the Advancement of Science, Washington, DC, https://doi.org/10.1126/science.1221840.

Harayama, Y. (2016), contribution to Blue Sky panel on "The Blue Sky Agenda", OECD Blue Sky Forum, Ghent, http://ocde.streamakaci.com/blueskyagenda/.

Heitor, M. (2016), "What do we need to measure to foster 'Knowledge as Our Common Future?'", position paper contributed to the OECD Blue Sky Forum, Ghent, http://oe.cd/heitorbluesky.

Hicks, D. (2010). "Systemic data infrastructure for innovation policy", NSTC's interagency task group on Science of Science Policy Science of Science Measurement Workshop, https://www.nsf.gov/sbe/sosp/tech/hicks.pdf.

Hicks, D. et al. (2015). "Bibliometrics: The Leiden Manifesto for research metrics" *Nature News*, Springer Nature. https://www.nature.com/news/bibliometrics-the-leiden-manifesto-for-research-metrics-1.17351.

Jaffe, A. (2016), "A Framework for Evaluating the Beneficial Impacts of Publicly Funded Research", Motu Note 15, Motu Publications, Wellington, https://motu.nz/our-work/productivity-and-innovation/science-and-innovation-policy/a-framework-for-evaluating-the-beneficial-impacts-of-publicly-funded-research/.

Krugman, P. (2013). "The New Growth Fizzle", *The New York Times*, New York, 18 August 2013, https://krugman.blogs.nytimes.com/2013/08/18/the-new-growth-fizzle/.

Lane, J et al. (2015), "New Linked Data on Research Investments: Scientific Workforce, Productivity, and Public Value", *NBER Working Paper*, No. 20683, National Bureau of Economic Research, Cambridge, MA, http://www.nber.org/papers/w20683.

Marburger, J. (2011), "Why Policy Implementation Needs a Science of Science Policy", in Fealing, K.H., J. Lane and S. Shipp (eds.), *The Science of Science Policy: A Handbook*, Stanford University Press, Redwood City, CA.

Marburger, J. (2007), "The Science of Science and Innovation Policy", in *Science, Technology and Innovation Indicators in a Changing World, Responding to Policy Needs*, OECD Publishing, Paris, https://doi.org/10.1787/9789264039667-en.

Martin, B. (2016), "Twenty challenges for innovation studies", *Science and Public Policy*, Vol. 43/3, pp. 432-450, Oxford University Press, Oxford, https://doi.org/10.1093/scipol/scv077.

Mazzucato, M. (2015), *The Entrepreneurial State: Debunking Public vs. Private Sector Myths*, Anthem Press, London and New York.

Moedas, C. (2016), "What new models and tools for measuring science and innovation impact?", keynote speech provided at OECD Blue Sky Forum, Ghent, https://ec.europa.eu/commission/commissioners/2014-2019/moedas/announcements/what-new-models-and-tools-measuring-science-and-innovation-impact_en.

National Research Council (2014), *Capturing Change in Science, Technology, and Innovation: Improving Indicators to Inform Policy*, The National Academies Press, Washington, DC, https://doi.org/10.17226/18606.

Nesta (2016), "Experimental innovation and growth policy: Why do we need it?", Nesta, London, https://www.nesta.org.uk/report/experimental-innovation-and-growth-policy-why-do-we-need-it/.

OECD/Eurostat (2018), *Oslo Manual 2018: Guidelines for Collecting, Reporting and Using Data on Innovation, 4th Edition*, The Measurement of Scientific, Technological and Innovation Activities, OECD Publishing, Paris/Eurostat, Luxembourg, https://doi.org/10.1787/9789264304604-en.

OECD (2016), Blue Sky Forum 2016, Ghent, general website, http://oe.cd/blue-sky.

OECD (2015), *Frascati Manual 2015: Guidelines for Collecting and Reporting Data on Research and Experimental Development*, The Measurement of Scientific, Technological and Innovation Activities, OECD Publishing, Paris, https://doi.org/10.1787/9789264239012-en.

OECD (2010a), The OECD Innovation Strategy: Getting a Head Start on Tomorrow, OECD Publishing, Paris, https://doi.org/10.1787/9789264083479-en.

OECD (2010b), Measuring Innovation: A New Perspective, OECD Publishing, Paris, https://doi.org/10.1787/9789264059474-en.

OECD (2007), *Science, Technology and Innovation Indicators in a Changing World: Responding to Policy Needs*, OECD Publishing, Paris, https://doi.org/10.1787/9789264039667-7-en.

Peltola, M.J. et al. (2016) "Value of official statistics", *OECD Statistics Newsletter*, Issue 65, November 2016, https://issuu.com/oecd-stat-newsletter/docs/oecd-statistics-newsletter-11-2016/2.

Polt, W (2016), "Scope and limits of indicator use by STI policy", contribution to OECD Blue Sky Forum panel discussion, Ghent.

Priem, J et al. (2011), *Altmetrics: A manifesto (v 1.01)*, http://altmetrics.org/manifesto.

Rafols, I. et al. (2016), "Towards more inclusive S&T indicators: A review on efforts to improve STI measurements in 'peripheral' spaces", paper submitted to the OECD Blue Sky Forum, Ghent, www.oecd.org/sti/086-2016-Rafols-OECDb.pdf.

Rozkrut, D. (2016), "New data and frontier tools: The challenge for official statistics in science and innovation", contribution to the OECD Blue Sky Forum panel discussion, Ghent.

Soete, L. (2016), "A sky without horizons. Reflections: 10 years after", keynote presentation at the OECD Blue Sky Forum, Ghent, https://www.slideshare.net/innovationoecd/soete-a-sky-without-horizons.

Stern, S. (2016), "Innovation-driven entrepreneurial ecosystems: A new agenda for measurement, policy, and action", keynote presentation at the OECD Blue Sky Forum, Ghent, https://www.slideshare.net/innovationoecd/stern-innovation-driven-entrepreneurial-ecosystems-a-new-agenda-for-measurement-policy-and-action.

Sugimoto, C. et al. (2016), "Social Media Metrics as Indicators of Broader Impact", paper submitted to the OECD Blue Sky Forum, Ghent, https://www.oecd.org/sti/172%20-%20SugimotoOECDaltmetrics.pdf.

Tett, G. (2018). "How to untangle the murky corporate web", *Financial Times*, September 12.

Varian, H. (2009), "How the Web challenges managers?", *The McKinsey Quarterly*, January 2009, McKinsey & Company, New York, https://www.mckinsey.com/industries/high-tech/our-insights/hal-varian-on-how-the-web-challenges-managers.

Thompson, W. (Lord Kelvin) (1883), "Electrical Units of Measurement" (3 May 1883), *Popular Lectures*, Vol. I, p. 73, https://archive.org/stream/popularlecturesa01kelvuoft#page/73/mode/1up%7C.

United Nations (2015), *Sustainable Development Goals*, United Nations, New York, https://sustainabledevelopment.un.org/?menu=1300.

ORGANISATION FOR ECONOMIC CO-OPERATION AND DEVELOPMENT

The OECD is a unique forum where governments work together to address the economic, social and environmental challenges of globalisation. The OECD is also at the forefront of efforts to understand and to help governments respond to new developments and concerns, such as corporate governance, the information economy and the challenges of an ageing population. The Organisation provides a setting where governments can compare policy experiences, seek answers to common problems, identify good practice and work to co-ordinate domestic and international policies.

The OECD member countries are: Australia, Austria, Belgium, Canada, Chile, the Czech Republic, Denmark, Estonia, Finland, France, Germany, Greece, Hungary, Iceland, Ireland, Israel, Italy, Japan, Korea, Latvia, Lithuania, Luxembourg, Mexico, the Netherlands, New Zealand, Norway, Poland, Portugal, the Slovak Republic, Slovenia, Spain, Sweden, Switzerland, Turkey, the United Kingdom and the United States. The European Union takes part in the work of the OECD.

OECD Publishing disseminates widely the results of the Organisation's statistics gathering and research on economic, social and environmental issues, as well as the conventions, guidelines and standards agreed by its members.

OECD PUBLISHING, 2, rue André-Pascal, 75775 PARIS CEDEX 16
(92 2018 04 1 P) ISBN 978-92-64-30756-8 – 2018